Asset Analytics

Performance and Safety Management

The main aim of this book series is to provide a floor for researchers, industries, asset managers, government policy makers and infrastructure operators to cooperate and collaborate among themselves to improve the performance and safety of the assets with maximum return on assets and improved utilization for the benefit of society and the environment.

Assets can be defined as any resource that will create value to the business. Assets include physical (railway, road, buildings, industrial etc), human and intangible assets eg software, data etc. The scope of the book series will be but not limited to:

- Optimization, modelling and analysis of assets
- Application of RAMS to the system of systems
- Interdisciplinary and multidisciplinary research to deal with sustainability issues
- Application of advanced analytics for improvement of systems
- Application of computational intelligence, IT and software systems for decisions
- Interdisciplinary approach to performance management
- Integrated approach to system efficiency and effectiveness
- Life cycle management of the assets
- Integrated risk, hazard, vulnerability analysis and assurance management
- Integration of data-information-knowledge for decision support
- Production rate enhancement with best practices
- Optimization of renewable and non-renewable energy resources

More information about this series at http://www.springer.com/series/15776

P. K. Kapur · Yury Klochkov
Ajit Kumar Verma · Gurinder Singh
Editors

System Performance and Management Analytics

Springer

Editors
P. K. Kapur
Center for Interdisciplinary Research
Amity University
New Delhi
India

Ajit Kumar Verma
Western Norway University of Applied Sciences
Bergen
Norway

Yury Klochkov
Department of Economics and Management in Machine Building
Peter the Great St. Petersburg Polytechnic University
Saint Petersburg
Russia

Gurinder Singh
Amity University
Noida, Uttar Pradesh
India

ISSN 2522-5162 ISSN 2522-5170 (electronic)
Asset Analytics
ISBN 978-981-10-7322-9 ISBN 978-981-10-7323-6 (eBook)
https://doi.org/10.1007/978-981-10-7323-6

Library of Congress Control Number: 2018942925

This Springer imprint is published by the registered company Springer Nature Singapore Pte Ltd.
The registered company address is: 152 Beach Road, #21-01/04 Gateway East, Singapore 189721, Singapore

Contents

Use of Bayesian Networks for System Reliability Assessment 1
Vipul Garg, M. Hari Prasad, Gopika Vinod and A. RamaRao

Predicting Code Merge Conflicts and Selecting Optimal Code Branching Strategy for Quality Improvement in Banking Sector 15
Viral Gupta, P. K. Kapur, Deepak Kumar and S. P. Singh

Testing the Effects of Agile and Flexible Supply Chain on the Firm Performance Through SEM 35
Mohammad Hossein Zavvar Sabegh, Aylin Caliskan, Yucel Ozturkoglu and Burak Cetiner

Analysis and Countermeasures for Security and Privacy Issues in Cloud Computing 47
Abdul Raoof Wani, Q. P. Rana and Nitin Pandey

An Efficient Approach for Web Usage Mining Using ANN Technique 55
Ruchi Agarwal and Supriya Saxena

Natural Language Processing Approach to Identify Analogous Data in Offline Data Repository 65
Nidhi Chandra, Sunil Kumar Khatri and Subhranil Som

Multi Release Reliability Growth Modeling for Open Source Software Under Imperfect Debugging 77
Diwakar and Anu G. Aggarwal

Barriers to Agile Adoption: A Developer Perspective 87
Noshiba Nazir, Nitasha Hasteer and Rana Majumdar

Toward Analysis of Requirement Prioritization Based Regression Testing Techniques 97
Varun Gupta, Yatin Vij and Chetna Gupta

Formulation of Error Generation-Based SRGMs Under the Influence of Irregular Fluctuations 103
Adarsh Anand, Deepika and Ompal Singh

Decision Aspect Prioritization Technique for Incremental Software 119
Varun Gupta, Siddharth Sethi and Chetna Gupta

Reliability Growth Analysis for Multi-release Open Source Software Systems with Change Point 125
Anu G. Aggarwal, Vikas Dhaka, Nidhi Nijhawan and Abhishek Tandon

Improvisation of Reusability Oriented Software Testing 139
Varun Gupta, Akshay Mehta and Chetna Gupta

Water Treatment System Performance Evaluation Under Maintenance Policies 147
Anshumaan Saxena, Sourabh Kumar Singh, Amit Kumar and Mangey Ram

Prediction of El-Nino Year and Performance Analysis on the Calculated Correlation Coefficients 167
Malsa Nitima, Gautam Jyoti and Bairagee Nisha

Performance of Static Spatial Topologies in Fine-Grained QEA on a P-PEAKS Problem Instance 179
Nija Mani, Gur Saran and Ashish Mani

Android Malware Detection Using Code Graphs 203
Shikha Badhani and Sunil Kumar Muttoo

Security in ZigBee Using Steganography for IoT Communications 217
Iqra Hussain, Mukesh Chandra Negi and Nitin Pandey

Implementation of Six Sigma Methodology in Syrian Pharmaceutical Companies 229
Yury Klochkov, Bacel Mikhael Alasas, Adarsh Anand and Ljubisa Papic

Developing Plans for QFD-Based Quality Enhancement 241
Dmitriy Aydarov, Yury Klochkov, Natalia Ushanova, Elena Frolova and Maria Ostapenko

Modified Counting Sort 251
Ravin Kumar

Analysis of Existing Clustering Algorithms for Wireless Sensor Networks 259
Richa Sharma, Vasudha Vashisht, Ajay Vikram Singh and Sushil Kumar

Process Mining for Maintenance Decision Support 279
Adithya Thaduri, Stephen Mayowa Famurewa, Ajit Kumar Verma and Uday Kumar

Software Release Time Problem Revisited 295
Nitin Sachdeva, P. K. Kapur and A. K. Shrivastava

Diffusion Modeling Framework for Adoption of Competitive Brands 307
Adarsh Anand, Gunjan Bansal, Arushi Singh Rawat and P. K. Kapur

Two-Dimensional Vulnerability Patching Model 321
Yogita Kansal and P. K. Kapur

A Hybrid Intuitionistic Fuzzy and Entropy Weight Based Multi-Criteria Decision Model with TOPSIS 333
Nitin Sachdeva and P. K. Kapur

The Role of Website Personality and Website User Engagement on Individual's Purchase Intention 347
Kokil Jain and Devnika Yadav

Impact of Social Media on Society—Analysis and Interpretation 361
Gurinder Singh, Loveleen Gaur and Kumari Anshu

Conceptual Framework of How Rewards Facilitate Business Operations in Knowledge-Intensive Organizations 373
Shweta Shrivastava and Shikha Kapoor

Tunnel QRA: Present and Future Perspectives 387
Jajati K. Jena, Ajit Kumar Verma, Uday Kumar and Srividya Ajit

Software Vulnerability Prioritization: A Comparative Study Using TOPSIS and VIKOR Techniques 405
Ruchi Sharma, Ritu Sibal and Sangeeta Sabharwal

About the Editors

Prof. P. K. Kapur is the Director, Center for Interdisciplinary Research, Amity University, Noida, and Former Dean of the Faculty of Mathematical Sciences and Head of the Department of Operational Research, University of Delhi, India. He has supervised over 40 Ph.D.s and 25 M.Phil. dissertations in the areas of innovation diffusion in marketing, software reliability, reliability-based optimization, and multi-criteria decision-making and other areas of management. He is the author of two world-renowned books "Software Reliability Assessment with O.R. Applications", Springer UK, 2011, and "Contributions to Hardware and Software Reliability", 1999, World Scientific, Singapore. He has been the President of the SREQOM (Regd.) since 2000 and Former President of ORSI. He is the Editor-in-Chief of IJSAEM, Springer, and has published over 300 papers in Indian journals and abroad in the areas of marketing, multiple-criteria decision-making (MCDM), optimization, hardware and software reliability.

Prof. Yury Klochkov is the Director, Monitoring Center for Education and Research, and Professor, Department of Economics and Management in Mechanical Engineering, St. Petersburg Polytechnic University, St. Petersburg, Russia. He was awarded the title of an honorary professor of Amity University, Noida, India, in September 2016. He was awarded the Ph.D. in standardization and quality management in 2006, and the postdoctoral degree, doctor of sciences in standardization and quality management in 2012. He is the author of about 100 publications in the areas of quality management and has implemented the results of his research in over 50 Russian enterprises.

Ajit Kumar Verma is a Professor (Technical Safety) of Engineering, Western Norway University of Applied Sciences, Haugesund, Norway (since March 2012), and was a Professor (since February 2001–December 2012) and Senior (HAG) Scale Professor (January 2013–January 2016) in the Department of Electrical Engineering at IIT Bombay, India, with a research focus on reliability and safety engineering. He has supervised/co-supervised 37 Ph.D.s and 95 Masters' theses in the area of electronic systems reliability, software reliability, reliable computing, power systems reliability (PSR), reliability-centred maintenance (RCM), RAMS in complex engineering systems and probabilistic safety/risk assessment (PSA) in power plants. He has over 250 publications in various journals and conferences. He is Senior Member of the IEEE and Life Fellow of the IETE and Editor-in-Chief of OPSEARCH, IJSAEM and Journal of Life Cycle Reliability and Safety Engineering.

Prof. (Dr.) Gurinder Singh is Group Vice Chancellor, Amity Universities, Director General, Amity Group of Institutions, India, and Vice Chairman, Global Foundation for Learning Excellence, and has an extensive experience of more than 21 years in institutional building, teaching, consultancy, research and industry. He is a renowned scholar and academician in the area of international business; he holds a prestigious doctorate and a postgraduate degree from IIFT, Delhi. He has spoken at various international forums which include prestigious Million Dollar Round Table Conference, at Harvard Business School, Thunderbird Business School, NYU, University of Leeds, Loughborough Business School, Coventry Business School, Rennes Business School, Essex University, UK, University of Berkeley, California State University, USA, NUS, Singapore and many more. He has received more than 25 international and national awards and has graced a host of talk shows on various TV channels.

Use of Bayesian Networks for System Reliability Assessment

Vipul Garg, M. Hari Prasad, Gopika Vinod and A. RamaRao

Abstract Probabilistic Safety Assessment (PSA) is a technique to quantify the risk associated with complex systems like Nuclear Power Plants (NPPs), chemical industries, aerospace industry, etc. PSA aims at identifying the possible undesirable scenarios that could occur in a plant, along with the likelihood of their occurrence and the consequences associated with them. PSA of NPPs is generally performed through Fault Tree (FT) and Event Tree (ET) approach. FTs are used to evaluate the unavailability or frequency of failure of various systems in the plant, especially those that are safety critical. Some of the limitations of FTs and ETs are consideration of constant failure/repair data for components. Also, the dependency between the component failures is handled in a very conservative manner using beta factor, alpha factors, etc. Recently, the trend is shifting toward the development of Bayesian Network (BN) model of FTs. BNs are directed acyclic graphs and work on the principles of probability theory. The paper highlights how to develop BN from FT and how it can be used to develop a BN model of the FT of Isolation Condenser (IC) of the advanced reactor and incorporate the system component indicator status into the BN. The indicator status would act like evidence to the basic events, thus updating their probabilities.

Keywords Fault tree · Bayesian networks · Fault detection

1 Introduction

PSA is a technique to quantify the risks associated with complex systems like NPPs, taking their design and operation aspects into account [1, 2]. PSA starts with the identification of the applicable Initiating Events (IEs). ETs are then developed that analyze the sequence of events from the IE to its final state. The purpose of

V. Garg (✉) · M. Hari Prasad · G. Vinod · A. RamaRao
Reactor Safety Division, Bhabha Atomic Research Centre,
Trombay, Mumbai 400 085, India
e-mail: vipulgarg@barc.gov.in

P. K. Kapur et al. (eds.), *System Performance and Management Analytics*,
Asset Analytics, https://doi.org/10.1007/978-981-10-7323-6_1

constructing ETs is that in the absence of safety systems how likely the IE will lead to an accident kind of situation [1]. In Level-1 PSA, the objective is to calculate the Core Damage Frequency (CDF). In calculating CDF, one needs information regarding IE frequency and unavailability of the different safety systems. The IE frequency in general can be obtained from operating experience, fault tree approach, and expert judgement. Similarly, safety systems unavailabilities can be obtained by performing system reliability analysis (e.g., Fault tree approach). PSA provides information about how safe the plant/system is, e.g., in Level-1 PSA, the lower the value of CDF, the better it is [1].

From the above discussion, it is clear that in PSA, system reliability analysis plays a major role. In general, FT approach is used in system reliability analysis. However, there are certain limitations in FTs such as taking a constant failure/repair data and consideration of independence among the basic events of the FT.

However, trend is now shifting toward the development of BN model of FTs. BNs apart from performing the FT analysis also offer certain other advantages. An inherent feature of a BN is diagnosis. This inherent feature of diagnosis can be utilized for fault diagnosis in a system, in which FTs are not equipped for. In a complex and critical system like an NPP, diagnosis of a faulty component or subsystem can assist the operator to take the necessary actions within the limited time.

2 System Reliability Analysis

System reliability analysis is performed to find the unavailability of different safety systems in a plant. There are different methods to perform the system reliability analysis. Two of those methods FT and BN techniques have been mentioned in the subsequent sections.

2.1 Fault Tree Analysis

FT is basically a top-to-bottom approach in which the top event represents the effect, e.g., failure of a system or subsystem. Below the top event are connected gates and events which specify the possible causes that could invoke the top event [3, 4]. These gates can be further expanded with the causes that could make the gate conditions as true and could thus lead to the top event. This expansion process continues till the bottom-most layer can no longer be expanded and only consists of the basic events. Events in an FT are assumed to be binary, i.e., a component can only have two states “working” and “not working” [3, 4]. FTs generally use AND/OR logic gates to represent the cause–effect relationships. An OR gate invokes the top event if at least one of the causes connected to it is true. An AND gate invokes the top event only if all the causes connected to it are true [3, 4]. Once the failure data is assigned to the basic components, the failure probability/failure frequency of the top event can be found.

Based on the information provided in the basic events, the FT provides the unavailability or the failure frequency of the top event. The information fed to the basic events is usually the constant failure rates, repair rates, etc. As a result, the output of the top event is also a constant value. Recently, the trend is shifting toward the development of BN model of FTs.

2.2 *Bayesian Networks*

BNs are directed acyclic graphs. Nodes in a BN represent the random variables. Arcs or links provide the cause and effect relationship between the nodes. Node with an incoming arc is called as the child node, and the node with the outgoing arc is called as the parent node [6]. A node that does not have any input arc is called as the root node. All the root nodes are provided with a prior probability data. A simple BN is shown in Fig. 1. Here, A and B are parent nodes while C is a child node.

All the child nodes have a Conditional Probability Table (CPT) associated with them. The CPT is indicative of the extent to which the linked nodes influence each other. According to the probability theory, the joint probability distribution of "M" nodes is given by the chain rule of probability as

$$P(N_1, N_2, N_3, \ldots, N_M) = P(N_1) * P(N_2/N_1) * \cdots * P(N_M/N_1, N_2, N_3, \ldots, N_{M-1}) \quad (1)$$

In BN, each node is conditionally independent of its non-descendants, given its parents, which transforms Eq. (1) as shown below [6]:

$$P(N_1, N_2, N_3, \ldots, N_M) = P(N_1/pa(N_1)) * P(N_2/pa(N_2)) * \cdots * P(N_M/pa(N_M)) \quad (2)$$

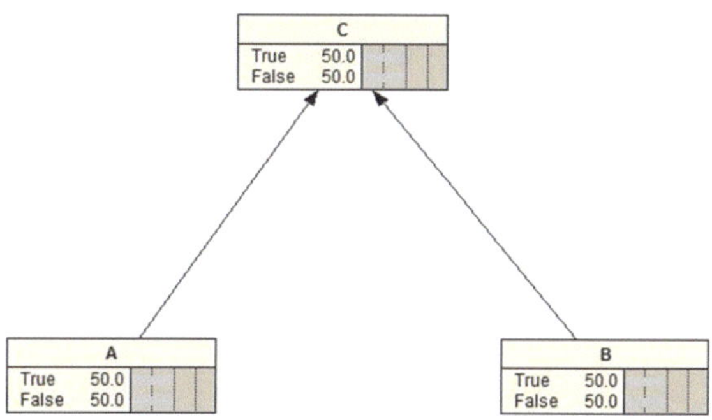

Fig. 1 A simple three-node BN

The joint probability distribution for Fig. 3 will be

$$P(A, B, C) = P(A) * P(B) * P(C/A, B) \tag{3}$$

where *pa* (*N*) represents parent of node "*N*".

The core of the BN lies in the Bayes theorem. It states that

$$P(A/B) = \frac{P(B/A) * P(A)}{P(B)} \tag{4}$$

where

P(*A*)	Prior probability of node "*A*",
P(*B*)	Evidence,
P(*B*/*A*)	Conditional probability of *B* given "*A*", and
P(*A*/*B*)	Posterior probability/updated probability of "*A*".

BNs also have the ability to reason backward, i.e., given the top event failure, which are the predominant nodes contributing to it.

2.3 Conversion of FT to BN

Seeing the strength of Bayesian network, various methods are investigated for assessing system reliability using Bayesian network. One of the popular methods is focussed on converting traditional FT to BN [5–7], using the steps given below:

1. All the basic events in the FT are drawn as root nodes in the BN.
2. The top event of the FT is drawn as the top node in the BN.
3. Every node in the BN has two states: True, i.e., failure state and False, i.e., working state.
4. All the root nodes, which are also the parent nodes, are fed with prior data which is same as that of the basic events in the FT.
5. All the remaining events in the FT are represented by the corresponding nodes in the BN, connected between the top node and the root nodes.
6. Links between the nodes are drawn in the same manner as represented by the FT. The direction of the links in the BN is from the cause toward the effect.
7. Top node which is a child node and has a CPT associated with itself.

A typical case study is selected to demonstrate the approach for conversion of fault tree to Bayesian network, which is discussed in subsequent section.

3 Case Study

3.1 Isolation Condenser of Advanced Reactor

Advanced reactor employs Isolation Condensers (IC) for decay heat removal during plant shutdown. The schematic diagram of IC is given (Fig. 2).

ICs are submerged in a pool of water. The flow is established from the steam drum to IC and then back to steam drum through the natural circulation. The decay heat in the form of steam enters from the steam drum into the ICs from the top, condenses in the tubes of the ICs, and the condensate returns by gravity to the steam drum. For this flow to establish under the cold shutdown conditions, the two Motor Operated Valves (MOVs) and the pneumatic/air-operated valve should be available, i.e., should remain in open state. Also, operator action is required to turn on the pneumatic/Air-Operated Valve (AOV). The details of these components are given in Table 1. The corresponding FT for an IC is as shown in Fig. 3.

The unavailability for the top event, i.e., IC comes out to be 4.118E−4 using the FT analysis.

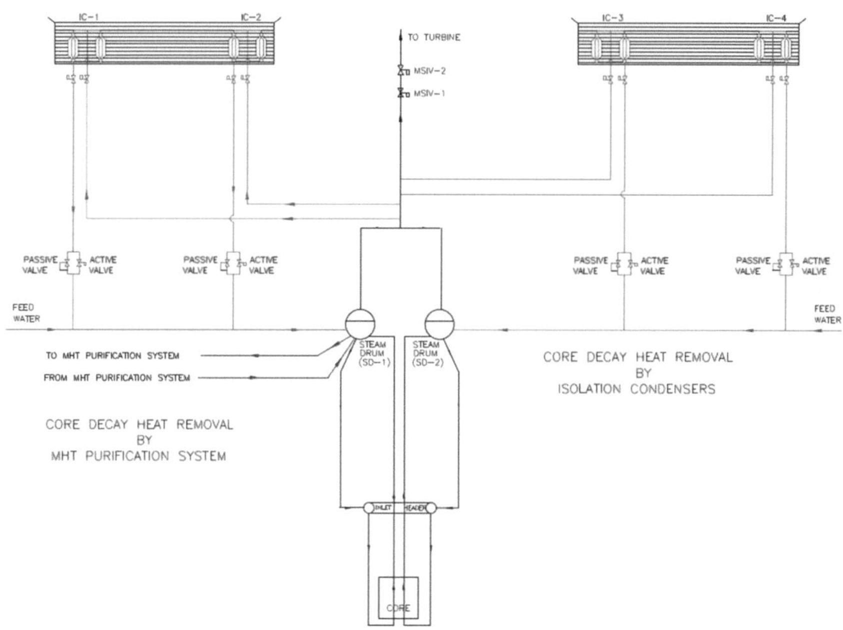

Fig. 2 Schematic diagram of IC

Table 1 Components/factors that influence the establishment of flow through IC

S. No.	Factor/component	FT denotation	Unavailability
1	MOV 1 fails to remain open	IC-MOV212D	3.1E−6
2	MOV 2 fails to remain open	IC-MOV211D	3.1E−6
3	Pneumatic valve/AOV	IC1-AOV1O	2.7E−4
4	Tube leakages in IC	IC1-TUBELEAK	5.63E−6
5	Human error in opening AOV	HE-IC-AOV	1E−4

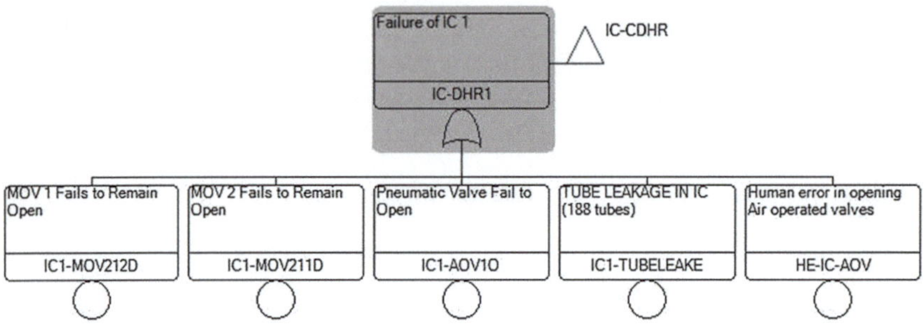

Fig. 3 Fault tree for an isolation condenser

3.2 *BN Model of the FT of IC*

The BN developed is as shown in Fig. 4. All the root nodes have been assigned the prior probabilities given in Table 2.

Similarly, a CPT has been assigned in the BN for the top event node, i.e., failure of isolation condenser. Since it is a series system, failure of any single component will lead to the IC failure. This has been implemented using the OR gate through the CPT.

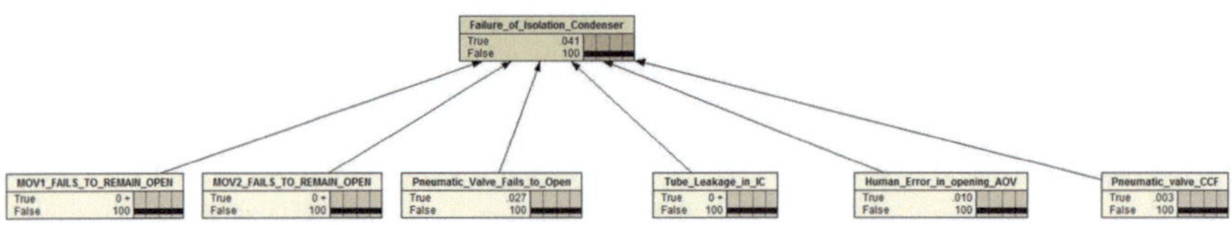

Fig. 4 Top event unavailability of the IC from the BN (0.041% = 4.1E−4)

Table 2 Prior probabilities of the root nodes in the BN of IC

S. No.	BN Node	State 1: true (%)	State 2: false (%)
1	MOV1 fails to remain open	0.00031	99.99969
2	MOV1 fails to remain open	0.00031	99.99969
3	Pneumatic valve fails to open	0.027	99.973
4	Tube leakages in IC	0.000563	99.999437
5	Human error in opening AOV	0.01	99.99

3.3 Similarities Between the FT and BN

FT calculates the unavailability or the frequency of occurrence of the top event. A BN developed from an FT can also find the unavailability or the frequency of occurrence as given in Figs. 4 and 5.

FT calculates the Minimal Cut Sets (MCS) which indicates the vulnerability of the system, i.e., the minimum number of the system components that could fail the system [3, 4], as shown in Fig. 5. BN, on the other hand, does backward reasoning to identify the predominant nodes contributing to the system failure, given system failure, as shown in Fig. 6.

The predominant MCS found from FT as shown in Fig. 5 is "ICI-AOV1O", i.e., pneumatic valve fails to open contributing 65.57% to the total unavailability.

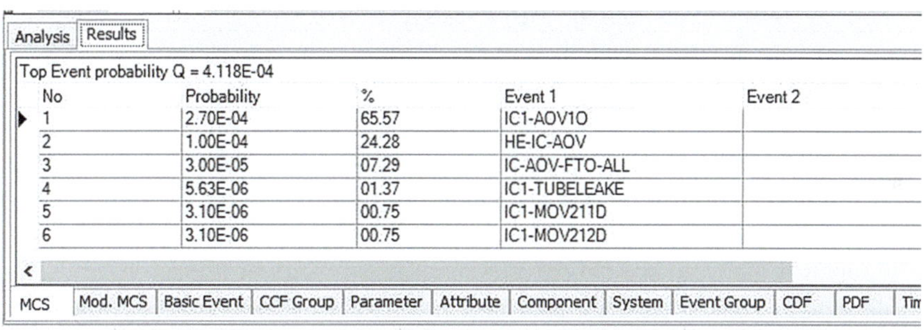

Fig. 5 Top event unavailability of the IC from the FT (4.118E−4)

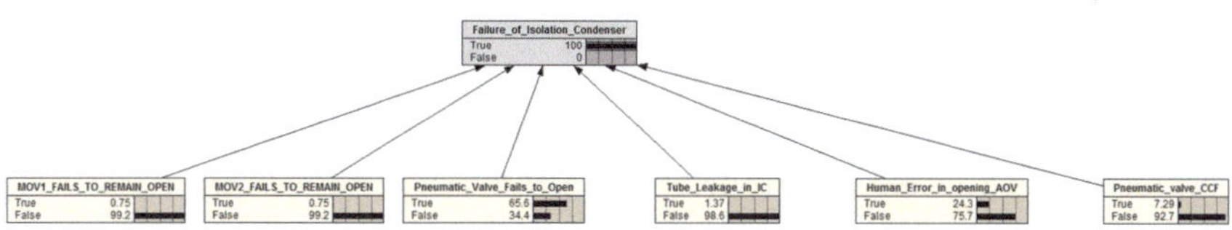

Fig. 6 Predominant nodes given system failure

- The predominant node found from the BN through backward reasoning is "Pneumatic_Valve_Fails_to_Open" having a contribution of 65.6% in the unavailability of the child node "Failure_of_Isolation_Condenser", as shown in Fig. 6.
- The remaining MCS and predominant nodes too reflect the same contribution to the unavailability of the top event in the FT and BN, respectively.

4 Advantage of BN Over FTA

BNs work on the principle of Bayes Theorem. The root nodes of the BN are the basic events of the FT. These root nodes are assigned some prior probability. Based on this prior data, the top node would provide the failure probability/frequency of the system, as in an FT. This top node/event value could be considered as the reference or base value. If these root nodes are fed with the observations made through some instrumentation, it would update their prior probability. This will in turn update the result of the top node/event. Any departure of the top node/event probability from its reference value is indicative of the change in the state of the system. This property could then be used for fault detection and diagnosis [8–10].

5 Fault Diagnosis Using Bayesian Networks

In advanced reactor, the working status of MOV 1, MOV 2, and pneumatic valve (AOV) can be found by examining the corresponding status from the Work Station (WS) and Main Control Room Hard Wired Panel (MCR-HWP). Thus, WS and MCR-HWP let the operator make certain observations. These observations can then be incorporated in the BN for fault detection and diagnosis as shown in Fig. 7.

Equipment status indicators get the requisite information regarding the equipment through some sensors. In case the sensor itself malfunctions or fails, it would have a direct impact on the status of the equipment as indicated by the indicators.

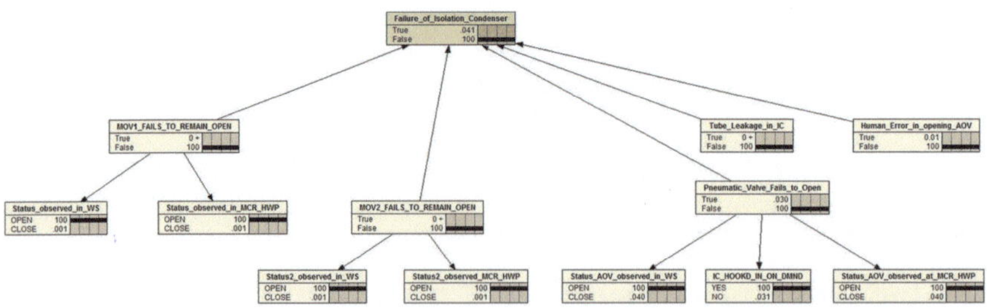

Fig. 7 BN of the IC with incorporation of the equipment status indicators observations

In order to take the effect of sensors into account, corresponding sensor nodes have been introduced in the BN as parents to the indicator nodes as shown in Fig. 8.

The position of the valves, i.e., open or close, is provided by the limit switches. These limit switch nodes have been provided with prior failure probability data [11].

The CPTs of the equipment status indicators will get modified due to the addition of sensor nodes [10]. Table 3 shows the CPT for the "Status_observed_in_WS" node.

The CPT value of 1.54E−4 has been taken from IAEA-TECDOC-478 [11]. However, when the limit switch is unavailable/non-functional, a probability of 0.5 or 50% has been assumed for the indicator to be indicating open or close state of the valve.

Under normal operating conditions, with all the equipment status indicators exhibiting the valve open state, the failure probability of the IC is 0.00014 or 0.014%, as shown in Fig. 8. This is an ideal case and therefore, 0.014% has been taken as the reference or base value. Fault detection tells whether the system is normal or not, and fault diagnosis identifies the root cause of the fault after it is detected [12]. If the failure probability of the top event, i.e., IC failure, exceeds its reference value of 0.014% by the virtue of some new observations, it would indicate some fault in the system. The new observations would act like new evidences, thus updating the posterior probabilities of their parent nodes or root nodes of the system.

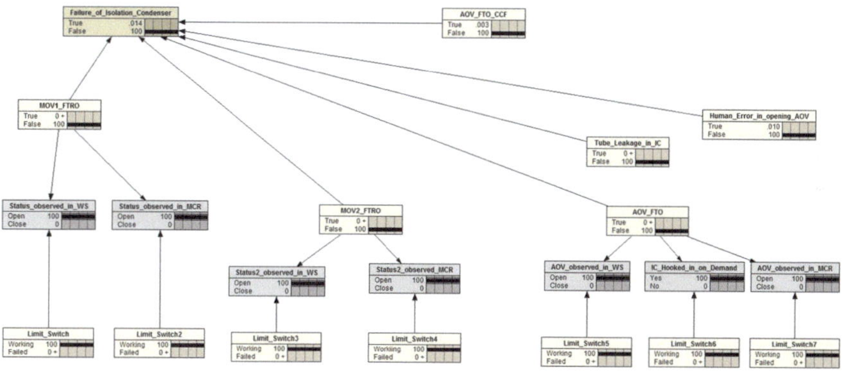

Fig. 8 BN of the IC along with the sensors or limit switch information

Table 3 CPT for the "Status_observed_in_WS" node

MOV1_Fails_to_remain_open	Limit_Switch	Status_observed_in_WS	
		Open (%)	Close (%)
True	Working	1.54E−4	99.999846
True	Failed	50	50
False	Working	99.999846	1.54E−4
False	Failed	50	50

Also, from the principle of d-separation, the child nodes of the diverging parent node are dependent on each other, if the status of the parent node is not known [13]. It may so happen that on the basis of the equipment status indicators, fault may be detected. However, there may not be enough rise in the posterior probability of any root node to diagnose the fault. If the system is still healthy and performing its task, which could be found using other parameters, e.g., temperature sensors installed at the inlet and outlet of the IC indicating requisite steam flow through IC—steam drum loop, then the incorrect fault detection may be due to a faulty indicator or sensor.

Generally, the higher the difference between the prior and posterior probabilities of the faulty states of the root nodes, the higher the chances of the corresponding fault occurrence [9]. Incorporation of the status indicators can only update the prior probabilities of the root nodes. The BN cannot perform fault diagnosis on its own. Some rules should be defined such that the faulty state of the indicator root node beyond a certain threshold value would be considered as successful diagnosis [8, 9] or if the difference between the prior and posterior probability is greater than a certain threshold value, then it would be considered as successful diagnosis [8]. This paper follows the first approach stated above [8].

Three typical cases have been given below.

5.1 *Case 1*

This case shows that failure or malfunctioning of an indicator due to some reason will not lead to spurious detection. Figures 9 and 10 show that despite one of the equipment indicators indicating a faulty status, no fault has been detected. This is because of the coupling present between the variables in the CPTs.

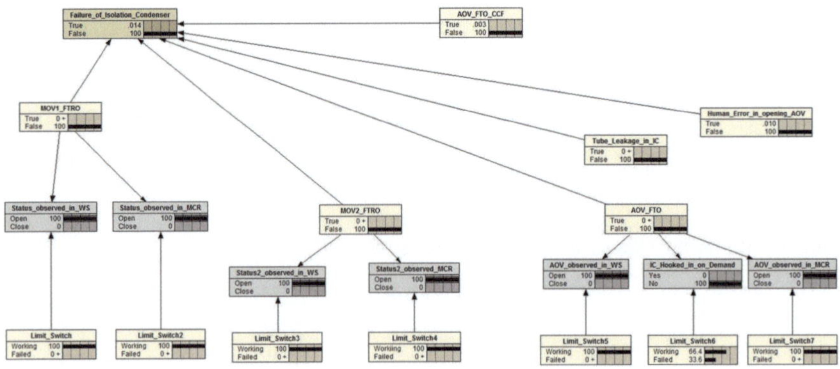

Fig. 9 IC BN with "IC_Hooked_In_on_Demand" indicator in faulty state

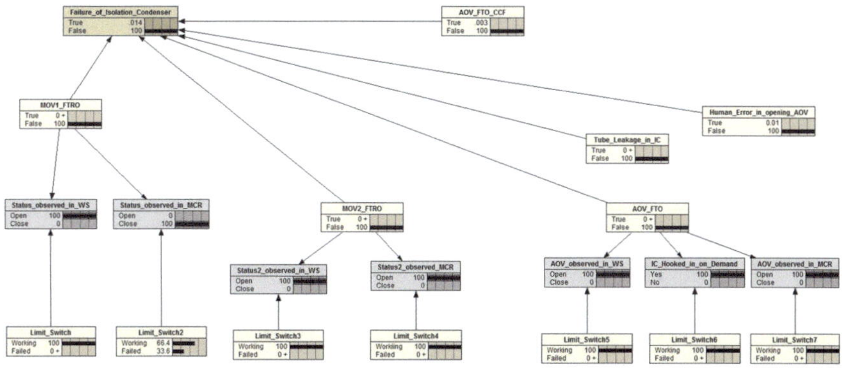

Fig. 10 IC BN with "Status_observed_in_MCR" indicator in faulty state

5.2 *Case 2*

This case considers a typical case that if a component of a system fails, it would be indicated by its respective indicators, thus leading to the fault detection and diagnosis as shown in Fig. 11.

The fault has been detected as the failure probability of the top node has increased from its normal conditions value of 0.014. Fault has been diagnosed as "MOV1_fails_to _remain_open".

5.3 *Case 3*

This case shows a situation where an indicator may have gone out of service due to some reason such that its output is not available. In Fig. 12, there is no status from

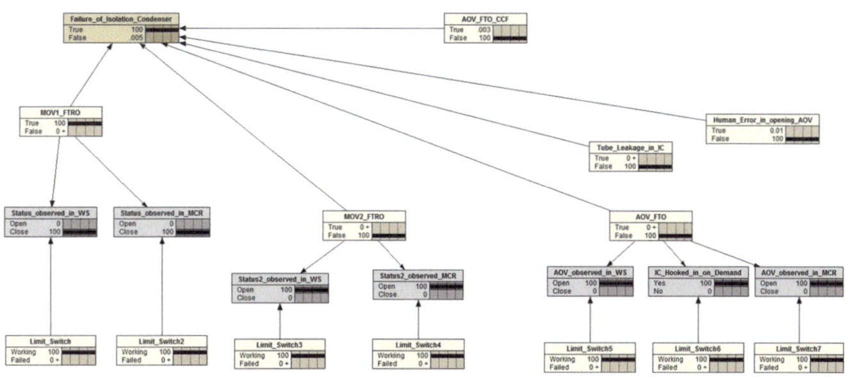

Fig. 11 IC BN with fault detection and fault diagnosis

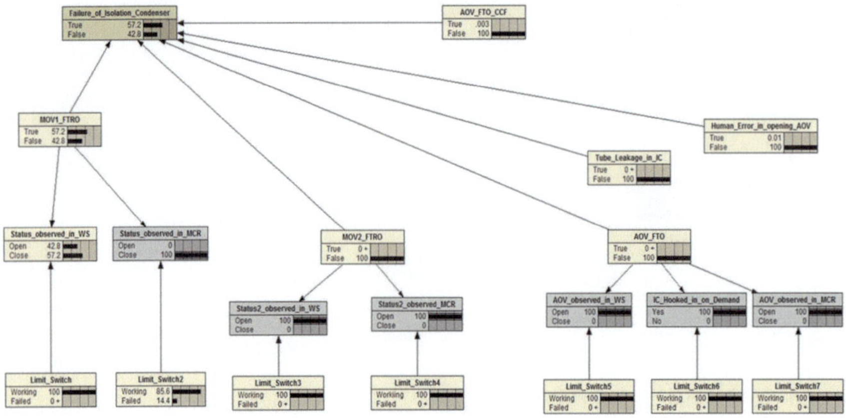

Fig. 12 IC BN considering unobserved state of indicator "Status_observed_in_WS"

"Status_observed_in_WS" indicator. However, the corresponding indicator node in the BN has the prior information and it is also dependent on "Status_observed_in_MCR_HWP" node, as the state of diverging parent node is not known. The fault has been diagnosed as "MOV1_fails_to _remain_open".

All the BNs have been developed in Netica tool [14].

6 Conclusion

This paper briefly tells about the importance of the FTs in the system reliability analysis. However, the capability of FTs is limited to finding the unavailability or failure frequency of the top event, which is a constant value. However, the same FT can be converted into a corresponding BN. BN, apart from calculating the top event unavailability, failure frequency, and predominant components that could cause the system to fail, can also facilitate fault detection and diagnosis. Incorporation of the status of the system components updates the prior probabilities of the root nodes of the BN, which are also the basic events of the FT. The updated root node probability further updates the failure probability of the top event node. This paper presents an application of fault detection and diagnosis using BN through the IC of the advanced reactor.

References

1. Development and Application of Level 1 Probabilistic Safety Assessment for Nuclear power plants, Specific Safety Guide, IAEA, 2010.
2. Cepin, M. (2015). Evolution of probabilistic safety assessment and its application in nuclear power plants. In *IEEE International Conference on Information and Digital Technologies* (pp. 53–60).
3. NUREG-0492. (1981). *Fault tree handbook*. USNRC.
4. Fault Tree Handbook with Aerospace Applications, NASA Publication, August 2002.
5. Przytula, K. W., & Milford, R. (2006). An efficient framework for the conversion of fault trees to diagnostic Bayesian network models. In *IEEE Aerospace Conference*.
6. Bobbio, A., Portinale, L., Minichino, M., & Ciancamerla, E. (2001). Improving the analysis of dependable systems by mapping fault trees into Bayesian networks. *Reliability Engineering and System Safety, 71,* 249–260.
7. Hamza, Z., & Abdallah, T. (2015). *Mapping fault tree into Bayesian network in safety analysis of process system*. In *IEEE International Conference on Electrical Engineering (ICEE)*.
8. Zhao, Y., Wen, J., Xiao, F., Yang, X., & Wang, S. (2017). Diagnostic Bayesian networks for diagnosing air handling units faults—Part I: Faults in dampers, fans, filters and sensors. *Applied Thermal Engineering, 111,* 1272–1286.
9. Cai, B., Liu, H., & Xie, M. (2016). A real-time fault diagnosis methodology of complex systems using object-oriented Bayesian networks. *Mechanical Systems and Signal Processing, 80,* 31–44.
10. Lampis, M., & Andrews, J. D. (2008). Bayesian belief networks for system fault diagnostics. *Quality and Reliability Engineering International, 25,* 409–426.
11. IAEA-TECDOC-478. (1988). Component reliability data for use in probabilistic safety assessment.
12. He, S., Wang, Z., Wang, Z., Gu, X., & Yan, Z. (2016). Fault detection and diagnosis of chiller using Bayesian network classifier with probabilistic boundary. *Applied Thermal Engineering, 107,* 37–47.
13. http://ai.stanford.edu/~paskin/gm-short-course/lec2.pdf.
14. Netica Application, Norsys Software Corporation (limited mode version). https://www.norsys.com/netica.html.

Predicting Code Merge Conflicts and Selecting Optimal Code Branching Strategy for Quality Improvement in Banking Sector

Viral Gupta, P. K. Kapur, Deepak Kumar and S. P. Singh

Abstract Code branching and merging plays a very critical role in the software development in an enterprise. Branching provides parallel development by enabling several development teams to work in isolation on multiple piece of code in parallel without impacting each other. Merging is a process to integrate the code of different teams together, which is achieved by moving the code around the branches. The process of merging can be very troublesome as it may contribute to enormous code merge or integration defects also known as code merge conflicts. One of the major problems faced by the practitioners is to predict the number of code merge conflicts and plan for the resolution of these conflicts. Another problem that is faced in an enterprise is to select an appropriate code branching strategy. Selection of a suitable code branching strategy is a multi-criteria decision making problem which involves multiple criteria and alternatives. This paper proposes a hybrid approach for predicting code merge conflicts and selecting the most suitable code branching strategy. Artificial neural network (ANN) is applied in a large enterprise to predict the code merge conflicts; thereafter analytic hierarchy process (AHP) is applied to select the most suitable code branching strategy. Total four code branching strategies have been considered in this paper. The outcome from the proposed approach successfully predicts the number of code conflicts and selects Branching Set-A as the most suitable code branching strategy with the highest priority weight

V. Gupta (✉) · D. Kumar
Amity Institute of Information Technology, Amity University, Noida 201313, Uttar Pradesh, India
e-mail: viralgupta@hotmail.com

D. Kumar
e-mail: deepakgupta_du@rediffmail.com

P. K. Kapur
Centre for Interdisciplinary Research, Amity University, Noida 201313, Uttar Pradesh, India
e-mail: pkkapur1@gmail.com

S. P. Singh
Department of Computer Science, BIT, Mesra, India
e-mail: sp.singh@bitmesra.ac.in

P. K. Kapur et al. (eds.), *System Performance and Management Analytics*, Asset Analytics, https://doi.org/10.1007/978-981-10-7323-6_2

of 0.287. The proposed methodology proved out to be very useful instrument for enterprises to quantitatively predict code merge conflicts and select the most suitable code branching strategy.

Keywords Code branching · Merging · Artificial neural network
Artificial hierarchy process · Quality · Enterprise applications

1 Introduction

One of the most prominent requirements of software industry is the capability of developing software by multiple teams simultaneously. The version control system [1] provides multiple code lines on which multiple teams can work in parallel [2, 3]. These code lines are known as branches. Branching provides ability to development teams to work on multiple pieces of code simultaneously without affecting each other [4]. Merging is conducted to integrate multiple code pieces, when they become ready for integration [5]. Code branching and merging process facilitates parallel development and is a critical component in the software development. There are complex problems associated with the code branching and merging process. One problem is that the merging process results into numerous code merge conflicts, software development teams are required to predict these code merge conflicts and plan the resources, efforts and cost to resolve these conflicts. Another problem that is faced by the practitioners is the selection of the most suitable code branching strategy. The major challenge that is faced by software development teams, after the application of merging process is the resolution of the code merge conflicts [3]. There can be numerous code merge conflicts and huge amount of efforts and cost is required to resolve these code merge conflicts. It becomes very critical for practitioners to predict these merge conflicts and plan for their resolution. Therefore, there is a requirement of a tool that can precisely predict the number of merge conflicts. Another prominent problem faced by large enterprise is to select the most suitable code branching strategy. Branching strategy governs the way in which the branches are created in the version control system and developers will use these branches in order to make their changes and track the code [6]. There are multiple branching strategies available like *Mainline/Trunk based, Release based, Feature based, team based, task based, component based, technology based, platform-based* etc. [6, 7]. The selection of the most suitable branching strategy is a multi-criteria decision problem, which involves multiple criteria and alternatives.

This paper addresses the problem of predicting the code merge conflicts and selecting the most suitable code branching strategy. In this paper a hybrid approach is adopted, code branching and merging processes are analysed for a large enterprise and ANN is applied for predicting the code merge conflicts, thereafter AHP is applied to select the most suitable code branching strategy. ANN is a family of model in the machine learning which is inspired by the biological neurons and are used to estimate functions that depends on large numbers of inputs [8]. In this paper, the input layer of ANN consists of seven inputs namely *Branch Pattern* [9], *Branch Topology* [9], *Branch Depth Level* [9], *Number of Components, Code Size, Release Efforts, Code Complexity.* These seven inputs collectively are used to predict the number of code merge conflicts. AHP is used to select the most suitable code branching strategy. AHP is a tool that enables decision makers to make decisions for complex, multi-criteria problems [10]. The framework enables the decision maker to gather the information regarding the problem, organize the information and analyze the gathered information by segregating it into hierarchy of criteria and alternatives. The five criteria considered in the decision making for this paper are *Ease of parallel development (EPD), Merging cost (MC), Propagation of features (POF), development efforts (DE) and Integration efforts (IE)* [11]. The four alternatives that are considered in this paper are *Branching Set-A, Branching Set-B, Branching Set-C and Branching Set-D.* The code branching and merging process along with the proposed model is applied in an enterprise application referred as Retail Banking Transaction System (RBTS) in remainder of the paper.

There has been very small amount of work done in the field of branching and merging in research community. The motivation behind this research work is the lack of a hybrid model in the literature and need of such model, which can predict the code merge conflicts and select the most suitable code branching strategy in an enterprise. As far as authors are concerned, they are not aware of any similar work done before. This paper contributes considerably to the field of code merging and branching in following three ways:

(1) The paper provides integrated approach comprising of ANN and AHP to predict the number of code merge conflicts and select the most suitable code branching strategy.
(2) The prediction of the code merge conflicts is performed using ANN considering seven input variables namely *Branch Pattern, Branch Topology, Branch Depth Level, Number of Components, Code Size, Release Efforts and Code Complexity.*
(3) The selection of most suitable code branching strategy is performed by utilizing the outcome from ANN and AHP.

The remainder of this paper is structured in the following way. Section 2 describes the review of literature conducted during this paper, Sect. 3 presents the proposed methodology in detail, and Sect. 4 illustrates the outcomes derived from the application of proposed methodology followed by the discussions and conclusion.

2 Literature Review

The concepts of code branching and merging have been established and discussed in the past. Many authors [9, 12, 13] have explicitly mentioned the importance of code branching and merging strategies and their impact on software quality, reliability [14, 15] and parallel development. Some authors [16] have examined the relationship between the goals of branching strategies and goals of software development teams. Few authors [17] have conducted the study to determine the code conflicts early so that these types of conflicts can be resolved earlier rather than delaying it towards the end of the project. Some authors [18] have investigated the impact of code branching and merging strategies along with the distributed versioning control system in an agile environment. In the past, various concepts of code branching and merging have been presented [12, 13, 16]. However, the field of prediction of code merge and selection of most suitable code branching strategy is under-represented in the literature. Arve [18] studied the impact of various code branching strategies with distributed version control system in agile projects. The paper describes various kinds of branching strategies those should be used in an agile workflow. Shihab et al. [9] in their paper investigated the empirical relationship between the code branching strategy and the software quality. The authors examined windows7 and windows vista and compared components that have different branch characteristics to quantify the impact of the code branching strategy on the software quality. Phillips et al. [3] in their study conducted the survey and examined the factors that define a successful branching and merging strategies. The survey was conducted on a diverse sample of 140 version control users. The key observation indicated that the continuous integration was typically not followed in practice. Data from most of the respondents demonstrated that the branches were long lived and 35–50% branches were staged rather than flat. Most of the respondents mentioned that code merge conflicts that generates from the code merging process was the biggest problem that was faced by the practitioners and the successful branching strategies were focused on reducing the frequency and complexity of code merge conflicts. Brun et al. [17] presented an approach that can help developers to identify the code merge conflicts early so that these can be resolved earlier before they turn into severe problem. The results of the study illustrated that 16% of the merge conflicts requires human efforts to resolve textual conflicts, 33% of the conflicts does not contain textual conflicts but higher order conflicts and finally conflict persists for at least average of 10 days. Bird et al. [16] developed a theory of relationship between goals and virtual teams on different branches. The study was conducted using the historical information from two releases of windows namely windows vista and windows7. The authors empirically investigated the relationship between the branches and the teams that were working on these branches. The results demonstrated the value of "p" to be less than 0.001, which signifies that the hypothesis was accepted. The empirical evidence indicated that the theory of branch similarity in terms of goals and teams were supported by development activity examined for the two releases of windows.

ANN is a model inspired by biological neural network. McCulloch and Pitts [19] in their study developed the first ANN that is based on the human neural network. The authors described the concept of neuron, which exists in a network of numerous neurons. This biological neuron receives input, performs some computation and produces the output. Post that many scientists and researchers have utilized the model of ANN in the field of pattern recognition, forecasting etc. Singh et al. [20] in their study utilized ANN to predict the testing efforts using object oriented metrics. The results demonstrated the ability of estimation of 35% of actual efforts in more than 72% of classes. Kaushik et al. [8] used ANN to estimate the software cost. The authors used back propagation learning algorithm and compared the results with the COCOMO model. The results demonstrated that the ANN model provided more accurate results compared with the COCOMO model. Many authors [8, 21–25] have predicted software development efforts, software-testing efforts, software lines of code, stock market rates, student's performance, fires, foreign exchange rates, election results etc. AHP is an analytic tool used to address complex decision-making problems by converting the qualitative values into quantitative values. Triantaphyllou and Mann [26] utilized AHP in addressing the decision problems in engineering applications. Kapur et al. [27] presented a methodology that was based on AHP to assess the health of ERP systems at various stages of implementation. The authors identified ten critical success factors of ERP implementation. Utility measures for these critical success factors were evaluated for all the five phases and the results were shared with the management team. Cheong et al. [28] developed an AHP tool that performs based on the fuzzy logic. The tool supported practitioners to take intricate decisions in the multi-criteria problem. The decisions can be taken in diverse kind of problem domains. Authors found that, there is a little amount of work done in research community on concepts of code branching and merging field in software development. However, there is lack of work that can predict code merge conflicts and select the most suitable code branching strategy in an enterprise.

3 Proposed Methodology and Experimental Set up

This paper proposes a hybrid model to predict the number of code merge conflicts using ANN and select the most suitable code branching strategy using AHP. ANN is capable of modeling complex nonlinear relationship between variables. The model can use to approximate any measurable function. The basic computational unit of ANN is neuron [29]. These neurons are present in three layers known as input, hidden and output layers. The neurons are connected with each other to form a network structure. Each connecting line has a weight associated with it. Each neuron has the net input function also known as activation function and output function. Based on the activation and output function, each neuron produces an output, which is sent as input to other neurons. The total output for the network structure is derived from the outputs of all the neurons present in the output layer and is compared against the target value. The differential of the output value and the target value is known as error. The objective of ANN is to adjust the weights of the network such that the total error is minimized. The process of adjusting weights and minimizing the error is known as training. The ANN is tested using the input data and the output value is predicted. The problem of selection of the most suitable code branching strategy is a complex multi-criteria decision problem, which is addressed using combined results from AHP and ANN. AHP is an analytic tool used to solve complex decision-making problems by allowing the conversion of the qualitative values into quantitative values [30]. In AHP, the decision problem is broken down into smaller components and is represented in a hierarchy or the top down flow of influence. “Criteria” form an important component of AHP structure. Various criteria and alternatives influencing the decision problem were identified. The experts conducted pairwise comparison and provided qualitative judgement data, which was converted into quantitative data. The proposed methodology was applied in a large enterprise application [31], which integrates disparate information technology systems in order to perform end-to-end business processes in an organization [4]. RBTS is a large enterprise application, which is a payments solution for a bank based out in United Kingdom. The proposed model was applied to predict the code merge conflicts and select the most suitable code branching strategy for the upcoming release of RBTS. RBTS is a set of applications, which process various payments flowing in and flowing out of the bank. These payments are also known as inbound and outbound payments. Our objective was to predict the code merge conflicts and select the most suitable code branching strategy.

3.1 Prediction of Code Merge Conflicts

This section explains the problem of prediction of code merge conflicts, various independent and dependent variables involved in the prediction of code merge conflicts, the setting up of ANN, the training and testing of the network and finally the prediction of code merge conflicts.

3.1.1 Define the Problem of Prediction of Code Merge Conflicts

The software development teams are required to predict the code merge conflicts and plan the resources, efforts and cost to resolve these conflicts. Code merge conflict is different from standard functional testing or unit testing defect, as it requires very careful inspection and detailed efforts from developers. If not planned carefully, these types of code conflicts can jeopardize the entire software development plan. It becomes critically important for the practitioners to understand the code branching strategy and predict the number of code merge conflicts in order to plan for resources required to address these conflicts. The dependent variable also known as output variable refers to the number of code merge conflicts. The independent and dependent variables are illustrated in Fig. 1.

The independent variables were extracted from the literature, and were validated by industry experts. These experts had more than fifteen years of experience in software development field and they have been working on code branching and merging strategies for the large enterprises. *Branching Pattern* [32] is a model that

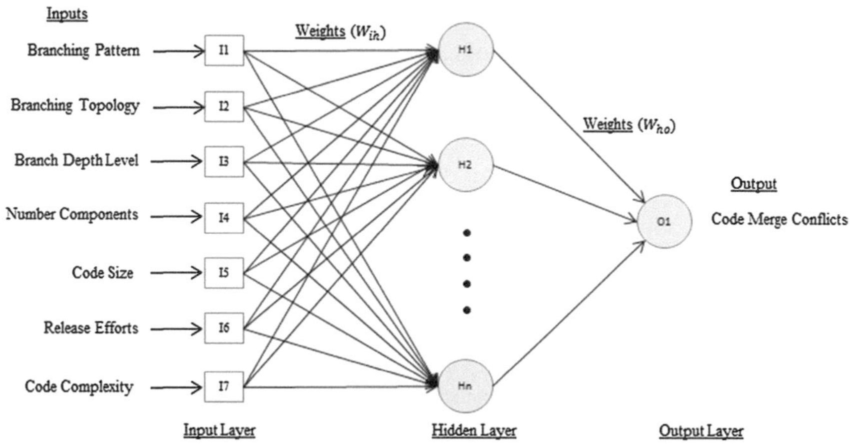

Fig. 1 Architecture of proposed ANN

represents the ways in which branching can be created in the versioning software. The most common types of branching patterns are mainline/trunk based, release based, feature based, team based, platform based etc. Mainline/trunk based branching pattern refers to a model having only one branch. Release based model refers to the creation of branches based on the number of releases. Feature based model refers to the creation of branches based on the features. Team based model refers to the creation of branches based on teams and platform based model refers to the creation of branches based on various platform. In this paper we have considered four types of branching patterns namely mainline/trunk based, release based, feature based and team based. Branching topology refers to the basic theme of branches that has been created in the versioning tool. There can be two types of branching topology. These are flat topology and stages topology, flat topology means only one or two layers of branches while staged topology refers to the multiple layers of branching. Branching depth level refers to the depth of the level of branches. Depth signifies to which level the child branches are created. Number of components refers to the total number of components that comes under the scope of the upcoming release. Code Size, release efforts and code complexity were also considered as input independent variables.

3.1.2 Empirical Data Collection

The data was collected from the RBTS enterprise application. The software development team was working on release number 47 and the problem was to determine the code merge conflicts for the upcoming releases. The data for the seven independent variables and one dependent variable was collected from all the previous releases. The data was collected from the version control system, which was Microsoft Team Foundation Server. The entire application was developed using 4600 components and the estimated code size was 1 million line of code (LOC). The values for independent and dependent variables were normalized using min-max normalization, which performed the linear transformation of the initial data.

$$I_n = \frac{I_o - \mathrm{Min}_I}{\mathrm{Max}_I - \mathrm{Min}_I} \tag{1}$$

Out of the seven independent and one dependent variable, there were few variables which could not be measured quantitatively. These variables were *branching pattern, branching topology and code complexity*. The values for these variables were transformed into quantitative values as per the input data transformation table given in Table 1. The transformed values of Table 1 were normalized based on Eq. (1). I_o represents *Original value of independent variable*. I_n represents *Normalized value of independent variable*. Min_I represents *Minimum value of independent variable* and Max_I represents *Maximum value of independent variable*. There are two types of learning or training that can be employed in ANNs. These leanings are supervised learning and unsupervised learning. In supervised learning, the output set of data is provided in the training set of data. In unsupervised learning no output data is provided in the training set of data. In supervised learning, the data can be classified into three categories commonly known as training, verification and testing set. The training set of data is used to observe the relationship between the input and output data sets. After the execution of training set of data, relationship between the input and output data set is established. Verification set of data is used to check whether the training of the network is being done as per expectations and the network is converging towards the target output values. The test set is used to evaluate the performance of the neural network. As per the literature, 60% of the data is considered as the training set, 10% of data is considered as the verification set and 30% of the data is considered as test set.

Table 1 Input data transformation

S. No.	Variables	Type	Domain	Final val.
1	Branch pattern	Input	Mainline/trunk	1
			Branch by release	0.75
			Branch by feature	0.5
			Branch by team	0.25
2	Branch topology	Input	Flat topology	1
			Staged topology	0
3	Code complexity	Input	Simple	1
			Medium	2
			Complex	3
			Very complex	4

3.1.3 Develop ANN Model

In this step, we configured various parameters of the network structure and developed the ANN. The values for various parameters of the ANN are specified in Table 2. The first parameter is the Network architecture. There are two types of architecture namely feed forward and recurrent/feedback networks [33]. In the Feed-forward network the signal travels in only one way while in the recurrent/feedback network the signal travels in both directions by inducting loops in the network structure. In this paper, feed-forward network structure was chosen. Multi-Layer Perceptron is a kind of feed forward network structure which has multiple layers namely input, hidden and output layers. Each layer can have multiple neurons. We considered there layers. Input layer had seven neurons, hidden layer has ten neurons and output layer had one neuron. The next parameter is the activation function, which is also known as transfer function. Activation function determines the relationship between the input and output node of a network. The activation function controls whether the neuron is active or inactive and it brings a level of non-linearity between the input and output. Various types of activation functions are threshold, logistic sigmoid, hyperbolic etc. In this paper, we used logistic sigmoid. There can be several learning algorithms like gradient descent [33], resilient back propagation, Bayesian regularization etc. In this paper, we used Levenberg-Marquardt algorithm as the training algorithm. The Levenberg-Marquardt algorithm updates the network weights and bias using Levenberg-Marquardt optimization. Levenberg-Marquardt is the fastest back propagation algorithm used in supervised learning. The last two attributes are performance and epochs. MSE measures the performance of the network using the mean of the squared errors. Epochs defines the number of times all the training set of data is used at least once to update the weights of the neural network. In this paper we considered 1000 epochs.

Table 2 ANN configuration

S. No.	Parameter	Value
1	Network architecture	Feed forward—multi layer Perceptron
2	Network layers	3
3	Input units	7
4	Hidden units	10
5	Output units	1
6	Activation function	Logistic sigmoid
7	Learning algorithm	Levenberg-Marquardt
8	Training function	Trainlm
9	Performance	Mean square error
10	Epochs	1000

3.1.4 Train Network

In this step, we trained the network using the configuration parameters that were set up in the previous step and the training set of data. The network model was executed iteratively based on the training set data. In each iteration, the weights of the network were updated and the error was minimized. Equation (2) illustrates the net input value for a particular neuron, Eq. (3) refers to the activation function used for the neurons in the neural network. Equation (4) signifies the total error of the network. Net input at *j*th Neuron is given below, where x_i is the input for *i*th input element and w_{ji} is the weight from *i*th input element to the *j*th neuron.

$$\mathrm{net}_j(x, w) = \sum_{i=0}^{n} x_i w_{ji} \tag{2}$$

$$\mathrm{out}_j = \frac{1}{\left(1 + \mathrm{e}^{\mathrm{net}_j(x,w)}\right)} \tag{3}$$

$$E_{\mathrm{total}} = \sum_{i=0}^{n} \frac{1}{2}\left(\mathrm{target}_j - \mathrm{output}_j\right)^2 \tag{4}$$

3.1.5 Test Network

The network was tested using the test data set. The training performance was evaluated using the measure mean squared error. The input data was provided for all the seven independent variables and the output value was determined. The results obtained from ANN were used as inputs in the determination of final prediction of the code merge conflicts in an enterprise application.

3.2 Selection of Most Suitable Code Branching Strategy

In this section, selection of most suitable code branching strategy is performed using AHP. AHP technique supports in prioritizing the alternatives by assigning the local priority vector to various elements of decision-making.

3.2.1 Formulate Decision Problem

The first step in AHP was to determine various elements of the decision problem. The decision problem attributes (also known as criteria) and alternatives for a decision problem were identified. The available code branching strategies were *Branching Set A, Branching Set B, Branching Set C and Branching Set D.*

Branching set A is a code branching strategy that is based on mainline or trunk based development with flat branch topology and only 1 level of branches. *Branching set B* is a code branching strategy that is based on release-based development with staged branch topology and two levels of branches. *Branching set C* is a code branching strategy that is based on feature-based development with staged branch topology and three levels of branches. *Branching set D* is a code branching strategy that is based on team-based development with staged branch topology and 4 levels of branches These four code branching strategies acts as four alternatives in AHP structure. Total five attributes were used to evaluate these four alternatives. These five attributes are *Ease of parallel development (EPD), merging cost (MC), propagation of features (POF), development (DE) and integration efforts (IE)*. The selection of the five attributes was based on the literature review and the expert opinion. EPD is the ability of large teams to work in parallel. MC is associated with the cost required to merge the code across various branches during the project execution phase. POF is a mechanism used to propagate the changes to all the software development teams. DI and EI refer to the efforts required to develop and integrate the software during a release.

3.2.2 Conduct Pairwise Comparison

Pairwise comparison is a mechanism using which all the elements of one level are compared among each other, two at a time, keeping one element as a control element. The control element is always from the higher level. There are two sub steps in this area. Firstly, all the attributes are compared among themselves keeping the decision problem as a control element. Secondly, all the alternatives are compared among themselves keeping one attribute as a control element. This is repeated by considering all the attributes as control element one at a time. During the pairwise comparison, experts compare two elements at a time and their qualitative judgement is converted into quantitative measure using Saaty's 1–9 Scale. Multiple judgement matrices are derived based on the selection of set of elements from the AHP hierarchy. Once the numerical values are assigned in the judgement matrix, the initial matrix was normalized and the row sums were calculated. Thereafter Eigen vector (λmax) was calculated by raising the matrix to the large power by successively squaring the matrix and calculating the row sums and normalizing them. The process was repeated until the Eigen values for the current matrix came closer to the values of previous matrix. For every judgement matrix (A) comprising of pairwise comparison scores, local priority vector (w) was calculated. The score of a_{ij} in the pairwise judgement matrix represents the relative importance of the element in the (i) row over the element in the (j) column. The local priority vector (w) for the matrix A can be calculated using the following equation, where λ_{max} is the largest Eigen value of matrix A. Multiple algorithms are available to calculate the local priority vector (w). Following algorithm was used in this paper to calculate the local priority vector (w). In this algorithm, "*J*" is the column number while "*I*" is the row number.

$$Aw = \lambda_{\max} \mathrm{w} \tag{5}$$

$$W_i = \frac{\sum_{i=1}^{I} \left(\frac{a_{ij}}{\sum_{j=1}^{J} a_{ij}} \right)}{J} \tag{6}$$

$$\mathrm{CI} = \frac{\lambda_{\max} - n}{n - 1} \tag{7}$$

$$\mathrm{CR} = \frac{\mathrm{CI}}{\mathrm{RCI}} < 0.1 \tag{8}$$

The following equations calculate the Consistency Index (CI) and Consistency Ratio (CR). RCI represents the random consistency index; the value of RCI was extracted from the literature. The consistency of the judgement data provided by the experts is validated for consistency by ensuring the value of Consistency Index to be less than 0.1 [34]. "*n*" in the below equations represent the number of elements in the pairwise matrix.

3.2.3 Determine AHP Global Priority Weights for Code Branching Strategy

After applying the AHP on the set of alternatives that constructs the problem domain, priority weights were calculated for each alternative. The association of these priority weights with the alternative was then used to solve the problems related to selection, prioritization, categorization etc.

3.2.4 Combine ANN and AHP Priority Weights and Determine the Final Weights

The weights received from ANN and AHP were combined together to determine the overall priority weights for these code branching strategies. This overall priority weight assigned to code branching strategies helped in determining the prioritization of the available code branching strategies using which the most suitable code branching strategy was selected.

4 Measures and Outcomes

This section explains the results obtained after applying all the steps of the proposed methodology to predict the number of code merge conflicts and select the most suitable code branching strategy. The first part of this section presents the results of ANN and presents the predicted number of code merge conflicts and the second part of this section presents the results from AHP.

4.1 Prediction of Code Merge Conflicts

This section presents the results obtained from the ANN and the predicted number of code merge conflicts and the code merge conflict score. MATLAB was used to develop model, train and test the ANN network structure. Total seven independent variables were considered and the best performance was achieved at the epoch 8 with the performance score of 0.0001812. The performance score signifies the MSE (mean squared error). The value of MSE close to 0 is considered to be very good and reflects that the error is very minimum. The performance score for Training, Testing and Validation is depicted in Fig. 2 that confirms the best score of 0.0001812 obtained during Validation. The value of "*r*" which is a coefficient of correlation is the most important metric in neural network. This coefficient of correlation signifies the relationship between the action and predicted value of number of code merge conflicts. The value of "*r*" greater than "0.95" and close to "1" considered very well, which reflects the high degree of convergence between predicted and actual values. The value of "*r*" for the training set was "0.99967"; the value of "*r*" for the validation set was "0.9957". The value of "*r*" for the testing set was 0.98548 and the final value of "*r*" was 0.99705. The predicted output value for the *branching set A* was 0.10136, which signifies the predicted number of code merge conflicts as 506. Based on the predicted output values for four branching sets, the final code conflict weights were derived. These results are illustrated in Table 3.

The results obtained from ANN will be combined with the results from AHP in the next section to get the final weights for the various branching strategies.

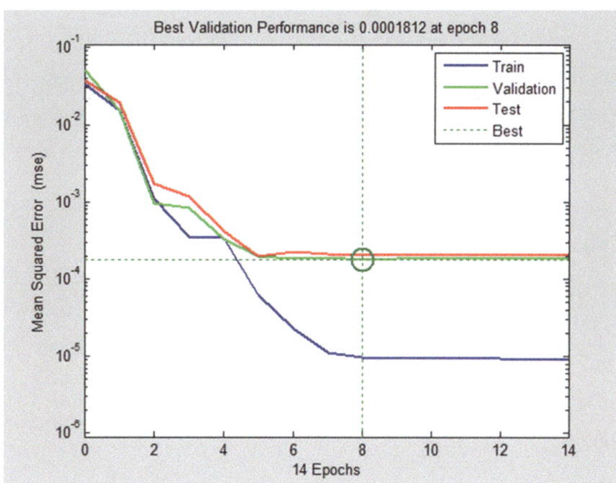

Fig. 2 Performance from ANN

Table 3 Outcome from ANN

Input/output variables	Branching Set A	Branching Set B	Branching Set C	Branching Set D
Branch pattern	1	0.75	0.5	0.25
Branch topology	1	0	0	0
Branch depth level	1	2	4	5
Number of components	0.252	0.252	0.252	0.252
Code size	0.006	0.006	0.006	0.006
Release efforts	0.205	0.205	0.205	0.205
Code complexity	1	1	1	1
Predicted output values	0.101	0.171	0.376	0.380
Predicted code merge conflicts	506.8	859.55	1882.4	1901.6
Local code conflict weight	0.098	0.166	0.365	0.369
Final code conflict weight	0.901	0.833	0.634	0.630

4.2 Selection of Most Suitable Code Branching Strategy

In this step, experts conducted pairwise comparison of various elements of AHP. Six individual judgement matrices were generated based on the selection of set of elements from the AHP network. One individual matrix was obtained, when all the attributes were compared among themselves keeping goal as a control element. The result from this comparison is illustrated in Tables 4 and 5. Table 4 represents the pairwise comparison, while Table 5 represents the derived data with the calculation of local priority vector. In Table 5, PEV represents local priority vector, PE represents principal or the largest Eigen value, CI represents Consistency Index CR represents Consistency Ratio. The value of P.E, CI and CR are 5.367, 0.0877 and 0.08 respectively.

Table 4 Pairwise comparison—goal

Goal	EPD	MC	POF	DE	IE
EPD	1	2	3	3	1/3
MC	1/2	1	2	2	1/3
POF	1/3	1/2	1	1/2	1/3
DE	1/3	1/2	2	1	1/2
IE	3	3	3	2	1
Total	5.167	7.00	11.00	8.50	2.50

Table 5 Pairwise comparison—goal—priority vector

Goal	EPD	MC	POF	DE	IE	PEV
EPD	0.194	0.286	0.273	0.353	0.133	0.212
MC	0.097	0.143	0.182	0.235	0.133	0.146
POF	0.065	0.071	0.091	0.059	0.133	0.095
DE	0.065	0.071	0.182	0.118	0.200	0.139
IE	0.581	0.429	0.273	0.235	0.400	0.407
Total	1.000	1.000	1.000	1.000	1.000	1.000

Five individual judgement matrices were obtained, when all the alternatives were compared among themselves keeping one of the attribute as a control element. The result from the pairwise comparison of all alternatives, keeping EPD as a control element is illustrated in Tables 6 and 7. Table 6 represents the pairwise comparison, while Table 7 represents the derived data with the calculation of local priority vector. In Table 7, PEV represents local priority vector, PE represents principal or the largest Eigen value, CI represents Consistency Index CR represents Consistency Ratio. The value of P.E, CI and CR are 4.242, 0.078 and 0.09 respectively.

Similarly, the result from the pairwise comparison of all alternatives, keeping MC, POF, DE and IE as a control element were determined. In Table 8, global priority weights were derived for various code branching strategies. The last column of this table represents the final global priority weight assigned to various code branching strategies using AHP. The priority weight derived from AHP was combined with the priority weight derived from ANN to determine the final consolidated priority weight for various code branching strategies. The consolidated

Table 6 Pairwise comparison—EPD

Control element—EPD	BS-A	BS-B	BS-C	BS-D
Branching Set A (BS-A)	1	1/3	1/2	1/3
Branching Set B (BS-B)	3	1	1/2	1/2
Branching Set C (BS-C)	2	2	1	1/3
Branching Set D (BS-D)	3	2	3	1
Total	9.000	5.333	5.000	2.167

Table 7 Pairwise comparison—EPD—priority vector

Control element—EPD	BS-A	BS-B	BS-C	BS-D	PEV
Branching Set A (BS-A)	0.111	0.063	0.100	0.154	0.118
Branching Set B (BS-B)	0.333	0.188	0.100	0.231	0.206
Branching Set C (BS-C)	0.222	0.375	0.200	0.154	0.217
Branching Set D (BS-D)	0.333	0.375	0.600	0.462	0.459
Total	1.000	1.000	1.000	1.000	1.000

Table 8 AHP—global priorities

Attributes	EPD	MC	POF	DE	IE	Weight
Priority weights	0.212	0.145	0.095	0.139	0.407	
Branching Set A	0.118	0.405	0.462	0.102	0.431	0.318
Branching Set B	0.205	0.288	0.287	0.221	0.293	0.263
Branching Set C	0.217	0.167	0.138	0.232	0.174	0.187
Branching Set D	0.458	0.137	0.111	0.443	0.100	0.230

Table 9 Final priority weight of code branching strategies

Alternatives	Priority weight	Code conflict weight	Final weight
	AHP	ANP	
Branching Set A	0.3183	0.9016	0.2870
Branching Set B	0.2636	0.8331	0.2196
Branching Set C	0.1873	0.6345	0.1188
Branching Set D	0.2308	0.6308	0.1456

priority weights for various code branching strategies are illustrated in Table 9. The results clearly indicate that the most suitable code branching strategy was *Branching Set A* with the highest priority weight of 0.287. The second most suitable branching strategy was *Branching Set B* with the priority weight of 0.219, which was followed by *Branching Set D* having priority weight of 0.1456 and *Branching Set C* with the priority weight of 0.1188. The AHP model has also provided the priority weight for the attributes. The top most attribute was *Integration Efforts* having priority weight of 0.4074, the second attribute was EPD with the priority weight of 0.212. The value of P.E, CI and CR are 4.269, 0.084 and 0.09 respectively.

The integrated model of ANN and AHP has proved to be very beneficial for the practitioners of code branching and merging as prediction of code merge conflicts is a very complicated task and selection of the most suitable code branching strategy is a very complex multi criteria decision-making problem. The result of ANN determined *Branching Set-A* as the best code merge option with code merge conflict weight of 0.901 which signifies lowest number of predicted code merge conflicts. The combined outcome from AHP and ANN demonstrated that the most suitable code branching strategy was *Branching Set-A* which received the highest priority weight of 0.287. The leadership team of the RBTS enterprise application was highly impressed by the results of the applied model. Using the model's results, the leadership team of the RBTS enterprise application selected Code Branching Set A as the most suitable code branching strategy for the upcoming release.

5 Conclusion

Code branching and merging are very crucial components of software industry, and these components influences the quality, schedule and cost parameters of any software development project. There are numerous code branching strategies available in the industry and careful selection of the code branching strategy becomes a very complex multi-criteria decision-making problem. The process of code merging may result into numerous code merge conflicts, which affects the quality, schedule and cost of an enterprise application. It becomes very important for an enterprise to predict the code merge conflicts and select the most suitable code branching strategy. This paper has proposed a hybrid model comprising of ANN and AHP to predict the code merge conflicts and select the most suitable branching strategy. The four available code branching strategies considered in this paper are *Branching Set A, Branching Set B, Branching Set C and Branching Set D*. To predict the merge conflict weight using ANN, the independent variables considered are *Branching Pattern, Branching Topology, and Branch Depth Level, number of components, code size, release efforts and code complexity*. To select the most suitable code branching strategy using AHP, various criteria considered are *ease of parallel development, merging cost, propagation of features, development and integration efforts*. The result of ANN allocated *Branching Set-A* the best code merge conflict weight of 0.901 which signifies lowest number of predicted code merge conflicts. The combined outcome from AHP and ANN demonstrated that the most suitable code branching strategy was *Branching Set-A* which received the highest priority weight of 0.287. The outcome of the methodology proved out to be very useful for the business organizations in decision making and deriving strategic objectives for code branching and merging. This paper has made significant contributions to the research community by proposing a hybrid model comprising of ANN and AHP in a large enterprise for predicting the code merge conflicts and selecting most suitable code branching strategy.

References

1. Baudiš, P. (2014). Current concepts in version control systems. arXiv preprint arXiv:1405.3496.
2. Buffenbarger, J., & Gruell, K. (1999). A branching/merging strategy for parallel software development. In *System Configuration Management* (pp. 86–99).
3. Phillips, S., Sillito, J., & Walker, R. (2011) Branching and merging: an investigation into current version control practices. In *Proceedings of the 4th International Workshop on Cooperative and Human Aspects of Software Engineering* (pp. 9–15). ACM.
4. Branching (version control). Website. https://en.wikipedia.org/wiki/Branching(version_control).
5. Merge (version control). Website. https://en.wikipedia.org/wiki/Merge_(version_control).
6. Jacob, J., Rodriguez, M., & Barry, G. *Team foundation server branching guidance*. Microsoft.

7. Somerville, I. (2001). *Software engineering* (6th ed.). Boston, MA: Addison-Wesley.
8. Kaushik, A., Soni, A. K., & Soni, R. (2013). A simple neural network approach to software cost estimation. *Global Journal of Computer Science and Technology, 13*(1).
9. Shihab, E., Bird, C., & Zimmermann, T. (2012) The effect of branching strategies on software quality. In *Proceedings of the 2012 ACM-IEEE International Symposium on Empirical Software Engineering and Measurement* (pp. 301–310). IEEE.
10. Saaty, T. L. (2008). Decision making with the analytic hierarchy process. *International Journal Services Sciences, 1*(1), 83–98.
11. Meier, J. D. (2009). *Microsoft application architecture guide* (2nd ed.). Website. https://www.microsoft.com/downloads/details.asp.
12. Appleton, B., Berczuk, S., Cabrera, R., & Orenstein, R. (1998). Streamed lines: Branching patterns for parallel software development. In *Proceedings of the Pattern Languages of Programs Conference, PLoP*, (Vol. 98).
13. Walrad, C., & Strom, D. (2002) The importance of branching models in SCM. *Computer, 35*, 31–38.
14. Kapur, P. K., Pham, H., Gupta, A., & Jha, P. C. (2011). *Software reliability assessment with OR applications*. London: Springer.
15. Kapur, P. K., Garg, R. B., & Kumar, S. (1999) *Contributions to hardware and software reliability*. World Scientific.
16. Bird, C., Zimmermann, T., & Teterev, A. (2011). A theory of branches as goals and virtual teams. In *Proceedings of the International Workshop on Cooperative and Human Aspects of Software Engineering* (pp. 53–56).
17. Brun, Y., Holmes, R., Ernst, M. D., & Notkin, D. (2011). Proactive detection of collaboration conflicts. In *Proceedings of the 19th ACM SIGSOFT Symposium and the 13th European Conference on Foundations of Software Engineering* (pp. 168–178). ACM.
18. Arve, D. (2010). *Branching strategies with distributed version control in Agile projects*. Website. http://fileadmin.cs.lth.se/cs/Personal/lars_bendix/research/ascm/in-depth/arve-2010.pdf.
19. McCulloch, W. S., & Pitts, W. (1943). A logical calculus of the ideas immanent in nervous activity. *The Bulletin of Mathematical Biophysics, 5*(4), 115–133.
20. Singh, Y., Kaur, A., & Malhotra, R. (2008). Predicting testing effort using artificial neural network. In *Proceedings of the World Congress on Engineering and Computer Science, WCECS*, San Francisco, USA.
21. Philip, A. A., Taofiki, A. A., & Bidemi, A. A. (2011). Artificial neural network for forecasting foreign exchange rate. *World of Computer Science and Information Technology Journal, 1*(3), 110–118.
22. Zhang, G., Patuwo, B. E., & Hu, M. Y. (1998). Forecasting with artificial neural networks: The state of the art. *International Journal of Forecasting, 14*, 35–62.
23. Aggarwal, K. K., Singh, Y., Chandra, P., & Puri, M. (2005). Bayesian regularization in a neural network model to estimate lines of code using function points. *Journal of Computer Sciences, 1*(4), 505–509.
24. Sarcià, S. A., Cantone, G., & Basili, V. R. (2007) A statistical neural network framework for risk management process. In *ICSOFT SE* (pp. 168–177).
25. Oladokun, V. O., Adebanjo, A. T., & Charles-Owaba, O. E. (2008) Predicting student's academic performance using artificial neural network: A case study of an engineering course, *The Pacific Journal of Science and Technology, 9*(1), 72–79.
26. Triantaphyllou, E., & Mann, S. H. (1995). Using the analytic hierarchy process for decision making in engineering applications: Some challenges. *International Journal of Industrial Engineering: Applications and Practice, 2*(1), 35–44.
27. Kapur, P. K., Nagpal, S., & Khatri, S. K. (2014). Critical success factor utility based tool for ERP health assessment—A general framework. *International Journal of System Assurance Engineering and Management, 5*(2), 133–148.

28. Cheong, C. W., Jie, L. H., Meng, M. C., & Lan, A. L. H. (2008). Design and development of decision making system using fuzzy analytic hierarchy process. *American Journal of Applied Sciences, 5*(7), 783–787.
29. Reddy, C. S., & Raju, K. (2009) A concise neural network model for estimating software effort. *International Journal of Recent Trends in Engineering, 1*(1), 188–193.
30. Zhu, L., Aurum, A., Gorton, I., & Jeffery, R. (2005). Trade off and sensitivity analysis in software architecture evaluation using analytic hierarchy process. *Software Quality Journal, 14*(4), 357–375.
31. Kalyani, K. (2012). Recent trends and challenges in enterprise application integration. *International Journal of Application or Innovation in Engineering & Management, 1*(4), 62–71.
32. Perry, W. E. (2006). *Effective methods for software testing* (3rd ed.). Indianapolis, IN: Wiley.
33. Khaze, A. R., Masdari, M., & Hojjatkhah, S. (2013). Application of artificial neural networks in estimating participation in elections. *International Journal of Information Technology, Modeling and Computing, 1*(3), 23–31.
34. Lai, V. S., Trueblood, R. P., & Wong, B. K. (1999). Software selection: A case study of the application of the analytic hierarchy process to the selection of a multimedia authoring system. *Information & Management, 36*(4), 221–232.

Testing the Effects of Agile and Flexible Supply Chain on the Firm Performance Through SEM

Mohammad Hossein Zavvar Sabegh, Aylin Caliskan, Yucel Ozturkoglu and Burak Cetiner

Abstract High competition, continuous, and rapid changing in consumer demands push companies finding differentiation ways to gain competitive advantage. Supply chain and logistics practices have been seen as the core strategic tools to survive for companies. In this research, the impacts of agile and flexible supply chain practices on the customer satisfaction, service quality, sales performance, and profitability are examined. As a research area, fast fashion industry was chosen. To the aim, a theoretical model was developed and tested through structural equation modeling (SEM). The results reveal that companies performing agile and flexible supply chains can reap the benefits in terms of service quality and customer satisfaction, and at the end can reap the resulting financial benefits in terms of increased sales and profits.

Keywords Supply chain management · SEM · Fast fashion · Firm performance

---

M. H. Z. Sabegh (✉)
ROES, Organizational Excellence Specialists, Courtenay, Canada
e-mail: mzavvar80@gmail.com

M. H. Z. Sabegh
QMF Review Board at American Society for Quality (ASQ), Milwaukee, USA

A. Caliskan · Y. Ozturkoglu · B. Cetiner
Faculty of Business, International Logistics Management, Yasar University, Izmir, Turkey
e-mail: aylin.caliskan@yasar.edu.tr

Y. Ozturkoglu
e-mail: yucel.ozturkoglu@yasar.edu.tr

B. Cetiner
e-mail: burak.cetiner@yasar.edu.tr

P. K. Kapur et al. (eds.), *System Performance and Management Analytics*, Asset Analytics, https://doi.org/10.1007/978-981-10-7323-6_3

1 Introduction

It is a well-known fact that supply chain management affects a firm's performance positively. In their research, Bayraktar et al. [1] indicated a positive correlation between a firm's performance and supply chain practices. Lenny Koh et al. [2] researched the effects of SCM practices on firm performance and indicated a meaningful and positive effect of SCM practices on firm performance. An effective and productive supply chain management includes many factors such as agility, simplicity, flexibility, proper use of information technologies, reliability, and information sharing. Fast fashion industry aims to offer more diversity to customers with the addition of interim seasons in between the existing seasons. To realize this aim, the concept of agility in supply chain management is inevitable in order to increase the frequency of new product entries, enhance customer service levels, and reduce the response time to changing market requirements. Also, the importance of a flexible supply chain management cannot be ignored due to its capability of adjusting to un-anticipated, long-term customer demands in fast fashion industry caused by high seasonality and changing order amounts, delivery times, delivery schedules, and production capacities based on market conditions. Consequently, in this study, flexibility factors, as the means to respond to the issues that can be encountered during agility and speed phases of the supply chain, are measured with their effects on firm performances in fast fashion industry.

1.1 Agile Supply Chain Management

For the firms that compete in an ever changing, dynamic supply chains, the motto has changed as "It's not the big that eats the small… it's the fast that eat the slow" [3]. Agility in a supply chain includes elements such as firm's new product frequency, the ability to reduce product development cycle time, production time reduction, customer service enhancement, and response time to changing market requirements [4]. We can consider agility in a supply chain as rearranging with haste. In literature, it is emphasized that the main element of agility is the ability to respond to changing market conditions and customer demands [5–8].

In previous researches, positive effects of agility on firms' performance were shown [9]. Chan et al. [48] empirically investigated the impact of supply chain agility on firm performance and found a positive and direct relationship. Yusuf et al. [10] indicated that an agile supply chain performance increases a firm's competitive advantage performance. Swafford et al. [4] in their experimental studies had found a positive correlation between compatibilities of information technologies, supply chain flexibility, supply chain agility, and competitive work performance. Gligor and Holcomb [11], as a result of their literature research, had found that the most important outcome of agility is the increasing sales in accordance with increasing customer ratio. In addition, several other authors had highlighted the close relationships between agility and fulfillment and satisfaction of customer demands [12–14]. Swafford et al. [4] only used financial performance elements while he was researching the effects of agile supply chain on a firm's performance and found a positive and direct correlation between them. On the other hand, Gligor et al. [12] emphasized the correlation but it is not direct.

1.2 Flexible Supply Chain Management

Flexibility in supply chain gives firms the ability to respond meaningfully to environmental changes, uncertainties, and contributes to its development of high-quality products and services [15, 16, 49]. Uncertainties may occur in many elements such as demand, supply, and cost, and flexibility in supply chain is must to respond to these uncertainties and changes effectively [17]. Flexibility in supply chain includes elements such as order amounts, delivery times, and ability to change production schedules [18]. Sanchez and Perez [18] had found a positive correlation between firm performance and flexible supply chain. Swafford et al. [4] in their experimental studies had found a positive correlation between compatibilities of information technologies, supply chain flexibility, supply chain agility, and competitive work performance. Supply chain flexibility concept has derived from flexible production literature; therefore, its dimension is usually related to production [19]. In this study, "ability to change production amount" is discussed as a dimension of flexibility. In addition, supply chain flexibility has a process-based perspective and consequently, it includes main processes such as purchasing/order placement and distribution/logistics [20, 21]. In accordance with the said processes, in this study, "the ability to change delivery schedule" and "the ability to change order amount" were discussed as two other dimensions.

1.3 Firm Performance Measures

As there is no agreement on specific firm performance measurement metrics in literature [22], it would cause no harm if the researchers were to use their own free will to determine performance measures. For example, Tan et al. [23] used market share, investment return, sales, active profitability, cost of production, customer service, product quality, and competitive advantage elements to measure the correlation between supplier and firm performances. Operation strategy and firm performance, gathered their performance elements into four groups: market, product, financial, and employee performance [24]. Some researchers [25–27] used service quality, customer satisfaction, productive internal processes, effective resource usage, fast service, growth rate, profitability, and productivity as their performance measures. Walker et al. [28] used profitability, productivity, growth, competitive advantage, customer satisfaction, job quit rates, investor relations, and environmental impact measures to measure firm performance. In order to measure a firm's performance, customer satisfaction focus is essential; otherwise, all supply chain efforts will be in vain and costly [29, 30]. Consequently, in this study, fast fashion supply chain's end customers' satisfaction is used for measurement. In addition, utilizing the researches mentioned above, profitability and sales numbers are used as financial performance indicators while service quality is used as a quality performance indicator.

2 Conceptual Development

The first two relationships we offer in this study are about the relationship between agile and flexible supply chain issues and customer satisfaction. As Gunasekaran et al. [30] indicated, flexibility and delivery performance to meet customer needs increase customer satisfaction. Inspired by Gunasekaran et al [30], we postulate that the speed of new product offering, the speed of customer services, and the speed of response to the changes and dynamics, also the abilities to change production amount, delivery schedule, and order amount effect satisfaction. Accordingly, we hypothesize:

Hypothesis 1: Agile supply chains have a positive effect on customer satisfaction.
Hypothesis 2: Flexible supply chains have a positive effect on customer satisfaction.

Service quality refers to ability of satisfying customer needs and wants. In highly competitive, turbulent, and volatile markets, the needs of customers have been changed so quickly and suddenly, so the agility and flexibility have become a must to meet the service quality requirements [17, 31]. Thus, we formulate the following hypotheses:

Hypothesis 3: Agile supply chains have a positive effect on service quality.
Hypothesis 4: Flexible supply chains have a positive effect on service quality.

In line with several authors and academic [32–35], we postulate the following hypothesis:

Hypothesis 5: Service quality has a positive effect on customer satisfaction.

Another relationship between constructs we offered exists between the customer satisfaction and sales performance and profitability. To support this idea, there have been several studies in the literature [36, 37]. How customer satisfactions increase company sales and profitability? it can be achieved through the reduced price elasticity of satisfied customers. Also, Reichheld and Sasser [38] indicated the tendency of satisfied customers to pay more for the products or services. Sales of the company increase in parallel with satisfied customers because satisfaction leads to loyalty and

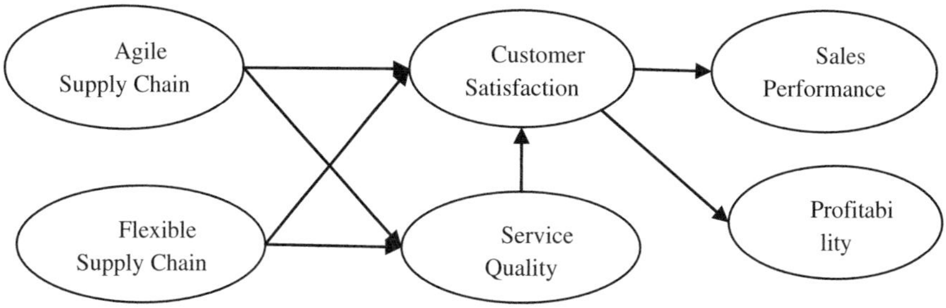

Fig. 1 Theoretical model of the research

for a firm, loyal customers mean more customers will purchase in the future [38]. The sixth and seventh hypotheses of the model can be stated as follows (Fig. 1):

Hypothesis 6: Customer satisfaction has a positive effect on sales.
Hypothesis 7: Customer satisfaction has a positive effect on profitability.

3 Methodology

Fast fashion also challenges existing traditional supply chain management techniques with its solutions for rapid changes in demand and high product diversity. This industry is well known with its characteristics of uncertainties and unstable environment [47]. In this study, instead of the traditional elements, the relationships between more flexible and agiler supply chain elements and firm performances of the brands operating in the textile industry are analyzed. There are 894 firms from 17th Occupation Group (Apparel Retail Commerce Group) registered to Izmir Chamber of Commerce to be included in the research. These firms include boutiques with one branch, private fashion houses, baby and kids clothing retailers, sports shops, readymade underwear, wedding dress shops, etc. Out of all officially registered 894 firms, firms that are eligible, operating in the fast fashion industry, have many branches, and thought be practicing supply chain management are selected. A total of 46 fast fashion brand brands are selected and due to the eligibility of access to the population, a sample is not created and the whole population is included in the research. Some of the firms that participate in this study, due to their firm policies, do not want their names to be revealed. Some of the firms included in our sample are Zara, Bershka, Mango, Pull and Bear, Loft, Mavi, Koton, Benetton, Stradivarius, and LC Waikiki. In this research, survey method is used for data collection. For hypothesis testing, five-point Likert scale is used. Due to the firms having difficulties with providing numeric values, subjective measures are utilized for performance questions. For the performance questions, "1 = too low" and "5 = too high" statements are used for the firms to compare themselves with their competitors. The answers for supply chain flexibility- and agility-related questions include consist of “1 = I absolutely agree” and “5 = I absolutely disagree” statements. As this survey was going to be conducted with the mid- and top-level managers of fast fashion brands located in Izmir, Istanbul, and Ankara, the possible difficulties that they would have in answering the questions regarding internal processes, cost of production, supplier performance, investment return, and active profitability were taken into consideration and instead the performance measures were handled as sales numbers, customer satisfaction, service quality, and profitability with the utilization of literature. The surveys were conducted in July 2015–November 2016 period with all chosen firms’ mid- and top-level managers in Izmir, Istanbul, and Ankara. 35% of the surveys were conducted face to face, while 62% of them were conducted via e-mail.

4 Analysis and Results

The LISREL 8.51 package was used to test the proposed research model. The two-stage testing process suggested by Anderson and Gerbing [39] was performed to apply structural equation modeling (SEM). In the first stage, the measurement model’s reliability and validity were tested. In the second stage, structural model was tested in terms of examining the hypothesized paths among the constructs.

4.1 Measurement Model

Calculating standardized loadings in CFA, average variance extracted (AVE) values and composite reliabilities (CR) are the common measurement ways of validity [40, 41]. For CFA, LISREL 8.51 package was used. Both the CR and AVE cannot be computed by LISREL, and therefore CR and AVE are computed manually in spreadsheet software Microsoft.

Hair et al. [41] suggest that all standardized loadings which are above the cut-off point of 0.70 are adequate for validity. Bagozzi and Yi [42] state that standardized loadings greater than 0.60 are adequate. For the measurement model, the standardized loadings are between 0.77 and 0.88, providing adequate evidence of validity (Table 1). Table shows that all latent variables show high composite reliabilities (CR) (between 0.71 and 0.88), well above the accepted 0.60 value [40]. Also as can be seen that all the latent variables’ AVE scores are well above 0.50 (between 0.55 and 0.69). So, according to the scores of CFA, CR, and AVE, it can be said that the measurement model validity is achieved. Finally, the overall fitness between the collected data and the measurement model was examined.

Table 2 lists the main fit indices outputted from LISREL and their acceptance thresholds. As the fit indices, normalized $\chi^2 = 1.870$, RMSEA = 0.054, GFI = 0.952, AGFI = 0.890, CFI = 0.960, and NFI = 0.980, all meet suggested acceptable range. So, we can conclude that the fit of measurement model is acceptable [43–46].

Table 1 CFA factor loadings, composite reliability, and AVE

Measures	CFA, standardized loadings (*T*-values)	Composite reliability	AVE
Agile supply chain		0.80	0.59
ASC 1	0.67 (6.88)		
ASC 2	0.75 (9.73)		
ASC 3	0.94 (13.26)		
Flexible supply chain		0.86	0.67
FSC 1	0.79 (10.67)		
FSC 2	0.88 (12.64)		
FSC 3	0.79 (10.65)		
Customer satisfaction		0.75	0.60
CS 1	0.73 (9.32)		
CS 2	0.82 (10.81)		
Product/service quality		0.71	0.55
SQ 1	0.81 (10.21)		
SQ 2	0.68 (8.38)		
Sales performance		0.82	0.69
SP 1	0.79 (10.39)		
SP 2	0.88 (12.00)		
Profitability performance		0.88	0.57
PP 1	0.84 (11.65)		
PP 2	0.80 (10.83)		

4.2 *Structural Model*

First, the overall fitness between the sample data and the structural model was tested using the six goodness-of-fit indices as the same used in the measurement model. As can be seen from Table 3, a sound fit of the data to the structural model was obtained. All six GOF indices achieve their acceptance thresholds ($\chi^2 = 1.894$, RMSEA = 0.048, GFI = 0.939, AGFI = 0.910, CFI = 0.961, and NFI = 0.978).

Table 4 contains the detailed results related to the structural model. All the hypotheses are accepted with positive directions. The expected relationships from agile and flexible supply chains to customer satisfaction and service quality were supported. We found strong relationship from service quality to customer satisfaction and from customer satisfaction to sales performance and profitability. Although agile and flexible supply chain plus service quality totally explain customer satisfaction 53%, the highest contribution is made by service quality. This result proves the mediator role of service quality. Also, agile supply chain affects customer satisfaction substantially more than flexible supply chain. The results

Table 2 Goodness-of-fit measures of the measurement model

Goodness-of-fit measure	Recommended value	Value of this study
χ^2/df	$\leq$2.00 (good fit) $\leq$3.00 (acceptable fit)	1.870
Root means square error of approximation (RMSEA)	$\leq$0.05 (good fit) 0.05–0.08 (acceptable fit) 0.08–0.10 (mediocre fit)	0.054
Goodness-of-fit index (GFI)	>0.95 (good fit) >0.90 (acceptable fit)	0.952
Adjusted goodness-of-fit index (AGFI)	>0.90 (good fit) >0.85 (acceptable fit)	0.890
Comparative fit index (CFI)	>0.97 (good fit) 0.95–0.97 (acceptable fit)	0.960
Normed fit index (NFI)	>0.95 (good fit) 0.90–0.95 (acceptable fit)	0.980

Table 3 Goodness-of-fit measures of the structural model

Goodness-of-fit measure	Recommended value	Value of this study
χ^2/df	$\leq$2.00 (good fit) $\leq$3.00 (acceptable fit)	1.894
Root means square error of approximation (RMSEA)	$\leq$0.05 (good fit) 0.05–0.08 (acceptable fit) 0.08–0.10 (mediocre fit)	0.048
Goodness-of-fit index (GFI)	>0.95 (Good fit) >0.90 (Acceptable fit)	0.939
Adjusted goodness-of-fit index (AGFI)	>0.90 (good fit) >0.85 (acceptable fit)	0.910
Comparative fit index (CFI)	>0.97 (good fit) 0.95–0.97 (acceptable fit)	0.961
Normed fit index (NFI)	>0.95 (good fit) 0.90–0.95 (acceptable fit)	0.978

Table 4 Results of the hypothesized model

Hypothesized path	Hypothesis	Standardized path coefficient	*T*-value
Effects of agile supply chain			
Agile supply chain → customer satisfaction	H1(+)	0.57	7.93
Agile supply chain → service quality	H3(+)	0.55	7.72
Effects of flexible supply chain			
Flexible supply chain → customer satisfaction	H2(+)	0.15	1.98
Flexible supply chain → service quality	H4(+)	0.48	6.51
Effects of service quality			
Service quality → customer satisfaction	H5(+)	0.77	10.51
Effects of customer satisfaction			
Customer satisfaction → sales performance	H6(+)	0.35	3.53
Customer satisfaction → profitability	H7(+)	0.23	3.73
Structural relationships	R^2		
Satisfaction = 0.57 * agility + 0.15 * flexibility + 0.77 * service quality	0.53		
Service quality = 0.55 * agility + 0.48 * flexibility	0.46		
Sales = 0.35 * satisfaction	0.31		
Profitability = 0.37 * satisfaction	0.36		

convincingly supported that satisfied customers with agile and flexible supply chain operations so, with service quality, effects positively the sales and profitability performance of the company.

5 Conclusion

The agile and flexible supply chain operations that stem from the rapidly changing needs of customers center on the supply chain management of companies especially in business-to-customer (B2C) markets. The main idea of this research is to reveal if agile and flexible supply chain operations do really pay off in terms of company financial performance. The main conceptual model comprising six constructs was proposed to examine the relations among agile, flexible supply chain, service quality, customer satisfaction, sales performance, and profitability. With the typical techniques and procedures of SEM, all proposed hypotheses were validated. Our study suggests that companies performing agile and flexible supply chains can reap the benefits in terms of service quality and customer satisfaction, and at the end can reap the resulting financial benefits in terms of increased sales and profits. In this study, it is demonstrated that both the agile and flexible supply chain benefits have a differential impact on the service quality and customer satisfaction. Even though the roles of the agile supply chain on customer satisfaction and service quality almost equally, flexible supply chain affects customer satisfaction statistically significant but much lower than it affects service quality. Also, according to the result of path analysis, the impact of service quality as a mediator has more effect on customer satisfaction than the direct effects of agile and flexible supply chain operations. Finally, the results convincingly supported that satisfied customers with agile and flexible supply chain operations so with service quality, impact positively the sales and profitability performance of the company.

References

1. Bayraktar, E., Demirbag, M., Koh, S. L., Tatoglu, E., & Zaim, H. (2009). A causal analysis of the impact of information systems and supply chain management practices on operational performance: evidence from manufacturing SMES in Turkey. *International Journal of Production Economics, 122*(1), 133–149.
2. Lenny Koh, S. C., Demirbag, M., Bayraktar, E., Tatoglu, E., & Zaim, S. (2007). The impact of supply chain management practices on performance of smes. *Industrial Management & Data Systems, 107*(1), 103–124.
3. Jennings, J., & Haughton, L. (2002). *It's not the BIG that eat the SMALL... It's the FAST that eat the SLOW: How to use speed as a competitive tool in business*. Harper Collins.
4. Swafford, P. M., Ghosh, S., & Murthy, N. (2008). Achieving supply chain, agility through IT integration and flexibility. *International Journal of Production Economics, 116*, 288–297.
5. Van-Hoek, R. I. (2001). Epilogue-moving forward with agility. *International Journal of Physical Distribution & Logistics Management, 31*(4), 290–301.
6. Holsapple, C., & Jones, K. (2005). Exploring secondary activities of the knowledge chain. *Knowledge and Process Management, 12*(1), 3.

7. Ismail, H. S., & Sharifi, H. (2006). A balanced approach to building agile supply chains. *International Journal of Physical Distribution & Logistics Management, 36*(6), 431–444.
8. Jain, V., Benyoucef, L., & Deshmukh, S. G. (2008). What's the buzz about moving from 'lean' to 'agile' integrated supply chains? A fuzzy intelligent agent-based approach. *International Journal of Production Research, 46*(23), 6649–6677.
9. Tallon, P. P., & Pinsonneault, A. (2011). Competing perspectives on the link between strategic information technology alignment and organizational agility: Insights from a mediation model. *MIS Quarterly, 35*(2), 463–486.
10. Yusuf, Y. Y., Gunesekaran, A., Adeleye, E. O., & Svayoganathan, K. (2004). Agile supply chain capabilities: Determinants of competitive objectives. *European Journal of Operational Research, 159,* 379–392.
11. Gligor, D. M., & Holcomb, M. C. (2012). Antecedents and consequences of supply chain agility: Establishing the link to firm performance. *Journal of Business Logistics, 33*(4), 295–308.
12. Gligor, D. M., Esmark, C. L., & Helcomb, M. C. (2015). Performance outcomes of supply chain agility: When should you be agile? *Journal of Operations Management, 33*(34), 71–82.
13. Goldsby, T. J., Griffis, S. E., & Roath, A. S. (2006). Modeling lean, agile, and leagile supply chain strategies. *Journal of Business Logistics, 27*(1), 57–80.
14. Christopher, M. (2000). The agile supply chain: Competing in volatile markets. *Industrial Marketing Management, 29*(1), 37–44.
15. Allegre, J., & Sard, M. (2015). When demand drops and prices rise. Tourist packages in the Balearic Islands during the economic crisis. *Tourism Management, 46,* 375–385.
16. Sabegh, M.H.Z., Ozturkoglu Y., & Kim, T. (2016). Green Supply Chain Management Practices' Effect on the Performance of Turkish Business Relationships, in Proceedings of 12th International Conference on Industrial Engineering, p. 5232–5245, 25–26 January 2016, Kharazmi University, Tehran, Iran.
17. Swafford, P. M., Ghosh, S., & Murthy, N. (2006). The antecedents of supply chain agility of a firm: Scale development and model testing. *Journal of Operations Management, 24*(2), 170–188.
18. Sánchez, M. A., & Pérez Pérez, M. (2005). Supply chain flexibility and firm performance: A conceptual model and empirical study in the automotive industry. *International Journal of Operations & Production Management, 25*(7), 681–700.
19. Stevenson, M., & Spring, M. (2007). Flexibility from a supply chain perspective: Definition and review. *International Journal of Operations & Production Management, 27*(7), 685–713.
20. Vickery, S. K., Calantone, R., & Dröge, C. (1999). Supply chain flexibility: An empirical study. *Journal of Supply Chain Management: A Global Review of Purchas and Supply, 35*(3), 16–24.
21. Merschmann, U., & Thonemann, U. W. (2011). Supply chain flexibility, uncertainty and firm performance: An empirical analysis of German manufacturing firms. *International Journal of Production Economics, 130*(1), 43–53.
22. Tan, K. C., Lyman, S. B., & Wisner, J. D. (2002). Supply chain management: a strategic perspective. *International Journal of Operations & Production Management, 22*(6), 614–631.
23. Tan, K. C., Kannan, V. R., & Handfield, R. B. (1998). Supply chain management: Supplier performance and firm performance. *International Journal of Purchasing and Materials Management, 34*(3), 2.
24. Dalgakıran, A.B., & Ozturkoglu, Y. (2017). Scale and Relationship Analysis for Turkish Furniture Sector, *Business and Management Studies: An International Journal*, *5*(1), 147–161.
25. Judge, W. Q., & Douglas, T. J. (1998). Performance implications of incorporating natural environmental issues into the strategic planning process: An empirical assessment. *Journal of Management Studies, 35*(2), 241–262.
26. Quinn, R. E., & Rohrbaugh, J. (1983). A spatial model of effectiveness criteria: Towards a competing values approach to organizational analysis. *Management Science, 29*(3), 363–377.
27. Cegarra-Navarro, J. G., Soto-Acosta, P., & Wensley, A. K. (2015). Structured knowledge processes and firm performance: The role of organizational agility. *Journal of Business Research, 69*(5), 1544–1549.

28. Walker, R. M., Chen, J., & Aravind, D. (2015). Management innovation and firm performance: An integration of research findings. *European Management Journal, 33,* 407–422.
29. Lee, H. L. & Billington, C. (1992). Managing supply chain inventory: Pitfalls and opportunities. *Sloan Management Review*, Spring, 65–73.
30. Gunasekaran, A., Patel, C., & Tirtiroglu, E. (2001). Performance measures and metrics in a supply chain environment. *International Journal of Operations & Production Management, 21*(1–2), 71–87.
31. Agarwal, A., Shankar, R., & Tiwari, M. K. (2006). Modeling the metrics of lean, agile, and leagile supply chain: An ANP-based approach. *European Journal of Operational Research, 173*(1), 211–225.
32. Anderson, E. W., & Sullivan, M. W. (1993). The antecedents and consequences of customer satisfaction for firms. *Marketing Science, 12*(2), 125–143.
33. Churchill, G. A., Jr., & Suprenant, C. (1982). An investigation into the determinants of customer satisfaction. *Journal of Marketing Research, 19*(November), 491–504.
34. Cronin, Jr., J. J., & Taylor, S. A. (1992). Measuring service quality: A reexamination and extension. *The Journal of Marketing, 56*, 55–68.
35. Oliver, R. L., & DeSarbo, W. S. (1988). Response determinants in satisfaction judgments. *Journal of Consumer Research, 14*(4), 495–507.
36. Rust, R. T., & Zahorik, A. J. (1993). Customer satisfaction, customer retention, and market share. *Journal of Retailing, 69*(2), 193–215.
37. Anderson, E. W., Fornell, C., & Lehmann, D. R. (1994). Customer satisfaction, market share, and profitability: Findings from Sweden. *The Journal of Marketing,* 53–66.
38. Reichheld, F. F. & Sasser, W. E. (1990). Zero defeofions: Quality comes to services. *Harvard Business Review, 68*, 105–111.
39. Anderson, J. C., & Gerbing, D. W. (1988). Structural equation modeling in practice: A review and recommended two-step approach. *Psychological Bulletin, 103*(3), 411.
40. Fornell, C., & Larcker, D. F. (1981). Evaluating structural equation models with unobservable variables and measurement error. *Journal of Marketing Research, 18,* 39–50.
41. Hair, J. F., Wolfinbarger, M. F., Ortinau, D. J. & Bush, R. P. (2008). *Essentials of marketing research*. New York: McGraw-Hill; Hines, T. & Bruce. M. (2007). *Fashion marketing: Contemporary Issues* (2nd. ed.). London: Butterworth-Heinemann.
42. Bagozzi, R. P., & Yi, Y. (1988). On the evaluation of structural equation models. *Journal of the Academy of Marketing Science, 16*(1), 74–94.
43. Bentler, P. M. (1988). *Theory and implementation of EQS: A structural equation program.* California: Sage.
44. Browne, M. W., & Cudeck, R. (1993). Alternative ways of assessing model fit. In K. A. Bollen & J. S. Long (Eds.), *Testing structural equation models* (pp. 136–162). Newbury Park, CA: Sage.
45. Marsh, H. W., & Grayson, D. (1995). Latent variable models of multitrait-multimethod data. In R. Hoyle (Ed.), *Structural equation modeling: Concepts, issues and applications* (pp. 177–198). Thousand Oaks, CA: Sage.
46. Schermelleh-Engel, K., Moosbrugger, H., & Müller, H. (2003). Evaluating the fit of structural equation models: Tests of significance and descriptive goodness-of-fit measures. *Methods of Psychological Research Online, 8*(2), 23–74.
47. Moon, K. K. L., Mo, P. L. L., & Chan, R. L. Y. (2014). Enterprise risk management: Insights from a textile-apparel supply chain. *International Journal of Risk and Contingency Management, 3*(2), 18–30.
48. Chan, A. T., Ngai, E. W., & Moon, K. K. (2017). The effects of strategic and manufacturing flexibilities and supply chain agility on firm performance in the fashion industry. *European Journal of Operational Research, 259*(2), 486–499.
49. Manders, J. H., Caniëls, M. C., & Paul, W. T. (2016). Exploring supply chain flexibility in a FMCG food supply chain. *Journal of Purchasing and Supply Management, 22*(3), 181–195.

Analysis and Countermeasures for Security and Privacy Issues in Cloud Computing

Abdul Raoof Wani, Q. P. Rana and Nitin Pandey

Abstract Cloud computing is having the capacity to dispose off the prerequisites for setting up high-cost computing framework and promises to provide the flexible architecture which is accessible from anywhere. The data in the cloud computing resides over an arrangement of network resources which enables position of the requirements for setting up costly data centers framework and information to be acquired to via virtual machines, and these serves might be arranged in any piece of the world. The cloud computing environment is adopted by a large number of organizations so the rapid transition toward the clouds has fueled concerns about security perspective. There are numbers of risks and challenges that have emerged due to use of cloud computing. The aim of this paper is to identify security issues in cloud computing which will be helpful to both cloud service providers and users to resolve those issues. As a result, this paper will access cloud security by recognizing security requirements and attempt to present the feasible solution that can reduce these potential threats.

Keywords Cloud computing · Cloud attacks · Security issues
Cloud security threats

A. R. Wani (✉) · N. Pandey
Amity University Noida, Noida, India
e-mail: wanirauf@gmail.com

N. Pandey
e-mail: Npandey@gmail.com

Q. P. Rana
Jamia Hamdard University, New Delhi, India
e-mail: qprana@jamiahamdard.ac.in

P. K. Kapur et al. (eds.), *System Performance and Management Analytics*,
Asset Analytics, https://doi.org/10.1007/978-981-10-7323-6_4

1 Introduction

Internet is the driving force behind various technologies but one of the discussed among all of them is cloud computing. It is still an advancing innovation technology that exchanges current innovating technology and figuring thoughts into utility like arrangements. The relocation diminishes time and cost of creation and offers better execution and unwavering quality [1]. Cloud computing is well defined as the convenient, on demand, and network access to the pool of resources like network servers, storage devices, and services that can quickly provisioned and released with nominal management effort [2]. The advantages of distributed computing incorporate diminishing the equipment and support cost, accessibility around globe, adaptability, and to a great degree mechanized process. It conveys unfathomable advantages to both individuals and ventures by decreasing the requirement for client association by concealing specialized points of interest, for example, updates, licenses, and support from its clients. Cloud can like wises provide improved safety over single-server arrangements and subsequently cloud totals resources and permits licensed security individual while as the typical organizations are restricted with system and network admin who will not be well learned about cybersecurity issues. Cloud computing can be stronger in distributive denial of service attacks in view of the availability of assets and flexibility of design.

2 Related Work

Analysts research on perceiving cloud issues, shortcomings, threats, and other security and protection matters to give countermeasures as plans, approaches, and architectures [3–5]. Various case studies [6–9] have led research on security in cloud computing and matters concerning single property, for example, information reconciliation, confirmation, shortcomings, and reviewing. Different scientists offer reviews [10–12] that cover the different zones and different security issues and resolutions. The joining of mobiles with cloud computing because of the utilization of cell phones has another security challenge identified with those that are related to ad hoc and sensor networks [13, 14]. The authors presented reviews on cloud security necessities like privacy, integrity, transparency, accessibility, and accountability.

3 Issues and Categories

This paper classifies the issues in the following categories (Tables 1 and 2).

Table 1 Cloud computing security categories

Category	Description
Standards	Criteria required to take safety efforts in cloud computing with a specific end goal to maintain safety and avoid attacks
Network issues	Incorporates issues in network, for example, connection accessibility, Denial of Service (DoS), flooding issues, web convention susceptibilities, and so on
Access control	Incorporates check and get to control and catches matters that influence confidentiality of client data and information storage
Data	Incorporates information linked to security matters including information development, quality, security, and warehousing
Cloud infrastructure	Incorporates issues that are precise to the cloud framework

Table 2 Cloud computing security issues and classifications

Category	Issues
Security standards	✓ Deficiency of security measures ✓ Compliance dangers ✓ Deficiency of looking into ✓ Lack of lawful components (service-level understanding) ✓ Trust
Network	✓ Appropriate establishment of system firewalls ✓ Security setups ✓ Internet protocol shortcomings ✓ Internet Requirements
Access control	✓ Accounts ✓ Malicious insiders ✓ Validation ✓ Private client access ✓ Browser Safety
Data	✓ Redundancy of information and data ✓ loss and data and information ✓ location of data and information ✓ Privacy of data and information ✓ Protection of information ✓ Data Availability
Cloud infrastructure	✓ Uncertain interface of API ✓ Quality of administration ✓ Allocation of technical defects ✓ Dependability of Suppliers ✓ Security misconfiguration ✓ Multi-occupancy ✓ Server Site and Backup

4 Attacks and Countermeasures

We have evaluated some of the known attacks in cloud computing and tried to find possible countermeasures to these known attacks.

4.1 Theft of Service

The theft of service attack [15] exploits the weaknesses in the scheduler of some hypervisor. This attack is recognized when scheduling mechanism is invoked by the hypervisor that fails to identify the account. The hacker guarantees that the process is certainly not scheduled. The common events of this attack are by means of cloud computing sources like human resources for a lengthy time and keeping it secreted from a dealer and using cloud computing means like storage or operating system platform for extended time without repeating it in billing cycle.

The countermeasures to this issue are given by Zhou et al. in [16] by altering the scheduling and changing the scheduling processes as well as checking policies and time intervals by means of exact scheduling, uniform scheduling, passion scheduling, and Bernoulli scheduling.

4.2 Denial of Service Attack

Out of the grave issues in the cloud security, denial of service attack is the most serious one. The attacks are at ease to perform and problematic for security professionals to deal with DDoS attacks that are more damaging than DoS attacks because there is no deterrence mechanism to avoid them.

Karnwal et al. [17] give a plan called cloud defender which deals with sensor filtering, hop count filtering, ip divergence filtering, puzzle resolving, and double signature filtering, yet the issue is that it needs an evidence and particle proof and is built on supposition.

4.3 Malware Injection

The malware injection issue accounts to a deployed replica of victims service instance uploaded by hacker; thus, the service requirements are processed within malicious instance. The hacker exports its private access to attack service security domain and acquires access to the customer data. The challenge is not only to identify malware injection but also ability to define the specific node on which hacker has loaded for malicious purpose [18].

The countermeasure is given by Oberheide et al. in [19] called Cloud Av which provides two features antivirus as a service and N-version defense. The authors prove the efficiency of Cloud Av by validating in cloud environment which provides improved detection of malicious software, improved forensic capabilities, and novel threat discovery approach.

4.4 Phishing Attacks

It is a type of social engineering attack often used to steal user data, including login credentials and credit card numbers. It occurs when an attacker, masquerading as a trusted entity, dupes a victim into opening an email, instant message, or text message. The recipient is then tricked into clicking a malicious link, which can lead to the installation of malware, the freezing of the system as part of a ransomware attack, or the revealing of sensitive information.

Cloud service alliance stated that CSP does not maintain adequate control over system in order to escape such attacks but CSA offers some precautionary measures such as registration procedure, security identity check technique, and improved monitoring skills [20].

4.5 Botnet Attacks

In this type of attack, the attackers do not reveal their identities to decrease the chance of discovery and traceback. This is accomplished by targeting victim by sequence of other hosts named stepping stone which is recruited through illegal botnets.

The countermeasures of stepping stone and botnet are by recognizing a specific host which is a stepping stone. The finding work is built on the hypothesis of relationship between licensing and outbound traffic of likely stepping stone host.

4.6 Audio Steganography Attacks

Audio steganography attacks are one of the grave attacks to cloud storage system. Audio steganography benefits customers to hide their top-secret information with normal audio records. The user communicates secret info via transferring media files which seem to be regular media records. Attackers are able to trick the present security mechanism by hiding their malicious cipher in sound records and direct it to target's server [21].

Liu et al. in [22] performed an investigation of audio steganography attacks on cloud storage system. The key is to investigate the hiding place of audio records beneath storage system by grayscale steganalysis technique.

4.7 VM Rollback Attacks

The VM part in cloud computing is most susceptible to issues. In VM rollback attack, an attacker takes benefit of prior snaps and run it without taking client into account and then erases history and again runs the similar or changed snap. The hacker launches brute force attack to give login and password for virtual machine and even if the guest operating system has restrictions on the amount of efforts such as blockade as user [22].

Szefer et al. [23] provided a design named hyperwall to cope with the hypervision susceptibilities. The hyperwall disables the suspended rescue functions of the hypervision.

5 Discussion

Out of the lots of challenges faced by cloud computing, security is still one of the biggest challenges introducing security resolutions like IDs, firewalls, contract out the personality supervision framework, and introducing antivirus, and so forth are costly and influence execution. The significant security research work lies in giving good security techniques in doing as such with minimal resources and without decline performance [23, 24].

This helps in providing the complete study of attacks in cloud, forming dependencies, and co-relating vulnerabilities across various cloud companies. It helps us to deliver protective measures as well as protection tools. This paper identifies few parts that are still not given attention in cloud computing such as checking and relocation of data from cloud to other. Security procedures must be dynamic and autonomous and should be implanted in cloud architecture for better results.

6 Conclusion

The adaption of clouds is rising day by day. With the gigantic evolution of cloud computing, the security of cloud remains still a big challenge and has not been addressed completely. In this work, we identified the security issues and tried to provide countermeasures and comparative analysis of effectiveness of the prepared solutions.

We identified the areas that are still unattended such as auditing and migration. We identified that emphasis should not only be given only on fast performance but quality of service should be considered seriously.

References

1. Tripathi, A., & Mishra, A. (2011). Cloud computing security considerations. In *Proceedings of the 2011 IEEE International Conference on Signal Processing, Communications and Computing (ICSPCC)* (pp. 1–8), Xi'an, China, 14–16 September 2011.
2. Mill, P., & Grace, T. (2011). The NIST definition of cloud computing, January 2011.
3. Wang, J.-J., & Mu, S. (2011). Security issues and countermeasures in cloud computing. In *Proceedings of the 2011 IEEE International Conference on Grey Systems and Intelligent Services (GSIS)* (pp. 843–846), Nanjing, China, 15–18 September 2011.
4. Houmansadr, A., Zonouz, S. A., & Berthier, R. (2011). A cloud-based intrusion detection and response system for mobile phones. In *Proceedings of the 2011 IEEE/IFIP 41st International Conference on Dependable Systems and Networks Workshops (DSN-W)* (pp. 31–32), Hong Kong, China, 27–30 June 2011.
5. Taifi, M., Shi, J. Y., & Khreishah, A. (2011). SpotMPI: A framework for auction-based HPC computing using amazon spot instances. In *Proceedings of the International Symposium on Advances of Distributed Computing and Networking (ADCN).*
6. Popovic, O., Jovanovic, Z., Jovanovic, N., & Popovic, R. (2011) A comparison and security analysis of the cloud computing software platforms. In *Proceedings of the 2011 10th International Conference on Telecommunication in Modern Satellite Cable and Broadcasting Services (TELSIKS)* (Vol. 2, pp. 632–634), Nis, Serbia, 5–8 October 2011.
7. Gul, I., ur Rehman, A., & Islam, M. H. (2011). Cloud computing security auditing. In *Proceedings of the 2011 the 2nd International Conference on Next Generation Information Technology (ICNIT)* (pp. 143–148), Gyeongju, Korea, 21–23 June 2011.
8. Kandukuri, B. R., Paturi, V. R., & Rakshit, A. (2009). Cloud security issues. In *IEEE International Conference on Services Computing, 2009. SCC'09* (pp. 517–520).
9. Chen, Z., & Yoon, J. (2010). IT auditing to assure a secure cloud computing. In *2010 6th World Congress on Services (SERVICES-1)* (pp. 253–259).
10. Kandukuri, B. R., Paturi, V. R., & Rakshit, A. (2009). Cloud security issues. In *Proceedings of the IEEE International Conference on Services Computing, 2009 (SCC'09)* (pp. 517–520), Bangalore, India, 21–25 September 2009.
11. Chen, Z., & Yoon, J. (2010). IT auditing to assure a secure cloud computing. In *Proceedings of the 2010* 6th *World Congress on Services (SERVICES-1)* (pp. 253–259), Miami, FL, USA, 5–10 July 2010.
12. Ryan, G. W., & Bernard, H. R. (2013). *Data management and analysis methods.* Available online: http://www.rand.org/pubs/external_publications/EP20000033.html. Accessed on 25 August 2013.
13. Khalil, I., & Bagchi, S. (2011). Stealthy attacks in wireless ad hoc networks: detection and countermeasure. *IEEE Transactions on Mobile Computing, 10*(8), 1096–1112.
14. Panta, R. K., Bagchi, S., & Khalil, I. (2009). Efficient wireless reprogramming through reduced bandwidth usage and opportunistic sleeping. *Ad Hoc Networks (an Elsevier Journal), 7*(1), 42–62.
15. Almorsy, M., Grundy, J., & Müller. I. (2016). An analysis of the cloud computing security problem. arXiv preprint arXiv:1609.01107.

16. Fangfei, Z., Goel, M., Desnoyers, P., & Sundaram, R. (2011). Scheduler vulnerabilities and coordinated attacks in cloud computing. In *Proceedings of the 2011 10th IEEE International Symposium on Network Computing and Applications (NCA)* (pp. 123–130), Cambridge, MA, USA, 25–27 August 2011.
17. Karnwal, T., Sivakumar, T., & Aghila, G. (2012). A comber approach to protect cloud computing against XML DDoS and HTTP DDoS attack. In *Proceedings of the 2012 IEEE Students' Conference on Electrical, Electronics and Computer Science (SCEECS)* (pp. 1–5), Bhopal, India, 1–2 March 2012.
18. Gruschka, N., Jensen, M. (2010). Attack surfaces: A taxonomy for attacks on cloud services. In *Proceedings of the 2010 IEEE 3rd International Conference on Cloud Computing (CLOUD)* (pp. 276–279), Miami, FL, USA, 5–10 July 2010.
19. Oberheide, J., Cooke, E., Jahanian, F., & Cloud, A. V. (2008). N-version antivirus in the network cloud. In *Proceedings of the 17th Conference on Security Symposium (SS '08)* (pp. 91–106) USENIX Association:Berkeley, CA, USA, 2008.
20. Top Threats to Cloud Computing V1.0; Cloud Security Alliance: March 2010.
21. Tupakula, U., Varadharajan, V., & Akku, N. (2011). Intrusion detection techniques for infrastructure as a service cloud. In *Proceedings of the 2011 IEEE Ninth International Conference on Dependable, Autonomic and Secure Computing (DASC)* (pp. 744–751), Sydney, Australia, 12–14 December 2011.
22. Liu, B., Xu, E., Wang, J., Wei, Z., Xu, L., & Zhao, B. et al. (2011). Thwarting audio steganography attacks in cloud storage systems. In *Proceedings of the 2011 International Conference on Cloud and Service Computing (CSC)* (pp. 259–265), Hong Kong, China, 12–14 December 2011.
23. Szefer, J., & Lee, R. B. (2012). Architectural support for hypervisor-secure virtualization. *SIGARCH Computer Architecture News, 40,* 437–450.
24. Motawie, R., El-Khouly, M. M., & El-Seoud, S. A. (2016). Security problems in cloud computing. *International Journal of Recent Contributions from Engineering, Science & IT (iJES),* 4(4), 36–40.

An Efficient Approach for Web Usage Mining Using ANN Technique

Ruchi Agarwal and Supriya Saxena

Abstract Web mining involves a huge variety of applications whose objective is to find and extract concealed information in web user data. It has provided an efficient and prompt mechanism for data access. Web mining enables us to extract out beneficial information from user's web access. Earlier studies on the subject are based on a concurrent clustering approach. In this approach, the clustering of the requests affected the performance results. In this paper, we have introduced the Enhanced Multilayer Perceptron (MLP) algorithm, a special technique of ANN (Artificial Neural Network) to detect patterns of use. The enhanced MLP technique is better than *K*-mean algorithm for web log data in terms of time efficiency. The aim of understanding the enhanced MLP technique is to improve the quality of e-commerce platforms, to customize the websites and improve the web structure.

Keywords Enhanced MLP · FCM · K-mean · Neural network
Web usage mining

1 Introduction

In today's world, the biggest influence on an individual's life is the Internet, because of its communication (audio and video), instant messaging, ordering food, purchasing clothes, reading the news, etc. All facilities and information are available to us at the click of a button on our mobile phones. E-commerce today has become a major distribution channel for goods and services. Access to product comparisons and rankings, user reviews and comments, and recommendations from bloggers with large followings have shaped a new scenario for consumer behavior, retail trade, and the economy in general [1]. Therefore, all major retail giants are

R. Agarwal · S. Saxena (✉)
Sharda University, Greater Noida, Uttar Pradesh, India
e-mail: supriya08.saxena@gmail.com

R. Agarwal
e-mail: ruchi.agarwal@sharda.ac.in

P. K. Kapur et al. (eds.), *System Performance and Management Analytics*,
Asset Analytics, https://doi.org/10.1007/978-981-10-7323-6_5

moving toward E-commerce and trying hard to provide customer delight by offering them easy access through relevant recommendations during their browsing sessions. Many websites utilize web usage mining for identification of user's habits. The web usage mining applies algorithms to identify patterns of web page usage. The web usage mining uses the log file data of web server for primary source. With its help, we can obtain four things namely, the user's habits, site customization data, make a tune-up server, and create business rules. Using it can lead a website/ company to explore new clients and better marketing campaigns. The user's web usage style detection has three steps: (i) Preprocessing, (ii) pattern discovery, and (iii) pattern analysis. It has become trivial for the webmaster to evaluate whether the products and services provided are catering to the need of the customers [2]. One effective solution to handle this issue is to provide personalized recommendation to individual user where he or she is interested in a product. A promising solution to overcome this issue is recommendation system. Recommendation system can be broadly classified into two: Content-based and collaborative filtering system [3, 4].

2 Web Usage Mining

Web usage mining aims at discovering a pattern of user's activity, so that it can provide user information in a better way. The ultimate objective of website is to provide the customer with more relevant information. Today's E-Commerce industry faces a cut-throat competition for enhancing user experience by providing them with better features and relevant results. Moreover, something extra must be available like dynamic content, links, etc. for suggesting the user to intrigue him/ her. Clustering of the user's log data is one of the main areas of concern for the web community. Log data is stored at three locations, namely, the server, the user end, and the proxy servers. As there are three places for data storage, analysis of browsing patterns of users for mining process has become tougher. Results are reliable only if data is available from the entire above said log file. Also, log data on the proxy servers provide other useful information. However, it is very difficult to collect information from the client side, e.g., Page requests, etc. Thus, many of the algorithm works depend on server data. Web mining is made up of three major steps [5, 6] as shown in Fig. 1:

(i) Preprocessing [7],
(ii) Pattern discovery, and
(iii) Pattern analysis.

During preprocessing, data is collected from the three stored locations: (i) client side, (ii) server side, and (iii) proxy servers. After user identification, user sessions are segregated using click streams, assuming the session to be 30 min each (approx.). The pattern discovery is also done by either of the three, clustering, association rule mining, or frequent pattern mining. Here, we are only clustering the web access log. Generally, in web usage mining, usage clusters and page clusters

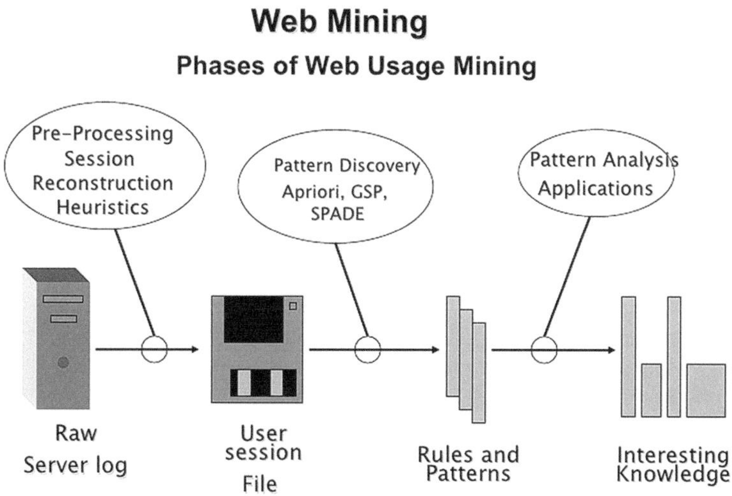

Fig. 1 Data processing in web usage mining

are used. Users having similar browsing patterns are grouped together into clusters. The users can be clustered on basis of number of parameters. On the one hand, we can request the user to fill a form stating their preferences such as registration on a portal. The user clustering can be obtained on the basis of the forms filled by them. On the other hand, during the customer's navigation, his log data can be collected to create the cluster. Various user information such as user's behavior pattern, his likes/dislikes, and his characteristics of the user, i.e., their personal information, are collected using these methods. Clustering web pages help in creating groups with content similar to each other. The last step is the analysis of the pattern found [8].

3 Related Work on Web Usage Mining

In data mining, huge data is analyzed to obtain hidden and unknown outcomes which are then used for better and more accurate decisions [9]. Real-time recommendation is provided to online users on the website irrespective of been registered or not. Rational recommendation technique is proposed that makes use of lexical patterns to generate item recommendation [10].

In a novel recommendation system for consumer products, here the system returns a ranked list of suitable brands, models available for a particular item as an output, and it performs rank aggregation to obtain a consensus ranking of products. The products are then returned in the order of the aggregated ranking. In this paper, it also presents a novel rank aggregation method for aggregation of partial lists [11].

The users are recommended a list of pages on the basis of their web history pattern and a list of unvisited pages [12].

A novel unified framework for data mining conceptualized through the composite functions. Real-life data is used in data mining algorithms [13] by increasing volume of data, and the variety of their formats have led to advance application for data analysis in order to transform it into relevant information. They study the perception of the specialists who experts that normally work in companies with these applications [14].

Current user behavior through his/her clickstream data is converted to a recommended system which is used to provide relevant information to the individual without explicitly asking for it [15]. Customer's behavior using the Web mining techniques and its application in e-commerce to explore customer behavior. The concept of Web mining describing the process of Web data mining in detail: source data collection, data preprocessing, pattern discovery, pattern analysis and cluster analysis [16].

4 Introduction to ANN

Artificial neural network (ANN) is a network which is adapted by the activity of nervous system, quite similar to the brain information processing function. The main characteristic of it is its architecture of the data processing system which is made up of many neurons working for solving the particular problem. However, the artificial neural network is also used in pattern recognition or data classification domain by means of learning process. Neural network has a capability of extracting meaning full information from the imprecise or complicated data which were further used in extracting the pattern and detecting the trends which were too complex and difficult for the computer or human to detect. The other advantages of neural network are as follows:

1. A neural network has a capability of adaptive learning which means the ability of the network to perform task on the basis of data giving for the training purpose.
2. An artificial neural network from the information it receives during learning process helps in creating its own organization and representation known as self-organizing map.
3. A neural network performs real-time operation which was carried out in parallel and also using this capability some special hardware devices can be designed and manufactured.
4. Fault tolerance via redundant information coding.

5 Web Usage Mining: Problem Faced

A. Logs processing—It is the process of cleaning server log files and user's session data [6].
B. Log files cleaning—Unnecessary files such as images and error logs need to be cleaned before processing data.

C. Users identifications—The data is then sorted user-wise, i.e., on the basis of cookies, user's IP address, and various other forms.
D. User session's identification—Once the user identification is done, the data is further divided into sessions, generally it is done by checking the time frame gap between two requests (clicks).
E. User's habit identification—It is a dynamic database which keeps updating as per the changing habit/behavior of the user.
F. Pattern discovery—All users having similar browsing patterns are grouped together.
G. Knowledge post-processing—The end result of the output can then be used by human/AI interface to create personalized suggestion/response database.

6 Neural Network Approach for Web Usage Mining

Here, we shall discuss a feedforward neural network. MLP is an extensively used supervised learning algorithm in ANN. It uses error-correction learning rule. A forward pass and a backward pass are used to calculate the error propagation. (i) Forward pass uses the input vector applied to the nodes of the network and it affects the network layer by layer. In the end, the output is determined by the response. When applying the forward pass, the synaptic weight of the networks is all fixed. When applying the back pass, the synaptic weights are all adjusted as per the error-correction rule. The error response is formed by subtracting the actual response from the desired response. This error response is then sent backward, i.e., in the opposite direction of synaptic conditions. Recalibration of the synaptic weights is done to make the actual output of the network similar to the desired output. The feedforward neural network architecture is able to approximate most problems with good accuracy and generalization ability.

As shown in Fig. 2, the input signal is fed to the first layer, the input layer. The output layer provides the output after the signal is processed via the hidden layer. The actual output is then compared with the expected output, and error value is

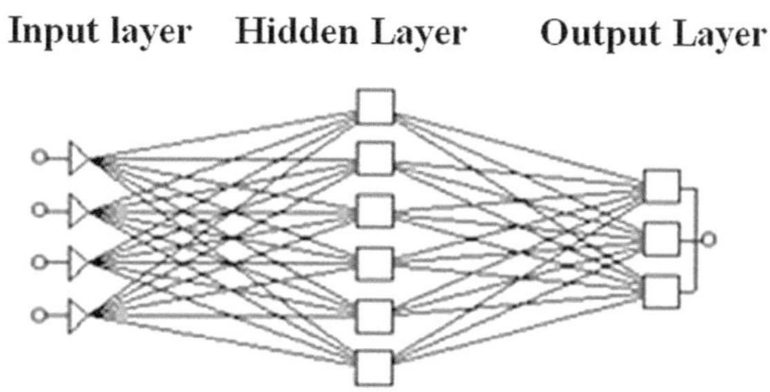

Fig. 2 Multilayer-perceptron architecture

calculated which in turn is used to adjust the weights. Then, from the output layer again, signals are transmitted backward to each unit in the intermediate layer which is connected directly from the output layer, and now every unit in the intermediate layer receives only error signal. Based roughly on the relative contribution, the unit made to the original output. This process repeats layer by layer, until each unit in the network has received an error signal that describes its relative contribution to the total error.

6.1 Multilayer Perceptron Algorithm

The multilayer perceptron algorithm can be used to create code for various languages. Here, we are assuming the use of the sigmoid function $F_{(net)}$.
Algorithm: [17, 18]

1. **Wt. value initiation**
   All wt. shall be set to random.
2. **Input feeding & Outputs**
   Present input $\boldsymbol{A_p} = \mathbf{a_0, a_1, a_2, \ldots, a_{x-1}}$ and target output $\boldsymbol{B_p} = \mathbf{b_0, b_1, \ldots, b_{y-1}}$ where x and y are the input/output nodes. Now we shall assume $\mathbf{w_0}$ to be $-\theta$, and $\mathbf{a_0} = 1$. The associated pattern is represented by A_p and B_p. We shall set $T_p = 0$, the only exception being A_p which is set to 1.
3. **Output Calculations**
   The following calculations are done for each layer:

$$\mathbf{T_{pj}} = \mathbf{f}[\mathbf{w_0 a_0} + \mathbf{w_1 a_1} + \ldots\ldots + \mathbf{w_x a_x}]$$

   The output $\mathbf{T_{pj}}$ is then transferred to the next layer for input. The final output is denoted by $\mathbf{o_{pj}}$.
4. **Adjusting the wt.**
   Once the output is achieved, we work back on the wt.

$$\mathbf{w_{ij}(b+1) = w_{ij}(b) + \tilde{n} Þ_{pj} o_{pj}},$$

   here, $\mathbf{\tilde{n}}$ = gain term, $\mathbf{Þ_{pj}}$ = error term **p represents the pattern and j = nodes.**

Output

$$\mathbf{Þ_{pj} = k o_{pj}(1 - o_{pj})(b - o_{pj})}$$

Hidden units are calculated as below:

$$\mathbf{Þ_{pj} = k o_{pj}(1 - o_{pj})[(Þ_{p0} w_{j0} + Þ_{p1} w_{j1} + \ldots + Þ_{pk} w_{jk})]}$$

6.2 Enhanced MLP Algorithm

In this section, the enhanced MLP shall be discussed which shall enhance the overall performance. Error adjustment shall lead to better learning. This shall be achieved due to the back transmission of the output signal to the intermediate signal. In other words, output error can be written as below:

$$\delta^0_{pk} = (Y_{pk} - O_{pk}).$$

Here, "p" is the pth training vector, "k" is the kth output unit, Y_{pk} is the output (ideal), and O_{pk} is the output (actual) at the k unit. Feedback δ_{pk} shall update the output signal wt. and the hidden signal weights.

Enhanced MLP Algorithm

1. Input the values.
2. Add the input values to the hidden layer.
3. Evaluate the output data from the hidden layer.
4. Find the errors from the output and replace updated $\delta^{\circ}{}_{pk}$ with old $\delta^{\circ}{}_{pk}$
5. Again calculate the output and the related error values.
6. In parallel, the wt. of the hidden and output layers need to be updated for convergence.
7. Repeat Steps 1-9 till error achieves an acceptable value.

7 Architecture of Proposed Recommender System

As shown in Fig. 3, we can see the split of architecture in two phases: (a) Back end phase and (b) front end phase. In the back end phase, the recommender system identifies the users browsing pattern using the five steps of data preprocessing, i.e., Data cleaning, user identification, session identification, and path completion identification. Using EMLP, the useful data is converted into user browsing pattern. This pattern is then added to the knowledge base. This knowledge base is then used to provide useful recommendations to the user during his browsing sessions. The front end can be briefly described as following. The user's active session is captured and compared with the already existing aggregate user profile from the knowledge base. The recommender system then searches for unvisited pages by comparing with other user's pattern and search algorithms. The best possible profile/results are achieved considering maximum similarity. Recommendation list as compared to other user's patterns is captured and added to original recommendation list.

Step 1: Collection of the data

- we will gather the data not from the Internet but from the live user's desktop. We will also gather data from the Internet from various sites which provide data.

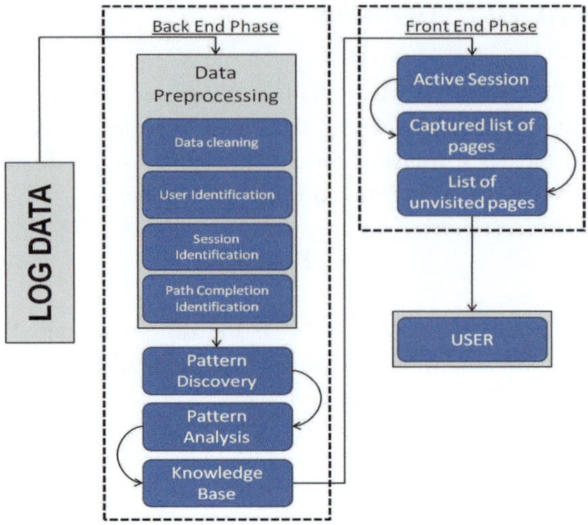

Fig. 3 Architecture of the proposed recommender systems

Step 2: Data analysis, i.e., Data preprocessing and pattern recognition

- In this phase, we clean the data cluster of the user session files having similar navigation patterns.

8 Conclusion

We have studied the possible use of ANN in web traffic data mining classification. The discovery of data patterns allows organizations to create customer-customized advertizing for a better outcome. To enhance user's web experience, the mining results must have these characteristics: the user's behavior model must accurately represent the user, the mining result must be quick, and recommendation should not include pages already visited. ANN is one of the most efficient soft computing approaches to making probabilistic models. Therefore, we have proposed an algorithm using ANN for web usage mining to overcome shortcoming in the traditional methods. The backpropagation method will readjust the weights and therefore reduce the error. After sufficient reiteration (till desired error threshold is achieved), the model can be accepted. Furthermore, a knowledge bank is proposed for better recommendation on the basis of behavior models of other users having similar habit patterns. The same can be validated by coding the abovementioned algorithm in any advanced programming language, e.g., Matlab, Java, .Net, etc.

References

1. Prajyoti, L., & Bidisha, R. (2015). Dynamic recommendation system using web usage mining for e-commerce users. *International Conference on Advanced Computing Technologies and Applications 2015.*
2. Huang, Z., Zeng, D., & Chen, H. (2007). A comparative study of recommendation algorithms in e-commerce applications. *IEEE Intelligent Systems, 22*(5), 68–78.
3. Lee, J., Sun, M., & Lebanon, G. (2012). PREA: Personalized recommendation algorithms toolkit. *The journal of Machine Learning Research, 13*(1), 2699–2703.
4. Kusmakar, A., & Mishra, S. (2003). Web usage mining: a survey on pattern extraction from web log. *International Journal of Advanced Research in Computer Science and Software Engineering, 3,* 834–838.
5. Mulvenna, M. D., & Buchner, A. G. (1997). Data mining and electronic commerce. In *Proceedings of of Overcoming Barriers to Electronic Commerce*, pp.1–7.
6. Keim, D. A. (2002). Information visualization and visual data mining. *IEEE Transactions on Visualization and Computer Graphics, 8*(1), 1–8.
7. Rajesh, S., Sanjay, S., & Chande, P. K. (2013). Web personalization systems and web usage mining: a review. *International Journal of Computer Applications, 72*(21), 0975–8887.
8. Kemi, D.A., & Ankerest, M. (2001). Visual data mining and Exploration of large database. *tutorial at ECML/PKDD* 01.
9. Mahendra, P. Y., Feeroz, M., & Yadav, V. K. (2013). Mining the customer behaviour using web usage mining In e-commerce. In *International Conference on Computing, Communication and Networking Technologies* (ICCCNT), pp. 462–466.
10. Song, Q., & Sheppered, M. (2006). Mining web browsing patterns for E-commerce. *Computers in Industry, 57*(7), 622–630.
11. Bhushan, R., & Rajender, N. (2012). Automatic recommendation of web pages for online users using web usage mining. *International Conference on Computing Sciences,* pp. 371–374.
12. Dost, M. K., Nawaz, M., Babajee D. R. K. (2013). A Unified Theoretical Framework for Data Mining. In *Information Technology and Quantitative Management* (ITQM2013), pp. 104–113.
13. Berta, D.-A. (2013). Study regarding the perception of the concept of business intelligence. In *Among Application Designers 2nd World Conference on Business, Economics and Management-WCBEM*, pp. 402–406.
14. Adeniyi, D. A., Wei, Z., & Yongquan, Y. (2014). Automated web usage data mining and recommendation system using K-Nearest Neighbor (KNN) classification method. *Applied Computing and Informatics.*
15. Pankaj, S., Deepshikha, P., & Amit, S. (2013). A neural network approach to improve the efficiency of image annotation. *International Journal of Engineering Research & Technology* (IJERT), *2*(1).
16. Rashid, A. (2013). Pro-Mining: Product Recommendation Using Web-Based Opinion Mining. *International Journal of Computer Engineering and Technology (IJCET), 4*(6), 299–313.
17. Jagadeesh, S., & Babu, P. K. (2012). Edge detection system using pulse mode neural network for image enhancement. *I.J. Image, Graphics and Signal Processing,* 42–48.
18. Richa, S., Mahendra, M., & Manish, S. (2014). To improve efficiency of intrusion detection system by using improved multilayer perceptron algorithm. *IJARCSSE* , 4(2).

Natural Language Processing Approach to Identify Analogous Data in Offline Data Repository

Nidhi Chandra, Sunil Kumar Khatri and Subhranil Som

Abstract There have been huge contributions to online communities and social media websites through posts, comments and blogs day in and day out. Some of this contribution is unstructured and unclassified. It is difficult to find similarities in terms of textual data in the posts as it comprises of mix of structured and unstructured data. The overall objective of this paper is to help identify similar text through natural language processing techniques. The approach has been demonstrated through linguistic features that points to similarity and use those features for the automatic identification of analogous data in offline data repository. To demonstrate the approach, we have used a collection of documents as an offline repository having similar text and a text corpus as a resource to identify analogous data. The proposed methodology processes a document against repository based on document preprocessing through lexical analysis, stop word elimination and synonym substitution check. Java data structure is used to hold and process data parsed from the file syntactic analysis is carried out with the help of WordNet™ database configured within the process. Part of speech (POS) and synonym check capabilities of WordNet API are being used in the process.

Keywords Document term matrix confusion matrix · Stemming
Scoring function · Linguistic feature analysis

N. Chandra (✉)
Amity School of Engineering and Technology, Amity University,
Noida, Uttar Pradesh, India
e-mail: nsrivastava5@amity.edu

S. K. Khatri · S. Som
Amity Institute of Information Technology, Amity University,
Noida, Uttar Pradesh, India
e-mail: skkhatri@amity.edu

S. Som
e-mail: ssom@amity.edu

P. K. Kapur et al. (eds.), *System Performance and Management Analytics*,
Asset Analytics, https://doi.org/10.1007/978-981-10-7323-6_6

1 Introduction

In the previous few decades, there has been huge demand from various businesses and organizations to make them access important relevant information more flexibly as mining such information from multiple disported sources has been a major area of exploration and concern. Once the solution approach to this problem has been text extraction, wherein data can be categorized based on similarity properties.

Natural language processing is an area that studies the interaction between human language and computer. This field unlocks the key to understand how human brain works. This field belongs to category that is intersection of computer science, artificial intelligence and computational linguistics. Natural language processing is a research domain that focuses on the machine level interpretation and translation of human understandable language. Natural language processing is a computational technique for computers to analyse, understand and derive meaning from human language in a meaningful way. By incorporating natural language processing techniques, programmers and domain expert can derive large number of application organizing and structuring knowledge to perform tasks such as free text classification by summarization, spam detection, Part of speech tagging, co-reference resolution, machine translation, information retrieval, text to speech translation, relationship extraction, sentiment analysis and topic segmentation. Major research area in natural language processing is paraphrasing detection and evaluations of subjective questions. By applying text mining techniques, text blocks can be summarized to extract the domain of the document by discarding or ignoring irrelevant information. Using natural language processing techniques, intelligent chatbot can be created using deep learning techniques.

Analogous text identification has many applications, for example, in content management system and the need is to categorize and put the proper documents in the proper categories in a proper and correct manner. It is widely used in many contexts, ranging from document categorization, web filtering, spam filtering, spam detection, fraud detection, and in any application that requires document organization and management.

Cavnar and Trenkle [1] stated text categorization and classification is also known as supervised text categorization as the datasets for classification and categorization used are predefined based on the domain-specific keywords. The automated programme then processes and compares the documents and accordingly assigns the documents into the predefined set of categories.

2 Related Work

'Automated building of domain ontologies from lecture notes in courseware' [2] highlights the automated domain-based ontology is being worked from course address notes. **The solution** developed in this paper helps provide pre-requisite and post-requisite terminology for specific term from specific area-based ontology when the user enters the term from courseware into the application. One of the limitations that have been identified in this approach is that it does not give dependency graph for pre-requisite and post-requisite of the submitted terms which could have helped imagine the aggregate intra-space relationship among the terms. One could have gained domain-based knowledge just by visualizing the ontology in terms of the dependency graph.

In this paper, [3] Naïve Bayes classifier, support vector machine, neural network, and KNN classifier are used for content portrayal and characteristic selection. These methods regard content order as a standard arrangement issue and along these lines that ease the learning procedure to two basic strides: (i) Feature engineering and (ii) classification learning over the component space, n-gram modelling is done in light of Markov's Model.

'R-tfidf, a Variety of tf-idf Term Weighting Strategy in Document Categorization' [4] lemmas define the basis of text representation method. Lemma demonstrates all word frames that lie under a similar gathering (networking, networks, network). In the Naïve Bayes technique, two speculations are chosen and the chosen classification corresponds to the maximum one. At that point, we calculate the likeliness. The features derived from the vector space model are used by the support vector machine (SVM) strategy. Thereafter, TF-IDF formula is used to find the weight of each term. Positive Z score which is above the threshold limit gives a result that the term has high density in the text, whereas negative Z score shows that the term has rarely been used.

The primary downside of this paper is that all the archives utilized were on two classifications divided by year. So, the proposed strategy was not randomly tried on any theme and the time period was also a helping factor.

3 Problem Analysis

Analogous data extraction is a difficult and time-consuming task as similarity of text needs to be accessed. Identifying relevant terms and comparing terms to access the similar behaviour can lead to large response time. Individual human has their own writing styles and speech can be active or passive add complexity to the process. Approach in this paper to identify analogous data is being derived by using WordNet database to identify permissible synonyms and their root words. We will be using domain-based ontology to provide the domain-specific keywords so that similar property data can be accessed. This paper is an approach toward the meaning of the text that needs to be classified as analogous data.

Domain-based ontology plays an important role in analogous data extraction here [5]. Ontology is a semantic structured approach that can convert unstructured data into a structured and meaningful format. Ontology helps in text classification in various domains such as textile, medical, agriculture, bioinformatics, and various other areas. Building ontology is the most difficult area as it requires in-depth knowledge of the respective domain. Ontology creation is also another area of research to automate respective domains documents. Ontology represents relationship among the domain-specific terms and their hierarchy.

To validate the approach, we are using an offline repository of text documents. The text can be stored in various forms, such as TXT, DOC, DOCX, XML, HTML, etc., before applying the text mining, and preprocessing steps need to be applied in the form of lexical text analysis, stop word removal, identify term frequency, and weighted term analysis. Synonym substitution check and ontology-based comparison lead this mechanism toward a text meaning-based approach and reduce the time of processing to identify analogous text.

4 Methodology

This research paper presents an improved methodology over term frequency as it is taking advantage of the semantic space and meaning of the term space against the WordNet API. Methodology is being proposed such that analogous data can be identified from the documents as well as those terms which has got the similar meaning can also be identified. By comparing it against the ontology, it can identify the relation among the named entity and provides the categorization (Fig. 1).

The steps to achieve the proposed methodology are shown in the figure below.

Fig. 1 Proposed methodology

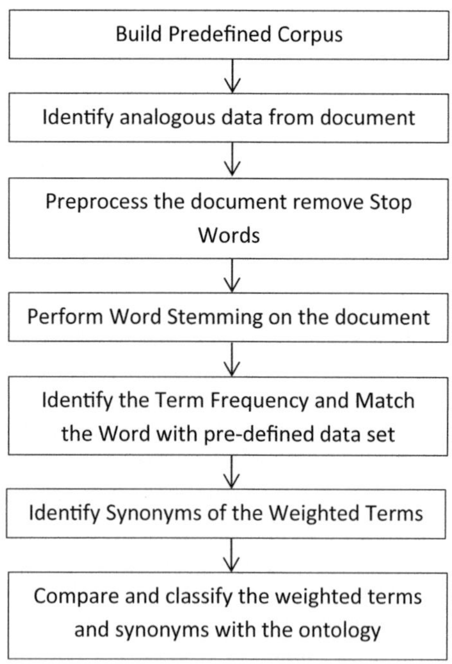

Document preprocessing can be categorized as four text operations (or transformations): lexical analysis, stop word removal, stemming of word, and selection of index term. Document preprocessing starts with the lexical analysis. Lexical analysis: Lexions extraction is where each and every word is being separated from the text document by Java String Tokenizer. Stop word elimination: Extracted tokens have been compared against the stop word database. Stop words are the words which have very less meaning in information retrieval, for example, conjunctions, prepositions and common words. Stop words have a large frequency of occurrence but very less significance and removal of those words from the document does not affect the information retrieval process. During search, string matching process discards these stop words only relevant keywords contribute in the process. In English Language, more than 400 stop words exist and if they are not being removed in the document preprocessing step, large amount of time spent on their comparison. Word stemming: It is the process where each word will be folded back to the root word by comparing it against the WordNet database. Weighted term left after the stop word removal process. Stemming process roots these weighted terms to their root keyword. For example, in fishing, fisher will root back to the word fish. Stemming of the words reduces the compilation and comparison time. Selection of index term is based on the term frequency calculation within the document and among the document space to find the highly occurred keywords. The term frequency (tf) is evaluated as inverse of document frequency (tf-idf).

Term frequency is defined as the frequency or the count of a word in a text or a corpus. This is the major step in the preprocessing of a natural language text. Inverse document frequency is a numerical value that determines the significance of a word in a text or a corpus. This is many of times used as weighting factor in text mining.

The tf-idf value grows relative to the quantity of times a word shows up in the archive and, however, is regularly balanced by the recurrence of the word in the corpus [6].

Term frequency was invented as a heuristic technique. In simpler word, various frequencies can be depicted as

Term Frequency = Bag of words,
Inverse Document Frequency = Weighing by how often word occurs in corpus,
Term Frequency absolute = Number of occurrence of terms,
Term Frequency Relative = Number of occurrence of term/Number of terms, and
Inverse Document Frequency = Log(1 + Number of Docs/Number of docs with term [6].

Weighted term frequency is to obtain an accurate output, and the next step is to determine the final score using a scoring mechanism. Weighted term frequency is evaluated as a count of the terms that appeared in a corpus.

A score is number of query terms over the matches of scores on each inquiry term. Based on the weight of query term '*t*' in a document '*d*', the score can be calculated. The most common approach, which is also known as term frequency, is to assign the weight of '*t*' as that of the number of appearances in the document. It is denoted by 'tft, d' [6].

Term frequency–inverse document frequency is derived from the basics of language modelling. Language modelling theory states that number of terms in a document gets split into property of eliteness, referring to term on topic in a document.

Term frequency–inverse document frequency has many disadvantages as it is on certain occasions considered as an ad hoc approach, because of issues related to derivation of mathematical model to perform relevance analysis and term distribution in context of dimensionality of input text data and vocabulary size across total text dataset [6].

Gather domain-based knowledge: Domain-based knowledge or knowledge base setup requires construction of ontology using domain-specific keywords. To construct ontology, one needs to have a domain expertise to help identify specific terms and their dependencies along with constraints. This is one of the important steps in the approach [7, 8].

Ontology is a semantic aspect of term space and is very difficult task to achieve. To build ontology of a term space, detailed knowledge and domain-specific term space with relations are required to formulate. Ontology requires classification and association details of the term space, how terms are derived and are related to each other. Dependency of one term on another term as well as the constraints need to be highlighted while drafting the ontology.

Creating domain-based ontology: Once the domain information is gathered, the ontology is constructed in a way so that each term will belong to a superclass, subclass and relationship among the terms. Once any term is being compared, it returns the complete classification of the term space.

Synonym substitution check: Synonym substitution and paraphrasing is the biggest problem in natural text processing. Synonyms are the similar words. Identification of synonyms in this approach is the strongest point to extract analogous data as data which can be substituted in a similar way will be identified from this approach. WordNet database is being used to identify the synonyms and their possible substitution.

5 Result Discussion

Lexical analysis involves the identification of Lexions from the input text document. Lexions will be extracted by identifying the space as the word separator, digits, hyphens, punctuation marks in case of letters (Fig. 2).

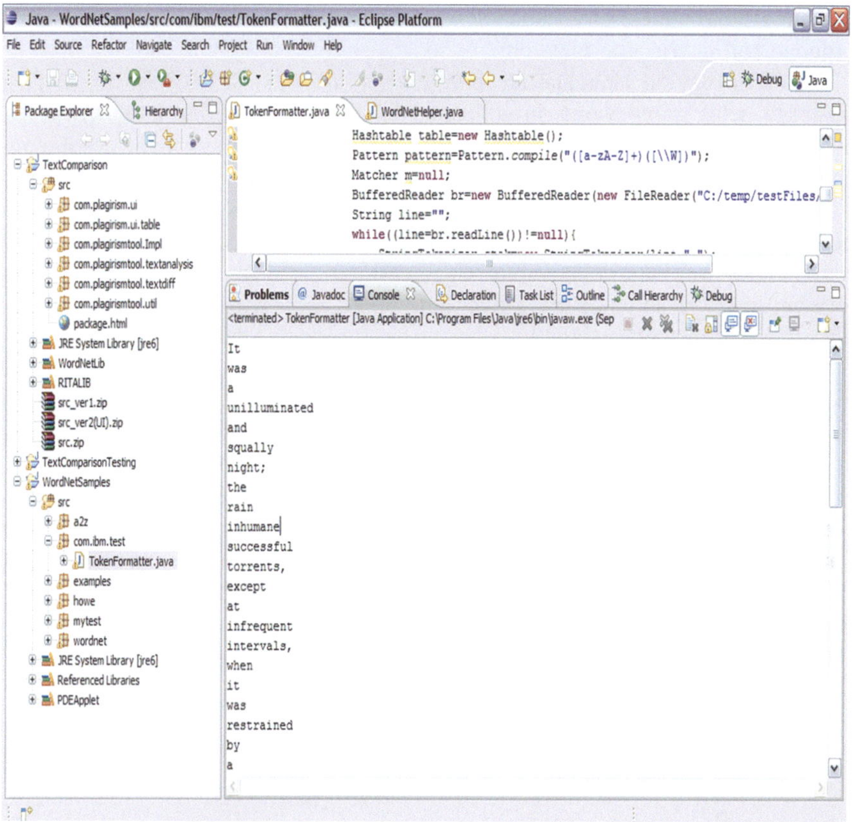

Fig. 2 Lexical analysis result

Synonym Substitution Check

The synonyms substitution could be prevented by WordNet API through replacing dictionary synonyms in suspected text. This method could be leveraged for advanced plagiarism detection.

WordNet

Thesauri are a very important in information extraction and WordNet is the biggest online thesauri. WordNet is a lexical knowledge based on conceptual lookup. This lexical knowledge base is widely used in identifying word sense ambiguities. It is a knowledge base where word is stored as a concept. The fundamental unit of word which is searched is concept not word and concept is meaning of the word. Lexical database structure is based on the psycholinguistic theory. In this information, base word is stored with the facts and arranged as a network of words linked by lexical and semantic relations. Words have been classified as content word and function word. Content word has been further divided into verb, noun, adjective and adverb and arranged as a hierarchy. Function word is divided into preposition, conjunction, pronoun and interjection hierarchy.

Words in natural language are polysemous. The lexical database for English language helps group words in collection of synonyms referred to as synsets. It helps provide semantic relationships between these words or synonym sets through short and general definitions [9, 10].

WordNet follows multiple different grammatical rules to help differentiate between nouns, verbs, adjectives and adverbs. All synset comprises collection of synonymous words or collocations where collocation refers to set of words going together to create some meaning like ‘auto pool’; each synset comprises of words with different meanings (Fig. 3).

The synsets are classified through short glosses (definitions or examples). For example—good, right, and ripe—(most suitable or right for a particular purpose; ‘a good time to plant tomatoes’; ‘the right time to act’; and ‘the time is ripe for great sociological changes’) [change example].

To form ontology, in-depth knowledge is required. Domain-based ontology could be constructed through knowledge acquired through domain expertise where ontology represents term dependencies, terms subclass, superclass, restrictions and constraints. It also represents domain and ranges of each class. The below figure depicts the term classification their associations and relationship among each other (Figs. 4, 5, 6, and 7).

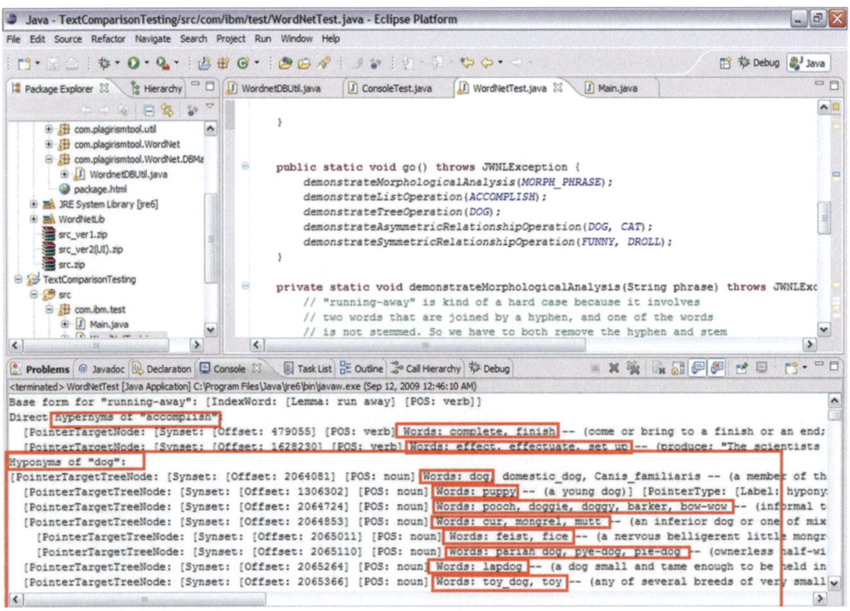

Fig. 3 Synonym substitution check

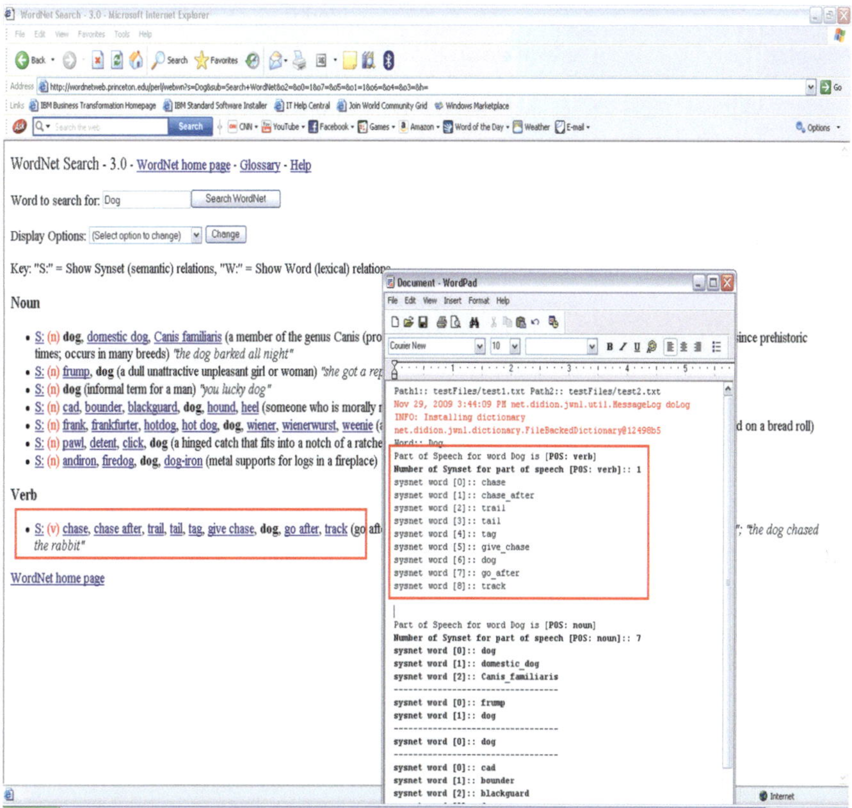

Fig. 4 Synonym check comparison with WordNet

6 Comparison Results and Future Work

The efficiency of proposed method could be evaluated through processing of multiple text documents referenced from certain source repository. The conclusions of the matching process and synonyms identified in the process are being projected in the previous figures. Comparison results show the total time spent in processing the document and providing the analogous synonym substitution check is being implemented by WordNet API and using the WordNet lexical database and the result is being provided that identifies the part of speech and the corresponding synonym for the word.

The future work will include weighed term evaluation and text classification and categorisation to preprocess the text. Similarity identification algorithm will be based on approximation similarity detection that will evaluate the approximate similarity in the repository documents with respect to the submitted document.

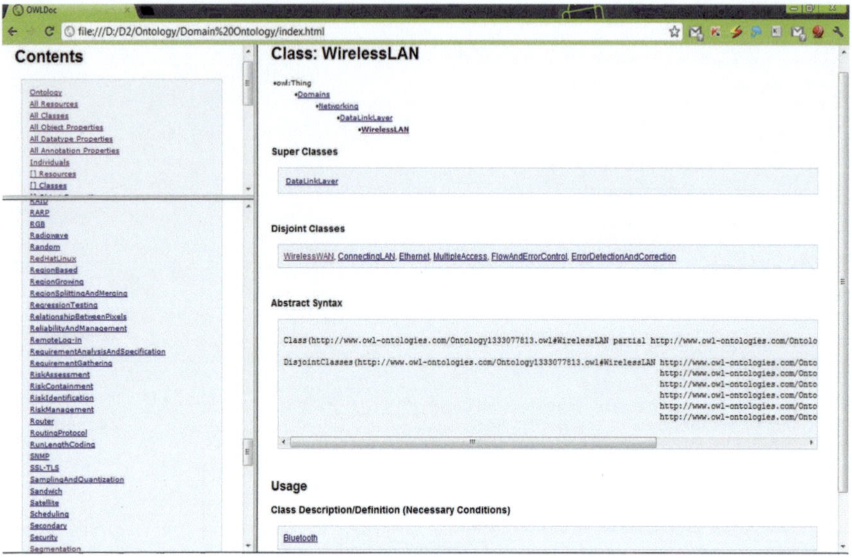

Fig. 5 Class hierarchy

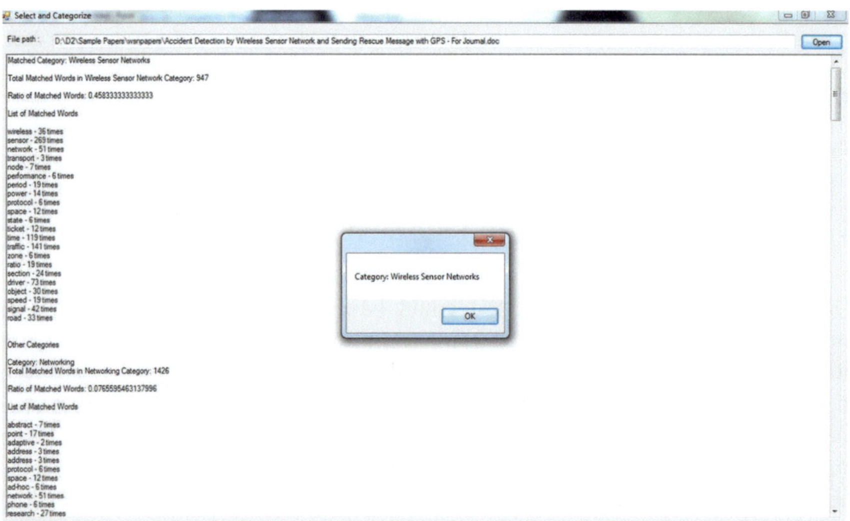

Fig. 6 Ontology showing class dependency [11]

Fig. 7 Response time

Serial Number	Size of Document (KB)	Execution Time (sec)
1	200	3.00
2	324	3.96
3	517	3.98
4	600	3.98
5	745	4.00
6	841	3.84
7	941	4.00
8	1028	5.74
9	1151	5.85
10	1243	6.60

References

1. Cavnar, W. B., & Trenkle, J. M. (1994). N-gram-based text categorization. In *Proceedings of SDAIR-94, 3rd Annual Symposium on Document Analysis and Information Retrieval.*
2. Gantayat, N., & Iyer, S. (2011). Automated building of domain ontologies from lecture notes in courseware. In *Proceedings of 2011 IEEE International Conference on Technology for Education.*
3. Farhoodi, M., Yari, A., & Sayah, A. (2011). N-gram based text classification for persian newspaper corpus. In *Proceedings of 7th International Conference on Multimedia Technology and its Applications (IDCTA).*
4. Zhu, D., & Xiao J. (2011). R-tfidf, a variety of tf-idf term weighting strategy in document categorization. In *Proceedings of 2011 Seventh International Conference on Semantics, Knowledge and Grids.*
5. Savoy, J., & Zubaryeva, O. (2011). Classification based on specific vocabulary. In *Proceedings of 2011 IEEE/WIC/ACM International Conferences on Web Intelligence and Intelligent Agent Technology*. 978-0-7695-4513-4/11. IEEE.
6. Term frequency and weighting. Term frequency and weighting. Accessed March 2017. https://nlp.stanford.edu/IRbook/html/htmledition/term-frequency-and-weighting-1.html.
7. Mohamed, R., & Watada, J. (2010). An evidential reasoning based LSA approach to document classification for knowledge acquisition. In *Proceedings of the 2010 IEEE IEEM, IEEE International Conference on Industrial Engineering and Engineering Management (IEEM).*
8. Pennington, J., Socher, R., & Christopher, D. (2014). Manning "GloVe: Global vectors for word representation. In *Proceedings of the 2014 Conference on Empirical Methods in Natural Language Processing* (EMNLP). pp. 1532–1543.

9. Princeton. WordNet A Lexical database for English. [Online: accessed 10-March-2017]. [Online] Available: https://wordnet.princeton.edu/.
10. Blake C. (2010). A comparison of document, sentence, and term event spaces. In *ACL-44 Proceedings of the 21st International Conference on Computational Linguistics and the 44th annual meeting of the Association for Computational Linguistics*, 601–608.
11. Ray, S., & Chandra, N. (2012). Domain based ontology and automated text categorization based on improved term frequency and inverse document frequency. In *International Journal of Modern Education and Computer Science*.
12. Frakes, W. B. (1992). Stemming algorithms. information retrieval: data structures and algorithms. In: Frakes W. B. & Baeza-Yates R. (Eds.), (pp. 131–160). Engle-wood Cliffs, US: Prentice Hall.

Multi Release Reliability Growth Modeling for Open Source Software Under Imperfect Debugging

Diwakar and Anu G. Aggarwal

Abstract In recent years, Open Source Software have gain popularity in the field of the Information technology. Some of its key features like source code availability, cost benefits, external support, more reliability and maturity have increased its use in all the areas. It has been observed that that people interests are shifting from closed source software to open source software due to size and complexity of real life application. It has become impractical to develop a reliable and completely satisfied Open source software product in a single development life cycle, therefore, the successive improved version or releases are developed. These successive versions are designed to meet technological arrangements, dynamic customer needs and to penetrate further in the market. But it also give rise to new challenges in the terms if deterioration in the code quality due to modification/addition in the source code. Sometimes new faults generated due to add-ons and also the undetected faults from the previous release become the cause of difficulty in updating the software. In this paper, an NHPP based software reliability growth model is proposed for multi-release open source software under the effect of imperfect debugging. In the model, it has been assumed that the total number of faults depends on the number of faults generated due to add-ons in the existing release and due to the number of faults left undetected during the testing of the previous release. Data of the three releases of Apache, an OSS system have been taken for the estimation of the parameters of the proposed model. The estimation result for proposed model has been compared with the recently reported multi release software reliability model and the goodness of fit results shows that the proposed model fits the data more accurately and hence proposed model is more suitable reliability model for OSS reliability growth modeling.

Keywords NHPP · Multi-release · Open source software · Imperfect debugging

Diwakar (✉) · A. G. Aggarwal
Department of Operational Research, University of Delhi, Delhi, India
e-mail: Diwakar.du.aor@gmail.com

A. G. Aggarwal
e-mail: anuagg17@gmail.com

P. K. Kapur et al. (eds.), *System Performance and Management Analytics*,
Asset Analytics, https://doi.org/10.1007/978-981-10-7323-6_7

Notations

$m(t)$	Expected number of faults removed in the time interval (0, t]
a_i	Fault content at starting of ith release
α_i	Constant rate at which new faults are introduced in ith release
b_i	A constant in the fault detection rate for ith release
$F_i(t)$	Cumulative distribution function for testing phase of ith release
k_i	Shape parameter for Weibull cdf for ith release
τ_i	Time for the ith release

1 Introduction

Open source software (OSS) have become very popular nowadays. OSS are the software whose source code is freely available to user for use, distribution, reproduction and modification as per the user needs under the licensing policies of OSS [1]. Open source software are developed by a single developer or a group of software developers initially but as the attractiveness of the software increases its users and volunteers also increases throughout the whole world. In recent years, people have become more reliant on OSS for their need. The reliability of the software is defined as the probability of the failure free software for a given interval of time in a specific environment [2, 3]. Software reliability is a very important attribute of the software quality, together with functionality, usability, performance, maintainability, capability, installation and documentation [4]. The proponent of the closed source software believe that hackers can easily incorporate the malicious files in the OSS as the source code is freely and easily available [5] but it is for the same fact that the OSS are more reliable than closed source software as thousands of volunteer are involved in the testing process of the OSS.

In software development process, due to availability of limited time and resources, it is not possible to detect all the faults of the software or to develop complete and reliable software in single development cycle [6]. Then, there is a need of up-gradation of the software and develop successive release by adding new functionalities which also helps in competing with other projects and capturing market. But up-gradation of the software is a very difficult task because up-gradation leads to additional faults in the software therefore there is an increase in failure rate after the up-gradation which then decreases gradually due to fault debugging process [6]. To estimate the mean number of faults detected for closed source software, multi up-gradation SRGM was proposed earlier by Kapur et al. [7, 8]. Recently, a number of multi release SRGM's have been proposed in the litrature for OSS. In 2011, Li et al. [9] proposed a multi attribute utility theory based optimization problem to determine optimal time for releasing next version of OSS. Yang et al. [10] discussed a multi release SRGM for OSS by incorporating fault detection process and fault correction process. Aggarwal et al. [11] proposed a discrete model for OSS which has been released into the market a number of times.

In this paper, we proposed an imperfect debugging based SRGM for OSS model with multiple releases. In the proposed model, bugs introduced during the addition of new add-ons to the current release and some undetected bugs of previous release are considered. This paper is divided into four sections. In Sect. 2, we discuss a multi release SRGM for OSS under the effect of imperfect debugging. In Sect. 3, we present parameter estimation results corresponding to three release fault data sets of Apache project. Finally the conclusions have been drawn in Sect. 4.

2 Modeling Software Reliability

For last few decades, several mathematical models have been proposed that describe the reliability growth of the software during testing process such as Goel and Okumoto [12], Yamada et al. [13]. In most of the models the software failure occurrence has been represented by Non-Homogenous Poisson Process (NHPP). The main focus of NHPP models is to determine the mean value function or the expected number of failure occurrences during a time interval.

Most of the NHPP models are based on the following assumption

- Failure occurs independently and randomly over time.
- Initially fault content in the software is finite.
- The efforts to remove underlying faults once a failure has occurred starts immediately.
- During testing and debugging process no new faults are introduced (i.e. perfect debugging).

But in case of imperfect debugging, the last assumption does not hold good. There is a possibility that some new faults may be added when detected faults are removed\corrected.

2.1 *Model Development*

In this section, an NHPP based SRGM is proposed to model reliability growth phenomena for an OSS incorporating imperfect debugging.

a. **A general NHPP model**

Letus assume that the counting process $\{N(t),\ t \geq 0\}$ is a Non-Homogenous Poisson Process, under, these assumption, $m(t)$, the mean value function for the fault removal process may be represented by the following differential equation.

$$\lambda(t) = \frac{\mathrm{d}m(t)}{\mathrm{d}t} = b(t)[a(t) - m(t)] \tag{1}$$

The mean value function for cumulative number of failure, $m(t)$ can be represented as

$$m(t) = \mathrm{e}^{-B(t)}\left[\int_0^t a(x)b(x)\mathrm{e}^{B(x)}\mathrm{d}x\right] \tag{2}$$

where $B(t) = \int_0^t b(x)\mathrm{d}x$.

b. **Weibull model**

In the case of open source software, when new software is released over the market, the fault removal rate of the OSS is quite distinct from that of the closed source software. In contrast to the closed source software system, OSS are released over the internet with little testing. Once it is released large number of volunteers and enthusiastic testers report bugs through bug tracking system which affect the reliability and attraction of the OSS [14]. Therefore, the fault removal rate (FRR) for OSS initially increases due to growth in the users population but later on decreases as newer versions come into the market and the attractiveness of the present release decreases. Its users shift their loyalty to the other versions/OSS. In order to incorporate such type of increasing and decreasing FRR in the model building [15], we use Weibull distribution function to describe fault removal process (FRP). Weibull distrubution is flexible distribution which may change its shape depending upon different values of its shape parameter (here k) see in Fig. 1. For example,

- When $\mathbf{k > 0}$, the rate corresponding to weibull distribution is increasing. In the context of OSS it may represent the phenomenon when more and more users are getting attached to OSS system and as result increasing numbers of faults are being reported through its bug tracking system.

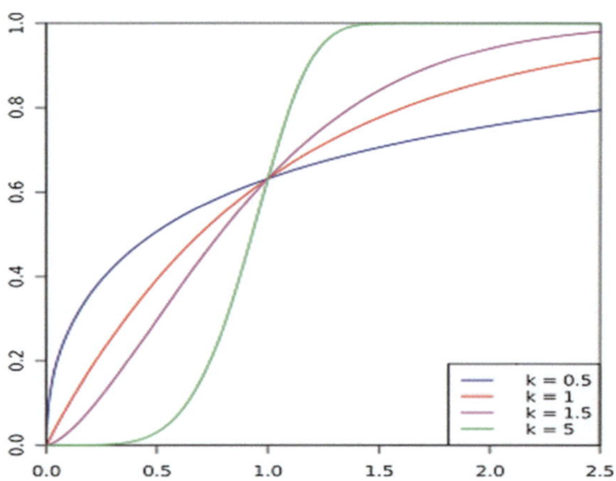

Fig. 1 CDF for weibull distribution

- When **k = 1**, this indicates constant rate of failure and it represent the case when an OSS has reached it maturity level with respect to its number of users.
- When **k < 1**, here failure rate is decreasing, this may occur when users are shifting due the availability of newer version on the internet.

The following differential equation using Weibull model may be formulated to measure the expected number of faults removed,

$$\frac{\mathrm{d}m(t)}{\mathrm{d}t} = bt^k(a - m(t)) \tag{3}$$

Under the initial condition that, $m(t) = 0$ at $t = 0$, the above differential equation gives the following result

$$m(t) = a\left[1 - \mathrm{e}^{-b\frac{t^{k+1}}{k+1}}\right] = aF(t) \tag{4}$$

where $m(t)$ represents expected mean number of faults removed and $F(t) = \left[1 - \mathrm{e}^{-b\frac{t^{k+1}}{k+1}}\right]$ is CDF of weibull distribution with shape parameter $k(>0)$, also known as weibull slope. Let us assume that the debugging process is not perfect over t, some new faults are introduced in the code during correction efforts [5]. Therefore the fault content of the software at time t is given as…

$$a(t) = a + \alpha m(t) \tag{5}$$

Here α is the constant rate at which new faults are introduced. Then, the Weibull model for open source software under the effect of imperfect debugging will be

$$m(t) = \frac{a}{1-\alpha}\left[1 - \mathrm{e}^{-b(1-\alpha)\frac{t^{k+1}}{k+1}}\right]. \tag{6}$$

2.2 *Multi Release Model with Imperfect Debugging*

$$\text{Let us assume } a_i^* = \frac{a_i}{1-\alpha_i} \text{ and } F_i(t) = 1 - \mathrm{e}^{-b_i(1-\alpha_i)\frac{(t-\tau_{i-1})^{k_i+1}}{k_i+1}} \tag{7}$$

where $F_i(t)$ is the *CDF* of weibul model for *i*th release. The mathematical expression for fault removal under imperfect debugging for the *i*th release can be shown as.

$$m_i(t) = \left[a_i^* + \left[a_{i-1}^* - m_{i-1}(\tau_{i-1})\right]\right]F_i(t) \quad \text{for } \tau_{i-1} \le t < \tau_i \tag{8}$$

The mathematical expressions for the number of faults removed during different releases are given as follows:

For release 1

When first release of the software comes in the market at time τ_1, it is tested before being introduced into the market. In the testing process, testing team tries to detect and correct maximum number of the bugs of the software. But practically it is not possible to detect all the faults of software, so testing team can detect only a finite number of bugs in the software which are less than the total fault content of the software [7]. The following equation represents the number of faults removed during testing of release 1.

$$m_1(t) = a_1^* F_1(t) \tau_0 \leq t < \tau_1$$

For release 2

Improved technology, rising competition and dynamic nature of market makes rise to the need of software up-gradation. Addition of new features and functions to the existing version of software can increase the probability of survival and adoption in the market. When new code is added some new faults are introduced into the code. These additional faults along with the fault content of previous release are corrected during the testing of second release with a new FDR [7]. Considering $[\tau_1, \tau_2)$ is the time interval for testing and at time τ_2 testing of release 2 is stopped and launched into the market. Then, the number of faults removed can be represent as

$$m_2(t) = \left[a_2^* + \left[a_1^* - m_1(\tau_1)\right]\right] F_2(t) \tau_1 \leq t < \tau_2$$

For release 3

In this release, the faults due to add-ons and left over fault content of release 2 are considered for removal process. Here we assume that during the testing of release 3 the faults of current version and just previous version are removed, do not take into consideration the undetected faults of version 1, which may be present in the code of version 3. It help us to keep the model simple and easy for parameter estimation. Let τ_3 be launched time for release 3. Then FRP for release 3 is given by

$$m_3(t) = \left[a_3^* + \left[a_2^* - m_2(\tau_2)\right]\right] F_3(t) \tau_2 \leq t < \tau_3$$

In the same manner, we can model FRP for the subsequent releases of the OSS. In the next section we discuss how to validate model to the real life application.

3 Data Set and Analysis

Data sets for three versions of Apache are considered for the validation of the proposed model. The data sets of Apache 2.0.35 (first release), Apache 2.0.36 (second release) and Apache 2.0.39 (third release) are used for the estimation of model parameters [9]. During 43 days of testing for first release (Apache 2.0.35) 74 faults were detected. For the second releases (Apache 2.0.36) testing was carried out for 103 days and 50 faults were detected. For release third (Apache 2.0.39) during 164 days 58 faults were detected.

For estimation of the parameters of the proposed model, the Least Square Estimation Method is used. In the field of Software Reliability Least Square Estimation Method is one of the commonly used methods [3]. SPSS, 'The Statistical Package for Social Sciences' software is applied for estimation of parameters a_i, b_i, α_i and k_i of ith release from the data sets. The estimated parameters of each release are demonstrated in Table 1. The proposed model is then compared with the Amir Garmabaki et al. reliability model [6]. For comparison purpose we have selected Amir et al. reliability model [6] because it proposes a multi release open source software reliability model under perfect debugging conditions. In our model we have incorporated error generation in the modeling framework. By comparing these two models, we can analyze the benefit of imperfect debugging based models. For comparison we have used important criteria (Coefficient of Multiple Determination (R^2) and Mean Square Error (MSE)), the goodness of fit analysis results are given in Table 2. From the result it may be observed that the proposed model provides better fit to the data in comparison to Amir et al. reliability model. The values of MSE and Ad-R^2 corresponding to the proposed models are better than Amir et al. reliability model [6]. The goodness of fit of our model may be further judge by looking Figs. 2, 3 and 4. It may be observed that estimated value is quite near to actual value for all the three releases. From the Table 1 we may observed the value of parameter α is highest for release 1 of Apache software as compared to other two versions. It indicates higher rate of error generation for the initial release as compared to subsequent releases. It may occur due to the fact that when the project is new then chances of introducing additional faults during debugging efforts are higher.

Table 1 Parameter estimation results

Parameters	Releases		
	Apache 2.0.35	Apache 2.0.36	Apache 2.0.39
a	73.995	49.991	58.134
b	0.050	0.033	0.027
α	0.082	0.038	0.065
k	0.162	0.046	0.137

Table 2 Comparison criteria results

Releases	Models	MSE	Ad-R^2
Apache 2.0.35	Proposed model	3.62	0.993
	Amir et al. [6]	3.68	0.992
Apache 2.0.36	Proposed model	5.28	0.989
	Amir et al. [6]	5.45	0.986
Apache 2.0.39	Proposed model	0.84	0.995
	Amir et al. [6]	0.70	0.995

- Coefficient of Multiple Determination (R^2)

It shows how much proportion of the variation of the data get explained by the regression model and it measure of the goodness of fit of the model, higher the value of R-squared, More the model fits to data.

$$R^2 = 1 - \frac{\text{residual}\, SS}{\text{corrected}\, SS}$$

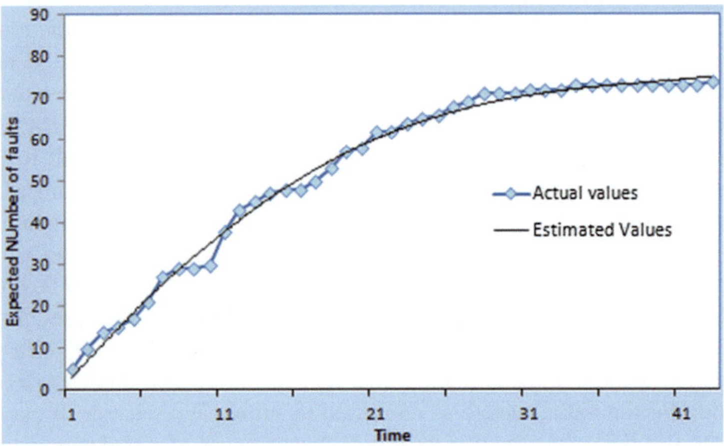

Fig. 2 Goodness of fit for first release

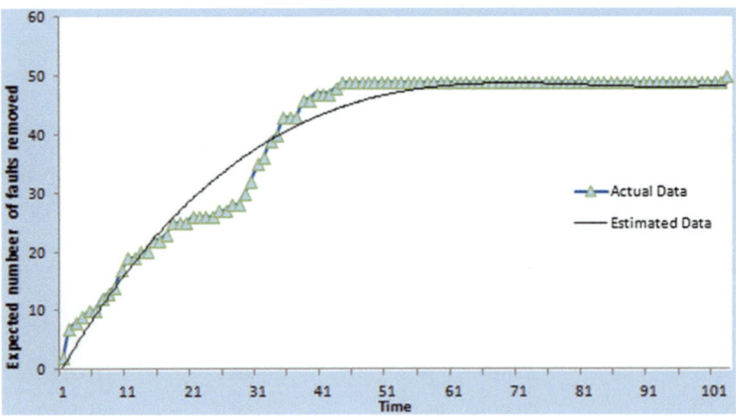

Fig. 3 Goodness of fit for release 2

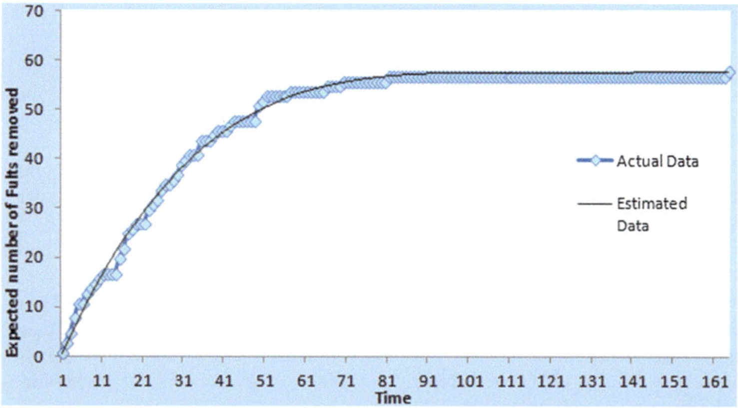

Fig. 4 Goodness of fit for release 3

- Mean Square Error

It is the mean of the square of the difference between the expected values and the observed value,

$$\mathrm{MSE} = \sum_{i=1}^{k} \frac{(m'(t_i) - y_i)}{k}$$

Here, k represents the number of observation. Better the goodness of fit to data if the MSE is lower. Figures 2, 3 and 4 shows the Goodness of fit for the three release of Apache respectively.

4 Conclusion

In this era of Information technology OSS represents a paradigm shift in the software development life cycle. Unlike in closed source software where testing is performed by a group of testers, OSS is tested by millions of spontaneous volunteers during its operational phase. In this paper, we use Weibull probability distribution function to model FRP of OSS so as to represents the initial increase and finally decrease in the bug reporting of OSS. As the entire bug reporting are not valid. Therefore the concept of imperfect debugging has been incorporated in model building. Proposed model has been validated on a 3 releases fault data sets of well-known OSS namely, Apache. The results on compared with other well-known model [6] to illustrate the accuracy of model and goodness of fit. In future we may extend the model to relate growth in the user population to the faults removed during debugging process.

References

1. Anant, K. S. & Still, B. (2009) Handbook of research on open source software technological. *Economic and Social Perspectives*.
2. Pham, H. (2006). *System Software Reliability*. Verlag.
3. Kapur, P. K., Pham, H., Gupta, A., & Jha, P. (2011). *Software reliability assessment with OR Application*. Berlin: Springer.
4. https://users.ece.cmu.edu/~koopman/des_s99/sw_reliability/.
5. Shyur, H. J. (2003). A stochastic software reliability model with imperfect–debugging and change point. *Journal of Systems and Software, 66,* 135–141.
6. Amir, S., Garmabaki, H., Barabadi, A., Yuan, F., Lu, J., & Ayele, Y. Z. (2015). *Reliability modeling of successive release of software using NHPP*. In Industrial Engineering and Engineering Management (IEEM), 2015 IEEE International Conference. pp 761–765.
7. Kapur, P. K., Singh, O., Garmabaki, A. S., & Singh, J. (2010). Multi Up-gradation software reliability model with imperfect debugging. *International Journal of System Assurance Engineering and Management 1*, 299–306. 2010/12/01.
8. Kapur, P. K., Tandon, A., & Kaur, G. (2010). *Multi Up-gradation Software Reliability Model.* In Reliability, Safety and Hazard (ICRESH), 2010 2nd International Conference. pp. 468–474.
9. Li, X., Li, Y. F., Xie, M., & Ng, S. H. (2011). Reliability analysisand optimal version-updating for open source software. *Information and Software Technology, 53,* 929–936.
10. Yang, J., Liu, Y., Xie, M., & Zhao, M. (2016). Modeling and analysis of reliability of multi release open source software incorporating both fault detection and correction processes. *The Journal of System and Software, 115,* 102–110.
11. Aggarwal, A. G., Nijhawan, N. (2017). *A discreate modeling framework for multi release open source software system. Accepted for publication in International Journal of Innovation and Technology Management*. World Scientific.
12. Goel, A. L., & Okumot, K. O. (1979) *Time-dependent error detection rat e model for software reliability and other performance measures*. In Reliability, IEEE Transactions on, Vol. 28. pp. 206–211.
13. Yamada, S., Ohba, M., & Osaki, S. (1983). S-shaped reliability growth modeling for software error detection Reliability. *IEEE Transactions, 32,* 475–484.
14. Raymond, E. S. (2001). *The Cathedral & then bazaar: Musings on Linux and open source by an accident revolutionary.* Sebastopol: O'Reilly Media, inc.
15. Garmabaki, A. H., Kapur, P., Aggarwal, A.G., & Yadaval, V. I. (2014) *The impact of bugs reported from operational phase on successive software releases.* In International Journal of Productivity and Quality Management, Vol. 14. pp. 423–440.

Barriers to Agile Adoption: A Developer Perspective

Noshiba Nazir, Nitasha Hasteer and Rana Majumdar

Abstract Agile methods are one of the most widely adopted methodologies for developing software. Agile methods refer to a family of lightweight methods that tend to favor working code over documentation, individuals over tools, and collaboration over negotiation. Agile methodology proves beneficial over conventional software engineering methods in terms of time and cost. However, apprehensions of developer community toward adopting agile are an area of concern that results in barriers toward complete agile adoption. In this work, we report the barriers identified through literature survey and results of investigating the relationship that exists between observed barriers. This paper focuses on structural equation modeling that utilizes different classes of modular approaches and establishes connections among identified variables, having the fundamental objective of providing a confirmatory test of a hypothetical model. Our work demonstrates a path model through the analysis of the identified barriers faced by developers during agile adoption.

Keywords Agile · Challenges · SEM methodology · Path analysis

1 Introduction

Various organizations use agile methodologies for software development. The main reason for the deployment of this method is assumed to be the effectiveness to create powerful applications. Agile consists of process models that have smaller

N. Nazir · N. Hasteer · R. Majumdar (✉)
Amity School of Engineering & Technology, Amity University,
Noida, Uttar Pradesh, India
e-mail: rmajumdar@amity.edu

N. Nazir
e-mail: noshibanazir@gmail.com

N. Hasteer
e-mail: nhaster@amity.edu

P. K. Kapur et al. (eds.), *System Performance and Management Analytics*,
Asset Analytics, https://doi.org/10.1007/978-981-10-7323-6_8

phases, where the end product is delivered in "sprints". Organizations have been using agile since a very long time, but there are still some issues especially the challenges that the development team faces that we need to address. The increase in the use of agile methodologies, leading to reduction in the usage of traditional methods, develops an extra pressure to take up agile. As we realize that agile has been very famous, there are numerous reasons why we ought to look at the different issues faced by the developers for its adoption. One of the issues being that the limit line of agile approach is currently changing and is no more restricted to little groups situated in the same spot. Some other issues being the quick pace of adoption, introducing new staff challenges, and to deal with the human assets [1].

Considering the present popularity of agile practices, it is critical to analyze the past work in this space and to portray out the different challenges faced by the group in an agile development environment. We in our previous work examined ten challenges through a literature review and discovered inter-relationship among them.

The 10 challenges identified were Re-organization of teams (ROT), distributed work being a continuous negative influence on performance of team (DWCNIPT), lack of business knowledge among developers (LBKD), requirement to be aware and learn beliefs and standards of Agile (RABSA), requirement for developers to be trained for all trades (RDTT), problem faced during individual performance evaluation (PIPE), absence of developer motivation to apply agile methods (ADMAAM), increased dependence on social skills (IDSS), developer panic created by translucence of weakness in skill (DPCTWS), and passing on the responsibility of decision-making (PRDM). A questionnaire-based survey was then conducted and the data collected was used to develop relationship matrix which is considered to be the first step toward the application of the interpretive structural modeling (ISM) methodology. The survey was structured on a 5-point Likert scale and among the ten identified challenges the respondents had to rate them according to their importance on the scale. In the 5-point scale, 1 correlates to "very low" and 5 to "very high". The next step was to administer the questionnaire. It was administered to a total of 50 software companies in India. Fifteen completely filled questionnaires were received. This gave us the response rate of 30%. ISM procedure has been given the name interpretive structural modeling as it depends on a gathering or individual understanding whether and how the variables might be identified with each other. ISM strategy is a procedure which includes illustrating the connections built up by gathering or individual translation and the general structure is set up into a digraph and lastly into a model [2].

In this work, we present a structured equation model of the challenges mentioned above. While ISM is more of a theoretical concept whose end product is a model, structured equation model (SEM) is a confirmatory concept. SEM utilizes the different sorts of modular approach to show connections among watched variables, having the fundamental objective of providing a confirmatory test of a hypothetical model guessed by the researcher. Particularly, different hypothetical models may be utilized in SEM giving a confirmation of the inter-relationship among the variables. The motivation behind the model is to provide representation of covariance matrix among the variables that are measured. It also helps in providing path analysis (e.g., regression) tests models and connections between the variables that are measured. One another very important parameter of SEM is confirmatory factor analysis that tests the interrelationships among variables that are measured and the ones that are latent. SEM also helps in finding latent growth curve models (LGM). Exceptional instances of structural equation modeling are path analysis, analysis of factors, and confirmatory factor analysis. In our work, we will be focusing on one significant model of SEM which is path analysis. To meet the objective of our study, we use Lisrel 8.80 (Student) software because it gives us statistical analysis of crude information (e.g., means, correlations, and missing information traditions), give schedule to take care of missing information and distinguishing exceptions, create the system's linguistic structure, graph the model, and accommodate import and fare of information and figures of a hypothetical model. The rest of the paper is organized as below. Section 2 is the literature review, and Sect. 3 outlines the methodology followed. In Sect. 4, we throw some light on the implementation of the methodology and in Sect. 5, we discuss the results and conclude the work.

2 Literature Review

The limits of agile are now transforming. There is no limitation restricted to small-scale assembled teams and is widely being used by companies coming out of their console [3], thus leading to latest issues faced in management of the human resource of the respective organizations. Now the big decision is finally on the shoulders of the team depending upon their requirements whether they want to adopt agile method of development or stick to the existing one. Progressively, more and more organizations and other stakeholders are pressurizing for the use of agile [4]. The need to use agile development has forced people from the industry to take up this task of identifying the challenges the people face for adoption. An analysis of the literature, allows us to see that agile environments are notably distinct in context to environments that use more traditional approaches although very often the distinction between the two is not so black and white [5]. The previous work focuses on the challenges that the developers face during Agile Software development, but they do not use the means to categorize these challenges [1], hence giving an incomplete idea of these challenges and their effects on development. From the overall literature review, it was found that although literature has the proof of mentioning various instances, wherein the Agile developers have faced challenges. But there is still the need of addressing these challenges by categorizing these challenges based on surveys conducted. We need to understand the relationship among these challenges, as to how one challenge affects other. This requires utilization of a proper methodology in establishing the relationship and developing a final model. We also require a confirmatory model that uses various types of models to depict relationships among observed challenges, with the goal of providing a quantitative test of a theoretical model. We in our work have attempted to establish a path model for analyzing the challenges in depth.

3 Methodology

Structural equation modeling (SEM) uses various types of models to depict relationships among observed variables, with the goal of providing a quantitative test of a theoretical model hypothesized by the researcher. More specifically, various theoretical models can be tested in SEM that hypothesize how sets of variables define constructs and how these constructs are related to each other [6]. For example, an educational researcher might hypothesize that a student's home environment influences her later achievements in school. A marketing researcher may hypothesize that consumer trust in a corporation leads to increased product sales for that corporation. A healthcare professional might believe that a good diet and regular exercise reduce the risk of a heart attack. In each example, the researcher believes, based on theory and empirical research, sets of variables define the constructs that are hypothesized to be related in a certain way.

The goal of SEM analysis [7] is to determine the extent to which the theoretical model is supported by sample data. If the sample data support the theoretical model, then more complex theoretical models can be hypothesized. If the sample data do not support the theoretical model, then either the original model can be modified and tested or other theoretical mode is needed to be developed and tested. Consequently, SEM tests theoretical models using the scientific method of hypothesis testing to advance our understanding of the complex relationships among constructs. SEM analysis consists of the following steps: Model selection, model recognition, model assessment, model examination, and model alteration.

3.1 Model Selection

In this step, we make use of the existing work, literature survey, research, and data to build up a hypothetical model. In this manner, before any information accumulation alternately examination, we indicate a specific model that is ought to be affirmed utilizing variance–covariance information. At the end, accessible data is utilized to choose which variables to incorporate into the hypothetical model. Model detail includes deciding the interrelationships and specifications within the model that is our priority. A hypothetical model is effectively resolved if the populace model is relatable with model that is hypothetical—which also means that the illustrated covariance matrix S is adequately replicated by the model that is hypothetical. Our objective was subsequently to find out the decision of model that is most appropriate and create an example covariance matrix. To illustrate the model selection, we took two-variable circumstance including watched variables P and Q. Knowing from earlier research that P and Q are exceedingly corresponded, the question that arises is why. Which hypothetical association is in charge of this connection? Does P impact Q, does Q impact P, or is there any other third variable R that impacts both P and Q? There can be numerous conceivable reasons why

P and *Q* are connected in a specific manner. We need literature review and speculations to pick among conceivable clarifications and in this way give the basis for determining a model—that is, trying an inferred hypothetical model [8].

3.2 Model Recognition

It is difficult that we solve the problem of recognition before identification of variables. But as soon as the model is selected and the variables are identified, the variables are connected to create a variance–covariance matrix. The three levels of model recognition are stated as below:

1. Under-recognized model: this happens if more than one variable is not individually recognized and the reason for this being that we are not having sufficient information inside matrix *S*.
2. Just recognized model: this happens when every variable is individually recognized and the reason for this being that we are having sufficient information in the matrix *S*.
3. Over recognized model: this happens when we have several different ways of recognizing each variable as we have more than sufficient information in *S* matrix [8].

3.3 Model Assessment

This particular step is to assess the variables of the populace in a basic structural equation model. There is a requirement for us to get estimations for the variables identified in the model which builds the advised matrix Ó, such that the values of the variables result in a matrix that is closely related to matrix *S*. The mechanism of assessment uses a definite quantity of fitting in order to reduce the difference between Ó and *S*. There are certain quantities of fitting or assessment methods that are available. Some of the early assessment techniques are summed up minimum squares (GLS), greatest probability (ML), and unweighted or common slightest squares (ULS or OLS).

There are various correlation methods such as polychoric, Pearson, and polyserial that are used by PRELIS to make an asymptotic covariance network for information into LISREL. There is a forewarning of not using the specifically utilized blended sorts of correlation matrices and the covariance matrices in a LISREL–SIMPLIS program; it rather utilizes an asymptotic variance–covariance network created by PRELIS alongside [8].

3.4 Model Examination

As soon as the variable assessment is complete, there is a need for a predetermined structured equation model, where we choose how well the information fits the model. Or we can also say to what degree is the hypothetical model backed by the acquired test information. If we want to consider model fit, there are two different ways to do that. One is to think of some as global sort test that tests the fittings for the whole model. The second is to analyze the fit of individual parameters in the model [8].

3.5 Model Alteration

In the event that the fitting of the inferred hypothetical model is not as we expected, it means as strong as we expect. If this is the case, then in this particular step we alter the model and consequently assess the changed model. With a specific end goal to decide in what way to change the model, there are various procedures accessible for determination errors, so that it becomes relatively easy to modify the model [8].

4 Implementation

4.1 Path Model

Path model is the consistent expansion of regression models. Path analysis makes use models having numerous identified variables; it may have any number of dependent and independent variables. Hence, in order to create a path model, we require assessing various equations and recognized variables. The data that was received during the survey acts as an input to create the path. The covariance matrix generated for the identified challenges is presented in Table 1. Figure 1 illustrates the path diagram, and the fittings of the parameters are given in Table 2.

5 Discussion and Conclusion

Finding a measurably huge hypothetical model that has useful, substantive significance is the essential objective of utilizing basic comparison demonstrating to test theories. We utilized the accompanying principles in deciding the statistical noteworthiness and significance of a hypothetical model. The primary premise is the nonstatistical centrality of the chi-square test and the root-mean-square error of

Table 1 Covariance matrix

	ID	ROT	DWCNIPT	LBKD	RABSA	RDTT	PIPE	ADMAAM	IDSS	DPCTWS	PRDM
ID	1.000										
ROT	0.349	1.000									
DWCNIPT	1.507	0.267	1.000								
LBKD	−0.818	0.810	−0.969	1.000							
RABSA	1.625	0.961	0.993	−0.932	1.000						
RDTT	−1.416	−0.963	−0.591	−0.945	−0.667	1.000					
PIPE	1.020	0.988	0.165	0.900	0.154	−0.591	1.000				
ADMAAM	−0.426	−0.857	−0.936	0.810	−0.985	0.974	0.267	1.000			
IDSS	1.294	0.938	0.343	0.969	0.302	−0.999	0.981	0.938	1.000		
DPCTWS	1.211	0.862	0.938	−0.811	0.974	−0.052	−0.267	−0.694	−0.936	1.000	
PRDM	0.433	0.916	0.190	0.979	−0.439	−0.551	0.972	0.916	0.765	−0.919	1.000

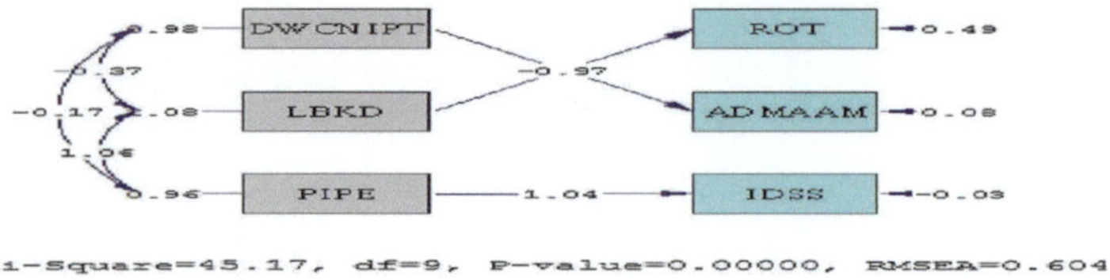

Fig. 1 Path diagram

Table 2 Model fittings

Degrees of Freedom	9	
Minimum Fit Function Chi-Square	0.39	(*P* = 1.00)
Normal Theory Weighted Least Squares Chi-Square	0.39	(I = 1.00)
Satorra–Bentler Scaled Chi-Square		45.17 (*P* = 0.00)
Chi-Square Corrected for Non-normality		467.24 (*P* = 0.0)
Estimated Non-centrality Parameter (NCP)		36.17 90
Percent Confidence Interval for NCP		(18.83; 61.03)
Minimum Fit Function Value		0.028
Population Discrepancy Function Value (F0)	3.29	90
Percent Confidence Interval for F0		(1.71; 5.55)
P-Value for Test of Close Fit (RMSEA < 0.05)		= 0.00
Expected Cross-Validation Index (ECVI)	6.29	90
Percent Confidence Interval for ECVI		(4.71; 8.55)
ECVI for Saturated Model	3	0.82
ECVI for Independence Model	1.19	
Chi-Square for Independence Model with 15 Degrees of Freedom	1	0.11
Independence AIC	1	3.011
Model AIC		69.17
Saturated AIC		42.00
Independence CAIC		23.36
Model CAIC		89.66
Saturated CAIC		77.87
Normed Fit Index (NFI)		−39.68
Non-normed Fit Index (NNFI)		5.34
Parsimony Normed Fit Index (PNFI)		−23.81
Comparative Fit Index (CFI)		0.0
Incremental Fit Index (IFI)		5.58
Relative Fit Index (RFI)		−66.80
Critical N (CN)		7.72
Root Mean Square Residual (RMR)		0.39
Standardized RMR		0.035
Goodness of Fit Index (GFI)		0.99
Adjusted Goodness of Fit Index (AGFI)		0.98
Parsimony Goodness of Fit Index (PGFI)		0.42

estimation (RMSEA) values, which are overall fit measures. Nonstatistically paramount chi-square esteem shows that specimen covariance grid and the delivered model derived covariance framework are tantamount. An RMSEA esteem not exactly or comparable to 0.05 is seen as commendable. The second premise is the measurable importance of individual parameter gauges for the ways in the model, where qualities are prepared by confining the parameter gauges by their different standard errors. This is alluded to as at esteem and is generally appeared differently in relation to a tabled t estimation of 1.96 at the 0.05 level of significance (two-tailed). The last principle is the size and course of the parameter gauges, giving watchful thought to whether a positive of course negative coefficient looks good for the parameter gauges.

In our case, we got the chi-square results to be 45.17, which still have got a scope to get refined, on modification of the model. Same is the case with RMSEA value wherein the value should always be 0.05 or less and we have got the value of 0.64 which is comparatively on the higher end. The future enhancements of the work would focus on optimization to improve the results.

References

1. Conboy, K., Coyle, K., Wang, X., & Pikkarainen, M. (2010). People over process: Key people challenges in agile development. *IEEE, 99*, 1–1.
2. Rajesh, A., Sandeep, G., Nikhil, D., & Deepak, K. (2012). Analysis of barriers of Total Productive Maintenance (TPM). In *International Journal System Assurance Engineering and Management*, https://doi.org/10.1007/s13198-012-0122-9.
3. Xiaofeng, W., & Kieran, C. (2009). Understanding agility in software development from a complex adaptive systems perspective.
4. Diane, J., Kevin, V., & Guy, C. (2006). Agile procurement and dynamic value for money to facilitate agile software projects. In *Proceedings of the Conference EUROMICRO*.
5. Luís, K., Roses, A., Windmöller, E., Almeida do, C. Favorability conditions in the adoption of agile method practices for software development in a public banking. JISTEM-Journal of Information Systems and Technology Management.
6. Tsun, C., & Dac, B. C. (2008). A survey study of critical success factors in agile software projects. *The Journal of Systems and Software*, *81*,961–971.
7. Warfield, J. W. (1974). Developing interconnected matrices in structural modeling. *IEEE Transactions on Systems Men and Cybernetics, 4*(1), 51–81.
8. Randall, E., Schumacker, R., & Lomax, G. (2010). A beginners guide to structural equation modeling. Taylor and Francis Group, LLC.

Toward Analysis of Requirement Prioritization Based Regression Testing Techniques

Varun Gupta, Yatin Vij and Chetna Gupta

Abstract The regression testing aims to validate the quality of successive software versions along with the validation of the existing functionality. The new functionality, change requests, and implementation of delayed requirements lead to the change in the source code, and it might be possible that existing functionality may malfunction as a result of such changes. Various regression testing approaches are proposed in the literature, and this paper tries to analyze the state of the art of requirement priority based regression testing approaches. Few requirement-based approaches are identified from the literature and were analyzed for their differences in functionality and other parameters that determine their applicability for doing regression testing. The results indicate that the existing techniques employ different parameters (with requirement priority as one of the parameters) and need validation on large dataset, and the applicability of particular technique as per circumstances is still uncertain. There is lack of consensus that helps the software tester to decide which technique is better as per existing scenarios.

Keywords Regression testing · Requirement prioritization · Testing

1 Introduction

Regression testing is aimed at the continuous testing of the newer versions of incremental software, in order to find errors in new addition or modification of the present parts of software and guarantees that no new errors have been presented in beforehand tested code. Software testers analyze software against the test cases and match the yield acquired with the normal result.

V. Gupta · Y. Vij (✉)
Amity University, Noida, India
e-mail: ytn.vj2@gmail.com

C. Gupta
JIIT, Noida, India

P. K. Kapur et al. (eds.), *System Performance and Management Analytics*, Asset Analytics, https://doi.org/10.1007/978-981-10-7323-6_9

Testing software is a costly and time-consuming affair if all the test cases are to be tested. This paper aims at minimizing the test cases that are required to be executed to test the incremental software; this can be achieved by prioritizing the test cases on the basis of requirement priorities.

Thus, the testers need selection of test cases which represent all the errors in the software. This paper is aimed at achieving this task by taking the priorities assigned requirements and the test suite as input and giving the prioritized test cases as the output.

The motivation behind this prioritization of test cases is that approx. 45% of the software functions are never used, 19% functionality is rarely used, and only 36% of the software functions are used always [1].

2 Related Work

Malhotra et al. [2] include successions of test suite choices and prioritization and is an augmentation of work completed by Aggarwal et al. [3]. Paper claimed to have 60% reduce in test cases for experiments subsequent to being connected to triangle classification as a case study.

Srikanth et al. [4] proposed test case prioritization strategy called prioritization of requirements for test (PORT), based on four factors, i.e., customer priority (CP), implementation complexity (IC), requirement volatility (RV), and fault proneness (FP). The values allotted to these four factors are mapped to single priority value associated with the test case. PORT calculation was utilized on understudy tasks of size 2500 LOC executed in Java.

In another paper, Srikanth and Banerjee [5] connected same PORT procedure on three complex projects which resulted in similar fault detection and percentage contributions.

Gupta et al. [6] proposed hybrid regression testing technique procedure that intends to diminish the quantity of experiments and all the while expanding error discovery rate. The test suite decrease was made conceivable by pruning the paths of the product to be tried. With a specific end goal to perform way pruning, those ways are chosen that contains proclamations that either influences or gets influenced by the (1) addition, or (2) deletion, or (3) modification. Resulting in ripple effects created because of the presence of either control or data dependency or both between variables in statements.

Gupta et al. [7] reported the direct relation between the requirement prioritization and regression testing. The authors reported that the test cases can be prioritized by clustering requirements on the basis of requirement priorities. This can fundamentally decrease the quantity of test cases without having bargain with fault detection rate. Hybrid regression testing technique as proposed by Gupta and Chauhan [1] was applied to a payroll management system project. The project was delivered in two increments in two forms, one employing clustering and another without clustering. The obtained results indicate that the clustering form is better than that without clustering. The results are for the project developed from scratch by employing decision aspect prioritization technique as proposed by Gupta et al. [8] and requirement prioritization technique as proposed by Gupta and Srivastav [9].

Arafeen et al. [10] in test case prioritization using requirement-based clustering proposed the use of requirement-based clustering for test case prioritization. They explored the grouping approach that joins traditional code analysis with enhancing test case prioritization systems. Their requirements clustering, requirements tests mapping, prioritization of test cases per cluster, cluster prioritization, and test case selection were the main elements of the proposal.

Siddik and Sakib [11] proposed that test cases can be prioritized on the basis of a framework consisting of requirements, design, and code collaboration.

Hettiarachchi et al. [12] in effective regression testing using requirements and risks showed that the test case can be prioritized using the requirement risks. It involved estimation of risks by correlating requirements, calculating risk weight for requirement and risk exposure value, and evaluating additional factors to prioritize requirement and test cases.

Wang and Zeng in history-based dynamic test case prioritization for requirement properties in regression testing (2016) proposed that test cases can be prioritized on the history of fault detection using the requirement classification and requirement importance.

Ansaria et al. [13] in optimized regression test using test case prioritization used the ant colony optimization technique to prioritize the test cases for regression testing. It will take a test case as input and choose one that covering maximum faults, and then it is checked to see fault coverage to see if all faults in it or not. If not, choose the next test case on the basis of remaining faults and repeat till all covered. Once all faults are covered, calculate the total number of faults covered by each test case which is stored in total fault test case matrix. All the combinations of test case called paths covering all faults are generated and the best path, from all paths, is selected on the basis of minimum execution time and pheromone value is updated as the best path.

3 Result Analysis

The proposed regression testing techniques employ various factors for prioritizing test cases including the requirement priorities, which is based on the logic that testing the test cases associated with higher priority requirements would be sufficient enough even if low priority requirements are neglected.

The proposed work requires more validations on varying complexities dataset to at least be sure about the suitability and scalability of the techniques. The automation tools must automate the proposed techniques so as to see which technique overweights others.

Further, the work performing the rigorous comparative analysis of the proposed techniques is missing from literature. Thus, which technique is suitable for which dataset and under what circumstances is still an unaddressed issue.

Last but not the least, the evaluation of accuracy of requirement priorities is still unaddressed issue, which may affect the test case prioritizations and hence the accuracy of the testing.

4 Conclusion and Future Work

The work highlights the need for comparative analysis and individual analysis of the regression testing approaches on different complexity datasets. The techniques can be automated which means that it provides more opportunities for technique improvements and reducing the efforts due to use of the tool.

More rigorous analysis of techniques will make it possible for the software tester to make a decision about the applicability of regression testing techniques as per the scenario and given conditions, which are ever changing.

References

1. Gupta, V., Singh, D., & Chauhan, K. (2011b). Hybrid regression testing technique: A multi layered approach. In *IEEE annual Conference "INDICON"*, Hyderabad, IEEE. https://doi.org/10.1109/indcon.2011.6139363.
2. Malhotra, R., Kaur, A., & Singh, Y. (2010). A Regression Test Selection and Prioritization Technique. *Journal of Information Processing Systems, 6,* 235–252.
3. Aggarwal, K. K., Singh, Y., & Kaur, A. (2004). Code coverage based technique for prioritizing test cases for regression testing. *ACM SIGSOFT, 29*(5), 1–4.
4. Srikanth, H.,. Williams, L., & Osborne, J. (2005). System test case prioritization of new and regression test cases. In *2005 International Symposium on Empirical Software Engineering*.
5. Srikanth, H., & Banerjee, S. (2012). Improving test efficiency through system test prioritization. *Journal of Systems and Software, 85,* 1176–1187.
6. Gupta, V., Chauhan, D. S., & Dutta, K. (2015). Hybrid regression testing based on path pruning. *International Journal of Systems and Service-oriented Engineering (IJSSOE).* IGI Global Publishers, 2015. *1*(5).
7. Gupta, V., Singh, D., Chauhan, K. D. (2012b). Impact analysis of requirement prioritization on regression testing. In *2nd World Conference on Innovation and Computer Sciences*.
8. Gupta, V., Chauhan, D. S., & Dutta, K. (2012). *Decision Aspect Prioritization Technique for Globally Distributed Developments: A Hybrid Approach*. Procedia Technology: Elsevier.
9. Gupta, V., & Srivastav, M. (2011). Web based tool supported requirement prioritization: Based on multiple stakeholder preferences. *International Journal on Computer Engineering and Information Technology (IJCEIT), 25*(1), 12–19.
10. Arafeen, M., & Do, H. (2013). Test case prioritization using requirements based clustering. In *International Conference of Software Testing, Verification and Validation (ICST)*.
11. Siddik, M. S., & Sakib, K. (2014). RDCC: An effective test case prioritization framework using software requirements, design and source code collaboration. In *2014 17th International Conference on Computer and Information Technology (ICCIT)*.

12. Hettiarachchi, C., et al. (2014). Effective regression testing using requirements and risks. In *2014 Eighth International Conference on Software Security and Reliability*.
13. Ansaria, A., Khanb, A., Khanc, A., Mukadamd, K. (2016). Optimized regression test using test case prioritization. In *7th International Conference on Communication, Computing and Virtualization 2016*. Procedia Computer Science Vol. 79 pp. 152–160.

Formulation of Error Generation-Based SRGMs Under the Influence of Irregular Fluctuations

Adarsh Anand, Deepika and Ompal Singh

Abstract Reliability growth models for software have been widely studied in the literature. Many schemes (like hazard rate function, queuing theory, and random lag function) have been proposed and utilized for modeling the fault removal phenomenon. Among these, hazard rate function technique has gained significant attention and has been excessively used for model debugging process. An essential aspect of modeling has been pertaining to reliability estimation under irregular fluctuations environment. Another major domain highlighted in Software Reliability Engineering (SRE) is that of error generation, which has been an important area of research up till now. This article shows how, using Hazard Rate Function approach, error generation concept can be studied in a fluctuating environment. The utility of the proposed framework has been emphasized in this paper through some models pertaining to different conditions. The applicability of our proposed models and comparisons in terms of goodness-of-fit and predictive validity has been presented using known software failure data sets.

Keywords Error generation · Hazard rate function · Irregular fluctuations
Nonhomogenous Poisson process (NHPP) · SRGMs

A. Anand · Deepika (✉) · O. Singh
Department of Operational Research, University of Delhi, Delhi,
New Delhi 110007, India
e-mail: deepika.sre@gmail.com

A. Anand
e-mail: adarsh.anand86@gmail.com

O. Singh
e-mail: drompalsingh1@gmail.com

P. K. Kapur et al. (eds.), *System Performance and Management Analytics*,
Asset Analytics, https://doi.org/10.1007/978-981-10-7323-6_10

1 Introduction

Software is practically inexorable in the modern era and since the time we have moved to the present decade, it has become the driver for almost everything surrounding us. Everyone is heavily affected by the speedy change of software technology. This heavy dependence of humans on software has enlarged the requirement for software firms to come up with software with desired level of reliability. Various SRGMs have been proposed in the past to help the firms to quantify the leftover faults in the system and majority of them are based on NHPP [1–7].

For the software product delivery, testing of the software is a leading ingredient of organizes. In this sight, many SRGMs can be employed for the evaluation of reliability. Software testing is a method for modeling the observation/correction of faults [8]. The very first article in software reliability was given by Hudson, where he talked about natural birth and death process [9]. After, Hudson, measurement of reliability was given by Jelinski and Moranda [10]. Many other researches tried to measure software reliability such as Moranda [11] and Schneidewind [12]. A methodical approach based on execution time is represented by Musa [13]. Goel and Okumoto [1] gave an SRGM based on the presumption that the fault causing a failure is immediately removed and is also called as exponential SRGM (G-O model). It is one of the most widely used models that very well exist in software reliability literature. A model with the concept of failure observation and corresponding fault removal phenomenon was given by Yamada et al. [7]. Researchers have also tried to organize the case of imperfect debugging environment under the idea of prolonged testing in which software firms issue patches in order to fix failures in operational phase [14]. Recently, features enhancement archetype SRGM was given by Anand et al. [15] in which they have discussed release time determination also.

First, SDE-based SRGM was introduced by Yamada et al. [16]. The model assumed that fault detection rate is constant along with the noise factor. Later, Yamada et al. [17] considered different types of fault detection rate to obtain different types of software reliability measures using the probability distributions. Then, Shyur [18] captured the behavior of stochastic-based SRGMs under the influence of imperfect debugging and change point. Moreover, Kapur et al. [19] employed logistic error detection rate in the modeling of SDE-based generalized Erlang model. Kapur et al. [20] gave an approach to formulate SDE-based SRGMs under unification scheme. Lately, Singh et al. [21] inculcated the impact of randomness in the formulation of multi-up-gradation software releases. Yamada et al. [16] proposed stochastic differential equation-dependent modeling for software reliability assessment. Like Yamada, many researchers [3] have proposed SRGMs inculcating the concept of Itô type SDE to measure software reliability. Later on, Tamura et al. proposed a flexible SDE model describing an FDP with distributed development environment [22]. An *S*-shaped SRGM that inculcates an irregular fluctuation has been represented by Lee et al. [23].

A significant number of researchers have also studied the impact of uncertain factors, and they suggested the influence of randomness in the fault detection rate [3]. During the debugging process, some changes that affect the whole testing environment are termed as randomness. Usually, random fluctuations arise due to numerous factors such as testing effort expenditure, testing skill, strategies, etc. [3]. The presence of these factors includes the noise factor in the fault removal phenomenon that represents the stochastic behavior in testing process.

Rest of the manuscript is prearranged as follows: First, modeling framework has been described in a section and further it comprises three subsections namely notations, assumptions, and SDE-based modeling. Then, we have discussed data analysis for all six SRGMs. Conclusion and acknowledgement have been provided at last, followed by references.

2 Modeling Framework

This section provides the illustrative description of the proposed SDE-based fault removal phenomenon. These are the set of notations and presumptions which have been used in this paper.

2.1 Notations

$N(t)$	Continuous random variable
$E(N(t)) = z(t)$	Expected number of faults observed or eliminated by time "t"
$a(t)$	Total fault content dependent on time "t"
ω	During debugging process, a rate at which the errors may be introduced
ρ	Error removal rate
$F(t)$	Cumulative distribution function
$f(t)$	Probability density function
v	Learning parameter in the logistic function
$\psi(t)$	Intensity function
s	A positive fixed value that symbolizes the scale of irregular fluctuations
$\zeta(t)$	Hazard rate function

2.2 Assumptions

During the testing phase, NHPP is used to illustrate the failure phenomenon. In nonhomogenous Poisson process, Poisson distribution has been used and it shows the probability of events occurring in a fixed interval of time. For continuous random variable, the counting process $\{N(t); t \geq 0\}$ of NHPP is represented in mathematically form as [3]

$$P[N(t) = k] = \frac{(z(t))^k}{k!} e^{-z(t)}, \quad k = 0, 1, 2\ldots\ldots; \quad e = 2.7181 \tag{1}$$

and

$$z(t) = \int_0^t \psi(x)dx \tag{2}$$

$\psi(x)$ is the intensity function in SRE literature and $z(t)$ shows the accumulation with definite limits (time interval) of intensity function. These are following postulates for proposed SRGMs:

(a) Failure phenomenon tends to follow NHPP.
(b) In the fault removal process, there is a probability that new faults may generate with a fixed rate ω.
(c) At any time, fault observation/elimination rate may transform accordingly.
(d) The rate of fault removal is represented as a stochastic process.
(e) It is assumed that the total amount of faults is a monotonically increasing function of time t.

In above presumptions, (b) captures the impact of error generation.

2.3 *Stochastic Differential Equation-Based Modeling*

In this segment, error generation-based SRGMs under the influence of irregular fluctuations has been formulated. In order to capture the randomness, we have stochastic differential equation that can be communicated analytically as an ordinary differential equation that includes irregular fluctuating function of time. This equation describes the Wiener process, and mathematically it is analyzed by Itô stochastic calculus [24]. In the software system, during testing process remaining number of faults steadily decreases. Under the common postulation of software reliability growth modeling, following first-order and first-degree differential equation can be considered as

$$\frac{dN(t)}{dt} = \rho(t)(a - N(t)) \tag{3}$$

where $\rho(t)$ refers to the fault detection rate dependent on testing time "t". It has been recurrently analyzed that $\rho(t)$ is not entirely identified and is subject to several environmental consequences. Therefore, it is considered that $\rho(t)$ to be stochastic parameter instead of deterministic which inculcates an arbitrary term in ordinary differential equation Eq. (3) to transform it into the SDE. The Stochastic parameter associates with a "noise" term and it takes the following form:

$$\rho(t) = \zeta(t) + {}'\text{noise}' \tag{4}$$

The accurate behavior of the noise is difficult to be known so the function $\zeta(t)$ is presumed to be stationary:

$$\rho(t) = \zeta(t) + s\,\gamma(t) \tag{5}$$

where $\gamma(t)$ is the standard Gaussian white noise and s is a nonnegative fixed value that symbolizes measure of the irregular fluctuations. Now the SDE given in (3) can be structured as

$$\frac{\mathrm{d}N(t)}{\mathrm{d}t} = (\zeta(t) + s\gamma(t))(a - N(t)) \tag{6}$$

$$\Rightarrow \frac{\mathrm{d}N(t)}{\mathrm{d}t} = \zeta(t)(a - N(t) + s\gamma(t)(a - N(t) \tag{7}$$

One-dimensional Wiener process $W(t)$ is calculative as the integration of random variable $\gamma(t)$, and Brownian motion is the best example of Wiener process, i.e.,

$$\frac{\mathrm{d}W(t)}{\mathrm{d}t} = \gamma(t) \Rightarrow W(t) = \int \gamma(t) \tag{8}$$

These are following basic axioms for Wiener process $W(t)$ [3]:

i. Wiener process is the continuous process.
ii. $\mathrm{W}(0) = 0$
iii. Wiener process follows a normal distribution with mean 0 and variance u, i.e., $W(t+u) - W(t) \sim N(0, u)$.

SDE given in Eq. (7) can be extrapolated to the following differential equation [16, 24–26]:

$$\mathrm{d}N(t) = (\zeta(t) - \frac{1}{2}s^2)(a(t) - N(t))\mathrm{d}t + s(a(t) - N(t))\mathrm{d}W(t) \tag{9}$$

Now in stochastic environment, it is considered that the deterministic term of detection rate, $\zeta(t)$, follows hazard rate [3]:

$$\zeta(t) = \frac{f(t)}{1 - F(t)} \tag{10}$$

Below is the basic mathematical derivation structure of hazard rate [3].

It is considered that in a time interval $[t_1, t_2]$, the probability of system failure can be expressed as

$$\left.\begin{aligned} &\int_{t_1}^{t_2} f(t) = \int_0^{t_2} f(t)\mathrm{d}t - \int_0^{t_1} f(t)\mathrm{d}t \\ &\int_{t_1}^{t_2} f(t)\mathrm{d}t = F(t_2) - F(t_1) \\ &\int_{t_1}^{t_2} f(t)\mathrm{d}t = R(t_1) - R(t_2) \qquad \because R(t) = 1 - F(t) \end{aligned}\right\} \tag{11}$$

The rate of failure is structured precisely as

$$\frac{\int_{t_1}^{t_2} f(t)\mathrm{d}t}{(t_2 - t_1)R(t_1)} = \frac{R(t_1) - R(t_2)}{(t_2 - t_1)R(t_1)} \tag{12}$$

If we redefined length of the time interval as $[t, t + \Delta t]$, the failure rate can be defined as

$$\frac{R(t) - R(t + \Delta t)}{\Delta t R(t)}$$

and hazard function $\zeta(t)$ can be obtained taking limit $\Delta t \to 0$, hence

$$\left.\begin{aligned} \zeta(t) &= \frac{R(t) - R(t + \Delta t)}{\Delta t R(t)} \\ &= \frac{1}{R(t)}\left[-\frac{\mathrm{d}}{\mathrm{d}_t}R(t)\right] \\ &= \frac{f(t)}{R(t)} = \frac{f(t)}{1 - F(t)} \end{aligned}\right\} \tag{13}$$

i.e., hazard rate is the ratio of probability density function $(f(t))$ and survival function $(R(t))$. There are two conditions which satisfy hazard rate [3]:

$$(a) \quad \zeta(t) \geq 0 \quad \forall t \geq 0$$

$$(b) \quad \int_0^{\infty} \zeta(t)\mathrm{d}t = \infty$$

In the testing process, the behavior of testing is influenced by the various factors like testing effort, expenditure, testing skill, method of testing, and strategy [3]; these major factors come randomly in the environment. In order to capture the stochastic nature (uncertainty) and the generic behavior of hazard rate (discussed above) in the proposed modeling framework, we have considered that Eq. (9) can be rewritten as

$$\mathrm{d}N(t) = \left(\frac{f(t)}{1 - F(t)} - \frac{1}{2}s^2\right)(a(t) - N(t))\mathrm{d}t + s(a(t) - N(t))\mathrm{d}W(t) \tag{14}$$

Further, integrate both sides to solve the above SDE,

$$\int \mathrm{d}N(t) = \int \left(\frac{f(t)}{1 - F(t)} - \frac{1}{2}s^2\right)(a(t) - N(t))\mathrm{d}t + \int s(a(t) - N(t))\mathrm{d}W(t) \tag{15}$$

Random variable $N(t)$ is presumed to be continuous and its expected value is given as $E(N(t)) = Z(t)$. Now taking expectation on both sides,

$$\int \mathrm{d}z(t) = \int \left(\frac{f(t)}{1 - F(t)} - \frac{1}{2}s^2\right)(a(t) - z(t))\mathrm{d}t + E\left[\int s(a(t) - N(t))\mathrm{d}W(t)\right] \tag{16}$$

Using the property of Itó Integral, the second component of the Eq. (16) is zero, i.e.,

$$E\left[\int s(a(t) - N(t))\mathrm{d}W(t)\right] = 0 \tag{17}$$

which implies that the nonanticipating function will be statistically independent in the future of "t "or mathematically we can say that if we take the expected value of any nonanticipating function then it vanishes the whole component, i.e., $E\left[\int_0^T J(t)\mathrm{d}W(t)\right] = 0$, This means that, by Ito convention, the integral has a constant mean for all "t" [27].

Therefore, SDE can be modeled as

$$\int \mathrm{d}z(t) = \int \left(\frac{f(t)}{1-F(t)} - \frac{1}{2}s^2\right)(a(t) - z(t))\mathrm{d}t \tag{18}$$

Now, we consider that faults can be introduced during the debugging phase with a constant fault introduction rate ω. Therefore, fault content rate function $a(t)$ is a linear function of the expected number of faults $z(t)$ detected by time t. This environment arises when the testing team might not be able to fix the bugs perfectly, that is, while performing the activity of removing the errors some new errors are also generated with a constant rate in the total fault span. It takes the following continuous algebraic functional form [3]:

$$a(t) = a + \omega z(t); \quad \omega > 0, \tag{19}$$

So Eq. (18) can be written as follows:

$$\int \mathrm{d}z(t) = \int \left(\frac{f(t)}{1-F(t)} - \frac{1}{2}s^2\right)((a + \omega z(t)) - z(t))\mathrm{d}t \tag{20}$$

The above differential equation is solved using the seed value $z(0) = 0$,

$$z(t) = \frac{a}{(1-\omega)}\left[1 - \{1 - F(t)\}^{(1-\omega)} \mathrm{e}^{(1-\omega)\frac{1}{2}s^2 t}\right] \tag{21}$$

Equation (21) represents the mean value function based on pure generation pedagogy and it inculcates the concept of irregular fluctuating environment.

To model the fault removal phenomenon, we assume that $F(t)$ follows different distributions in above differential Eq. (21) and making use of different distributions, some models are elaborated in the following SRGMs [28, 29].

SRGM-1

Exponential distribution function is used in SRGM-1. This distribution has a constant rate and it is broadly used in modeling of software reliability. It designates the uniform distribution of faults.

Let $F(t) \sim \exp(\rho)$

i.e., $F(t) = 1 - \exp(-\rho t)$

Substituting the value of $F(t)$ in Eq. (21), we have

$$z(t) = \frac{a}{(1-\omega)}\left[1 - \mathrm{e}^{(1-\omega)\left(-\rho + \frac{1}{2}s^2\right)t}\right] \tag{22}$$

In Eq. (22), the mean value function (MVF) has been represented for exponential distribution.

SRGM-2

Let $F(t)$ be two stages Erlangian distribution function, i.e., in this distribution, fault removal process can be done in two stages: first, faults are detected and finally those detected faults are removed by the testing team in the software.

$F(t) \sim$ Erlang $(2\ \rho)$
i.e., $F(t) = [1 - (1 + \rho t)\ \exp(-\rho t)]$

$$z(t) = \frac{a}{(1-\omega)}\left[1 - (1+\rho t)^{(1-\omega)} e^{(1-\omega)(-\rho + \frac{1}{2}s^2)t}\right] \tag{23}$$

in above Eq. (23), $z(t)$ shows expected number of faults using Erlang two-stage distribution.

SRGM-3

In SRGM-3, logistic distribution has been used. It has an S-shaped representation that is widely used in reliability. It looks like the normal distribution in contour.

Let $F(t) \sim$ logistic distribution $(\rho,\ v)$

$$F(t) \sim \left(\frac{1-\ \exp(-\rho t)}{1 + v\ \exp(-\rho t)}\right)$$

$$z(t) = \frac{a}{(1-\omega)}\left[1 - \left(\frac{1+v}{1+ve^{-bt}}\right)^{(1-\omega)} e^{(1-\omega)(-\rho + \frac{1}{2}s^2)t}\right] \tag{24}$$

Above Eq. (24) integrates learning phenomenon in MVF.

SRGM-4

Erlang 3-stage distribution has been used in SRGM-4. In this distribution, phenomenon of fault removal can be done in three stages: first, testing team observes the faults and then isolates and finally they remove the isolated faults in the software system.

$$F(t) \sim [1 - \left(1 + \rho t + \frac{\rho^2 t^2}{2}\right) \mathrm{e}^{-\rho t}]$$

$$z(t) = \frac{a}{(1-\omega)}\left[1 - \left(1 + \rho t + \frac{\rho^2 t^2}{2}\right)^{(1-\omega)} \mathrm{e}^{(1-\omega)(-\rho + \frac{1}{2}s^2)t}\right] \tag{25}$$

Equation (25) represents the MVF for Erlang 3-stage distribution.

SRGM-5

Weibull distribution (in SRGM-5) is much used in SRE. Due to its versatile nature, it can take the characteristics of other type of distributions. Thus, we can say that generalization of exponential distribution is Weibull distribution because of its flexible nature.

$$F(t) \sim (1 - \mathrm{e}^{-\rho t^k}); \quad k > 0$$

where k is the shape parameter (or slope).

$$z(t) = \frac{a}{(1-\omega)}\left[1 - \mathrm{e}^{(1-\omega)(\rho t^k + \frac{1}{2}s^2 t)}\right] \tag{26}$$

This expression (in Eq. (26) describes the MVF for Weibull distribution.

SRGM-6
In SRGM-6, Rayleigh distribution is the measure of a two-dimensional random vector whose coordinates are distributed identically.

$$F(t) \sim \left(1 - \mathrm{e}^{-\rho \frac{t^2}{2}}\right)$$

$$z(t) = \frac{a}{(1-\omega)}\left[1 - \mathrm{e}^{(1-\omega)(-\frac{1}{2}\rho t^2 + \frac{1}{2}s^2 t)}\right] \tag{27}$$

Using the Rayleigh distribution, we can see the MVF with stochastic environment in Eq. (27).

3 Model Validation, Comparison Criteria, and Data Analysis

To illustrate the application of proposed SRGMs, SAS [30] has been used and parameters have been calculated. The first dataset (DS 1) has been taken from Wood [31] that comprised of 12 weeks and 61 faults. The second dataset (DS-2) has been taken Kanoun et al. [32] that comprised a total of 461 faults removed in 81 weeks.

The parameter estimation is carried out using the least square estimation procedure of nonlinear regression method and in order to statistically infer the results of nonlinear regression, five types of goodness-of-fit measures SSE, MSE, Root MSE, R^2, and Adj. R^2 are applied. The parameter estimation and comparison criteria results for DS-1 and DS-2 of all models under consideration can be viewed through Tables 1, 2, 3, and 4. The performance of SRGMs is judged by their capability to fit the past software failure data (goodness-of-fit) and predicting the future performance of the faults (as shown in Figs. 1 and 2 for respective datasets).

From Tables 3 and 4, it can be keenly analyzed that the value of R^2 and adj. R^2 are higher for SRGM-5 (for Weibull distribution) and value of all statistical error is lower in comparison with other models (SRGMs) and provides fine goodness of fit

Table 1 Parameter estimates for DS 1

Models	a	ρ	ω	s	v	k
SRGM-1	100	0.0674	0.5239	0.0400	–	–
SRGM-2	70	0.3169	0.0600	0.1600	–	–
SRGM-3	70	0.3999	0.0600	0.5577	1.2803	–
SRGM-4	71	0.4670	0.0003	0.0003	–	–
SRGM-5	67	0.04114	0.0600	0.0030	–	1.6500
SRGM-6	69	0.06000	0.0069	0.2724	–	–

Table 2 Parameter estimates for DS 2

Models	a	ρ	ω	s	v	k
SRGM-1	510	0.02673	0.0661	0.400	–	–
SRGM-2	465	0.6555	0.0600	0.0030	–	–
SRGM-3	500	0.3951	0.0600	0.0033	1.1000	–
SRGM-4	476	0.1027	0.0003	0.003	–	–
SRGM-5	500	0.5086	0.01724	0.990	–	1.0052
SRGM-6	469	0.00199	0.0069	0.030	–	–

Table 3 Comparison criteria for DS 1

Models	SSE	MSE	Root MSE	R^2	Adj. R^2
SRGM-1	200.7	20.741	4.4804	0.961	0.957
SRGM-2	78.288	7.8288	2.7980	0.985	0.983
SRGM-3	65.851	7.316	2.705	0.988	0.984
SRGM-4	93.304	8.482	2.912	0.982	0.982
SRGM-5	71.122	7.112	2.666	0.986	0.985
SRGM-6	272.2	24.74	4.975	0.945	0.947

Table 4 Comparison criteria for DS 2

Models	SSE	MSE	Root MSE	R^2	Adj. R^2
SRGM-1	8089.0	103.7	10.183	0.994	0.994
SRGM-2	48415.2	605.2	24.600	0.966	0.966
SRGM-3	19384.4	242.3	15.566	0.987	0.984
SRGM-4	107072	1338.4	36.584	0.925	0.925
SRGM-5	6780.1	86.9246	9.323	0.995	0.995
SRGM-6	134,189	1677.4	40.955	0.915	0.915

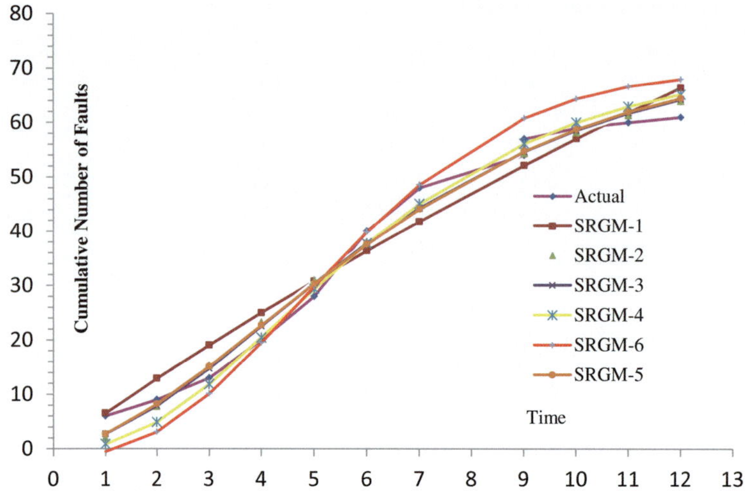

Fig. 1 Goodness-of-fit curve for DS1

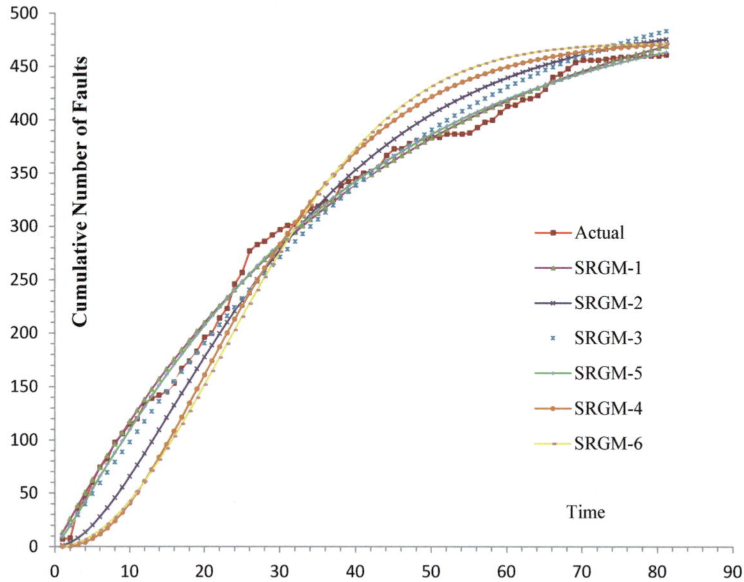

Fig. 2 Goodness-of-fit curve for DS 2

in graphically(as shown below). Finally, on the basis of performance analysis, we can say that SRGM-5 gives better results in all perspectives for both data sets.

4 Conclusion

Numerous SRGMs have developed to scrutinize the reliability growth during the testing phase and several are in pipeline. In order to capture the uncertainty in the environment, we have used the hazard rate technique for proposing all models and also discussed the different types of distribution such as exponential, Erlang 2-stage, logistic, Erlang 3-stage, Weibull, and Rayleigh. In this manuscript, we have formulated SRGMs under the influence of irregular fluctuations with the concept of error generation. Different comparison criteria are considered for the analysis and to compare the models. The proposal has been validated on software failure datasets. It is clear from the tables that the value of R^2 (which measures the percentage of total variation by the fitted curve) for SRGM-5 is higher and value of SSE, MSE, and root MSE is lower in comparison with other models and thus provides better goodness-of-fit curves for DS-1 and DS-2. Thus, SRGM-5 which is analyzed by Weibull distribution is most significant for both fault removal datasets.

Acknowledgements The research work presented in this chapter has been supported by grants to first and third author from Department of Science and Technology, via DST PURSE phase II, India.

References

1. Goel, A. L., & Okumoto, K. (1979). Time-dependent error-detection rate model for software reliability and other performance measures. *IEEE Transactions on Reliability, 28*(3), 206–211.
2. Kapur, P. K., Garmabaki, A. S., & Singh, J. (2011). Multi up-gradation software reliability model with imperfect debugging. In *Proceedings of the International Congress on Productivity, Quality, Reliability, Optimization and Modeling* (ICPQROM); Feb 7–8; New Delhi, India, 136.
3. Kapur, P. K., Pham, H., Gupta, A., & Jha, P. C. (2011). *Software reliability assessment with OR applications*. London: Springer.
4. Kapur, P. K., Tandon, A., & Kaur, G. (2010, December). Multi up-gradation software reliability model. In *Reliability, Safety and Hazard (ICRESH), 2010 2nd International Conference on* (pp. 468–474). IEEE.
5. Musa, J. D., Iannino, A., & Okumoto, K. (1987). *Software reliability: measurement, prediction, application*. McGraw-Hill, Inc.
6. Pham, H. (2006). Software Reliability Modeling. In *System Software Reliability* (pp. 153–177). London: Springer.
7. Yamada, S., Ohba, M., & Osaki, S. (1984). S-shaped software reliability growth models and their applications. *IEEE Transactions on Reliability, 33*(4), 289–292.

8. Anand, A., Deepika, Singh, N. & Dutt, P. (2016). Software reliability growth modeling based on in house testing and field testing. *Communication in dependability and quality management: An International Journal, 19*(1), 74–84.
9. Hudson, G. R. (1967). Program errors as a birth and death process. *System Development Corporation. Report SP-3011, Santa Monica, CA.*
10. Jelinski, Z., & Moranda, P. B. (1972). Software reliability research, Statistical Computer Performance Evaluation. In W. Freiberger (Ed.), 465–484.
11. Moranda, P. B. (1975, January). Prediction of software reliability during debugging. In *Proceedings Annual Reliability and Maintainability Symposium* (No. Jan 28, pp. 327–332). 345 E 47th St, New York, NY 10017-2394: IEEE-Inst Electrical Electronics Engineers Inc.
12. Schneidewind, N. F. (1972, December). An approach to software reliability prediction and quality control. In *Proceedings of the December 5–7, 1972, fall joint computer conference, part II* (pp. 837-847). ACM.
13. Musa, J. D. (1975). A theory of software reliability and its application. *IEEE Transactions on Software Engineering, 3,* 312–327.
14. Anand, A., Das, S., & Singh, O. (2016, September). Modeling software failures and reliability growth based on pre & post release testing. In *Reliability, Infocom Technologies and Optimization (Trends and Future Directions)(ICRITO), 2016 5th International Conference on* (pp. 139–144). IEEE.
15. Anand, A., Deepika, & Singh, O. (2016). Incorporating features enhancement archetype in software reliability growth modeling and optimal release time determination. *International Journal of Computer Applications (0975 – 8887), 139*(4) 1–6.
16. Yamada, S., Nishigaki, A., & Kimura, M. (2003). A stochastic differential equation model for software reliability assessment and its goodness of fit. *International Journal of Reliability and Applications, 4*(1), 1–11.
17. Yamada, S., Ohba, M., & Osaki, S. (1983). S-shaped reliability growth modeling for software error detection. *IEEE Transactions on Reliability, 32*(5), 475–484.
18. Shyur, H. J. (2003). A stochastic software reliability model with imperfect-debugging and change-point. *Journal of Systems and Software, 66*(2), 135–141.
19. Kapur, P. K., Anand, S., Yamada, S., & Yadavalli, V. S. (2009). Stochastic differential equation-based flexible software reliability growth model. *Mathematical Problems in Engineering*.
20. Kapur, P. K., Anand, S., Yadav, K., & Singh, J. (2012). A unified scheme for developing software reliability growth models using stochastic differential equations. *International Journal of Operational Research, 15*(1), 48–63.
21. Singh, O., Kapur, P. K., & Anand, A. (2011, December). A stochastic formulation of successive software releases with faults severity. In *Industrial Engineering and Engineering Management (IEEM), 2011 IEEE International Conference on* (pp. 136–140). IEEE.
22. Tamura, Y., & Yamada, S. (2006). A flexible stochastic differential equation model in distributed development environment. *European Journal of Operational Research, 168*(1), 143–152.
23. Lee, C. H., Kim, Y. T., & Park, D. H. (2004). S-shaped software reliability growth models derived from stochastic differential equations. *IIE Transactions, 36*(12), 1193–1199.
24. Øksendal, B. (2003). Stochastic differential equations. In *Stochastic differential equations* (pp. 65–84). Berlin Heidelberg: Springer.
25. Singh, O., Kapur, P. K., Anand, A., & Singh, J. (2009). Stochastic Differential Equation based Modeling for Multiple Generations of Software. In *Proceedings of Fourth International Conference on Quality, Reliability and Infocom Technology (ICQRIT), Trends and Future Directions, Narosa Publications* (pp. 122–131).
26. Singh, O., Anand, A., Kapur, P. K., & Aggrawal, D. (2012). Consumer behaviour-based innovation diffusion modelling using stochastic differential equation incorporating change in adoption rate. *International Journal of Technology Marketing, 7*(4), 346–360.
27. Gardiner, C. (1985). *Handbook of stochastic methods* (Vol. 4). Berlin: Springer.

28. Anand, A., Kapur, P. K., Agarwal, M., & Aggrawal, D. (2014, October). Generalized innovation diffusion modeling & weighted criteria based ranking. In *Reliability, Infocom Technologies and Optimization (ICRITO)(Trends and Future Directions), 2014 3rd International Conference on* (pp. 1–6). IEEE.
29. Deepika, Singh, O., Anand, A., & Singh J. N. P. (2017). Testing domain dependent software reliability growth models. *International Journal of Mathematical, Engineering and Management sciences*. 2(3), 140–149.
30. SAS, S. (2004). STAT user guide, version 9.1. 2. Cary, NC, USA: SAS Institute Inc,.
31. Wood, A. (1996). Predicting software reliability. *Computer, 29*(11), 69–77.
32. Kanoun, K., de Bastos Martini, M. R., & De Souza, J. M. (1991). A method for software reliability analysis and prediction application to the TROPICO-R switching system. *IEEE Transactions on Software Engineering, 17*(4), 334–344.

Decision Aspect Prioritization Technique for Incremental Software

Varun Gupta, Siddharth Sethi and Chetna Gupta

Abstract In incremental softwares, software is delivered incrementally where each increment implements some agreed high priority requirements. Priority of a requirement is decided by considering different aspects. A new technique has been proposed for the prioritization of decision aspects. The proposed technique prioritizes the decision aspects by using historical data thereby reducing the time taken in the prioritization and prioritization of decision aspects is done by the stakeholders. The technique aims to enhance the software success rate by optimal selection of decision aspects for prioritization of software requirements in an efficient manner which is not time-consuming and thus increases software's success rate.

Keywords Decision aspects · Decision aspect prioritization

1 Introduction

In incremental software, software is delivered incrementally where each increment implements some agreed high priority requirements [2]. Priority of a requirement is decided by considering different aspects. Prioritization must result in selection of optimal of requirements to enhance software success rate and as requirements are prioritized by aspects, it is important to choose the right aspect as the success of a software invariably depends on the selection of requirements which are prioritized by decision aspects [4].

As there are a large number of aspects, there is a need for prioritization of decision aspects. A new technique has been proposed for prioritization of decision

V. Gupta · S. Sethi (✉)
Amity University, Noida, India
e-mail: sethisiddharth504@gmail.com

V. Gupta
e-mail: vgupta7@amity.edu

C. Gupta
JIIT, Noida, India

P. K. Kapur et al. (eds.), *System Performance and Management Analytics*,
Asset Analytics, https://doi.org/10.1007/978-981-10-7323-6_11

aspects by drawing aspects from similar successful projects and the aspects considered important by the software developers. Criteria, factor, element, parameters are terms that express the same notion as aspects [3]. Decision aspects influence the priority of requirements which creates a win-win situation for the software projects.

The effectiveness of decision aspects depends on how the trade-off between aspects is negotiated and how the viewpoints of the stakeholders are handled [2].

In this paper, a new method for prioritization of decision aspects has been proposed. The proposed method uses historical data of decision aspects chosen before for similar projects thus thereby reducing the time taken in selection of decision aspects and thus not causing much delay in the project. The proposed approach can be used for prioritization of decision aspects by considering the aspects chosen from similar successful projects thereby knowing what aspects have more importance in the success of the project and as with time the aspects priority changes it allows the management to change its priority value. It also allows the management members to add the aspects that they are relevant to the project's success.

2 Related Work

Most of the work in the field of requirement prioritization focuses upon the various requirement prioritization techniques and the comparison between the various prioritization techniques. Only one paper was extracted from the literature survey and one research work [1] which was not identified at the time of literature survey due to non-indexing was also included for review. Both the papers talk about the importance of decision aspects and give a method for prioritization of decision aspects. The details of the extracted papers are as follows:

Gupta et al. [2] identified that if the right set of decision aspects are not chosen, then the efficiency of the prioritization technique will have no role to play in selection of optimal set of requirement, hence the need to prioritize the decision aspects. The paper gives a method for prioritization of decision aspects in a globally distributed environment. The proposed technique involves giving weights to development sites based on the experts and the kind of activity to be done at that site. Followed by prioritization of the aspect at every site and then generating the priority of a decision aspect by multiplying the values of the site with the priority value of the aspect at that site and summing all the values generated. If the priority values differ beyond the threshold value at a site, then local negotiations are done whereas if priority values differ beyond the threshold value at different sites then global negotiations are performed. Negotiations perform prioritization by drawing a common consensus among the stakeholders thus, creating a win-win situation. Berander et al. [1] highlight that most work has been done in finding approaches to compare different techniques, and less emphasis has been given on which decision aspects should be focused upon. In the literature study, it was found that different studies focus on different aspects, and it is highly unlikely that there exist an ultimate set of aspects. Thus, aspects vary for project to project, and the perspective of stakeholders should be considered, hence the need for decision aspect prioritization. The process adopted for the decision aspect prioritization firstly involved elicitation of decision aspects and defining of decision aspects followed by the prioritization of decision aspects and lastly, a feedback meeting is held. The overall literature doesn't really focus on decision aspects. The work that has been identified in literature doesn't get executed in the software engineering practices of selection of decision aspects.

3 Problem Statement

This paper uses historical data of decision aspects chosen before for similar successful projects and negotiations for the prioritization of decision aspects. This is achieved by gathering decision aspects used in successful similar aspects with their prioritization values and negotiations if there is a difference in opinion thereby creating a win-win situation.

> How the software organizations will optimally prioritize a large number of ever-changing aspects implying a large number of stakeholders.

4 Proposed Approach

As in incremental software, a different set of requirements are implemented in each version, and it is necessary to choose the right set of requirements for its success. As for example for gaming application if the graphics aren't proper it will ensure that the game isn't a success as it won't appeal to the gamers. Thus, to increase the success rate of the product following steps must be undertaken.

- First of all, similar successful projects should be seen and taken a cue from like which aspects have been considered important and check the priority value assigned to them and make a list of all the aspects by taking the average of the priority value of similar aspects.
- The list generated should be sent to all the members of the management team and each stakeholder/member should have assigned weightage as per their importance to the team.
- As the relevance of an aspect may increase or decrease with time as the requirements are ever changing each member can modify the value of an aspect by a range from −9 to +9. So, if $P_{Aspect1}$ is a value of a particular aspect then, its value is

$$P_{Aspect1} = [W_{s1} * (P + P_{UPDT1}) + W_{s2} * (P + P_{UPDT2}) + \cdots + W_{sn} * (P + P_{UPDTN})]/n \quad (1)$$

where W_{Si} the weight of a stakeholder, P is the average value as generated from previous successful similar projects and P_{UPDTi} is the value updated by the stakeholder.

- Following this, an empty list should be sent to all stakeholders ask them to send the aspects they consider to be important along with their priority value. The generated list should be merged and the average of similar aspects should be done as

$$P_{Aspect2} = [W_{s1} * (P_{S1}) + W_{s2} * (P_{S2}) + \cdots + W_{sn} * (P_{Sn})]/n \quad (2)$$

- Both the lists generated from Eq. (1)–(2) should be merged by taking the average of value of an aspect which occurs in both the lists. Based on the final list aspects should be ranked based on their priority value.
- After this, a feedback meeting should be held where if there is a difference of opinion, negotiations among the stakeholders should take place to draw a common consensus among them.

Algorithm: Decision Aspect Prioritization

STEP 1 Make a list of decision aspects which have been used in the similar successful projects and with their prioritization values.

STEP 2 In the list, take the average of the decision aspects which represent the same meaning for ex. Customer satisfaction and business value by taking average of those aspects.

STEP 3 Send the list generated from step 1 to the management team.

STEP 4 Assign weight W_s to each team member according to their importance to the team.

STEP 5 Ask each team member to update the priority value *P* of each aspect by adding value P_{UPDT} to the priority value, where P_{UPDT} ranges from −9 to 9.

$$P_{Aspect1} = [W_{s1} * (P + P_{UPDT1}) + W_{s2} * (P + P_{UPDT2}) + \cdots + W_{sn} * (P + P_{UPDTN})]/n \quad (3)$$

STEP 6 Calculate P_{aspect} for all aspects following step 4.

STEP 7 Send the stakeholders an empty list and ask them to give the aspects they consider important with their priority value P_{ASP}.

STEP 8 Each member should list all the aspects considered to be important and give their priority value.

STEP 9 Calculate priority value of each aspect P_{Aspect} as

$$P_{Aspect2} = [W_{s1} * (P_{S1}) + W_{s2} * (P_{S2}) + \cdots + W_{sn} * (P_{Sn})]/n \quad (4)$$

STEP 10 Merge both the list and rank aspects based on their priority value P_{Aspect}

$$P_{Aspect} = [P_{Aspect1} + P_{Aspect2}]/2 \quad (5)$$

STEP 11 Hold a feedback meeting to perform negotiations if there is difference in opinion to draw a common ground and create a win-win situation.

5 Research Questions

The foremost objective of this research is to provide a method for decision aspect prioritization which uses data from similar successful projects thus thereby reducing the reliance on gut feeling/personal opinions. The paper tries to answer the following questions:

RQ [1]: How to optimally prioritize a large number of aspects?
RQ [2]: How decision aspect prioritization will be able to satisfy all stakeholders?

6 Conclusion and Future Work

Paper proposes a new approach which uses data from similar successful projects for the optimal prioritization of decision aspects. Simple prioritization of aspects uses greatest number of votes or value [2]. As aspects are an important part of prioritization of requirement, selection of right decision aspects is important. The above-proposed algorithm reduces time consumption by taking data from similar successful projects available for prioritization of decision aspects. As aspects and priority value of aspects are also taken for similar software projects, it reduces the risk and over-reliance on guesswork as has been the case with the software's specially in the beginning of software development.

With respect to research questions given in the previous section is concluded as follows:

RQ1: To optimally prioritize a large number of aspects findings from similar related projects are used which will decrease the reliance on personal opinions and gut feeling of stakeholders involved thereby resulting in optimal selection of decision aspects as these aspects have been resulted in successful projects in the past.

RQ2: The decision aspect prioritization will be able to satisfy stakeholders as the priority values of aspects taken from previous successful projects can be modified by the stakeholders and they can give the aspects they consider to be important and if there is a conflict of interest negotiations can take place among the stakeholders.

This paper provides future work directions. Tools for prioritization of decision aspects can also be expected and techniques shall be made which shall reduce reliance on guesswork. This approach can further be extended for globally distributed environments.

References

1. Berander, P. (2007). Evolving Prioritization for software produce management. Ph.D Thesis.
2. Gupta, V., Chauhan, D. S., & Dutta, K. (2012). Hybrid decision aspect prioritization technique for globally distributed developments. *Procedia Engineering, Elsevier 38,* 3614–3627.
3. Sher, F., Jawawi, D. N. A., Mohamad R., & Babar, M. I. (2014). Requirements prioritization techniques and different aspects for prioritization a systematic literature review protocol. In *Software Engineering Conference (MySEC), 2014 8th Malaysian*, Langkawi, pp. 31–36.
4. Sher, F., Jawawi, D. N. A., Mohamad R., & Babar, M. I. (2014). Multi-aspects based requirements prioritization technique for value-based software developments. In *Emerging Technologies (ICET), 2014 International Conference on*, Islamabad, 2014, pp. 1–6.

Reliability Growth Analysis for Multi-release Open Source Software Systems with Change Point

Anu G. Aggarwal, Vikas Dhaka, Nidhi Nijhawan and Abhishek Tandon

Abstract Open source software has now become an essential part of the business for huge segment of developers to enhance their visibility in public. Many of the open source communities are continuously upgrading the software through series of releases to improve its quality and efficiency. Here in this paper, general framework is presented to model fault removal process (FRP) for multiple releases of OSS using the concept of change point on discrete probability distribution. To validate our model, we have chosen two successful open source projects-Mozilla and Apache for its multi release failure data sets. Graphs representing goodness of fit of the proposed model have been drawn. The parameter estimates and measures of goodness of fit criteria suggest that the proposed SRGM for multi release OSS fits the actual data sets very well.

Keywords Open source software (OSS) · Software reliability growth model (SRGM) · Multiple releases of OSS · Change point

1 Introduction

Over a few years, open source software has come a long way. It provides its users boundless liberty to freely use, examine and modify the source code. Large number of software firms is stepping up open source initiatives for the better stability, security, quality and the accessibility [1, 2]. However, developers as of today find software multi up gradation as almost inevitable so as to sustain dynamic com-

A. G. Aggarwal (✉) · N. Nijhawan
Department of Operational Research, University of Delhi, New Delhi, India
e-mail: anuagg17@gmail.com

V. Dhaka
Daulat Ram College, University of Delhi, New Delhi, India

A. Tandon
Shaheed Sukhdev College of Business Studies, University of Delhi, New Delhi, India

P. K. Kapur et al. (eds.), *System Performance and Management Analytics*,
Asset Analytics, https://doi.org/10.1007/978-981-10-7323-6_12

puting needs of their customers and extreme market challenges. To unveil the latest versions of their product with more features, software companies embrace the multi release development approach. These upgraded versions may lead to high fault count in the software. Software upgrade is worthy only if it leads to better reliability with less safety hazard.

Software reliability has been an important concern in IT industry and is used to appraise software quality relating to the residual fault content in the system. To estimate software reliability while testing phase of development process, SRGMs are very useful. Researchers have contributed their effort on the study of multi up-gradation for closed source software [3, 4, 5, 6, 7, 8] but the same approach is limited in OSS [9, 10, 11, 12, 13]. Due to distinct contributors with different level of skills and resources, it is imperative and realistic to consider the effect of change point during fault removal phenomena of open projects. During the testing process, sudden changes in debugging situation may results to change in fault removal rate (FRR) depending on the severity of faults detected, changing strategies adapted by testing team, program size, testing efforts and software testability. The time points at which change in FRR appear are known as change points. Lately, the concept of change point with respect to fault removal process has been widely discussed [14, 15] but the effect of up-gradations incorporating change point has been discussed by a few [16, 17, 18].

The proposed model considers the possibility of having multiple change points during reliability growth of OSS with successive releases under discrete time scale. Discrete models relate expected number of faults removed to number of test cases executed or number of testing periods for which testing continued and perform as good or sometimes better as compared to continues time models [19].

This paper is divided into eight sections as follows: Sect. 2 outlines few important notations followed by a list of assumptions stated in Sect. 3. Model development of fault removal process for an OSS is presented using hazard rate function in Sect. 4. Section 5 presents the generalized framework for modeling the multiple releases incorporating the effect of change point and in particular discrete Weibull function has been used as a hazard rate function for illustration. The parameters estimation of the proposed model (on considered change points) using real life fault count data sets for two OSS projects: Mozilla and Apache for three consecutive releases followed by the comparison analysis has been done in Sect. 6. Curves representing goodness of fit have been shownand data analysis has been performed in Sect. 7. Section 8 presents conclusion of the proposed work.

2 Notations

n	Number of testing periods
$m(n)$	Expected number of faults removed/corrected during n
$m_i(n)$	Mean value function (MVF) for FRP corresponding to releasei; $i =$ 1,2,…
$\lambda(n)$	Intensity functions for FRP
a	Initial number of faults present in the software just before the testing starts
a_i	Initial number of faults present in the ith release of software
b_i	Per fault FRR corresponding to release i
k_i	Shape parameter corresponding to release i
$F(n)$	Probability distribution function (PDF) for the number of faults removed/corrected during n
$f(n)$	Probability mass function (PMF) for the number of faults removed/corrected during n
$b_{i(j-1)}$	FRR per remaining fault before change point during n for ith release (j = 1, 2, … k + 1)
b_{ij}	FRR per remaining fault after change point for ith release
n_{ic}	Change point in ith release, $\tau_{i-1} < n_{ic} \leq \tau_i$; $c = 1, 2, \ldots .k$
τ_i	Release time for ith release

3 Assumptions

1	In the course of software testing, a failure is said to have occurred when test cases are run and the desired and actual output obtained do not match
2	During execution software is subject to failures due to leftover faults of the software
3	At any time failure rate is proportional to the leftover faults of the software
4	Fault content in the software before the testing starts is finite
5	Debugging process is perfect i.e., new faults do not occur in the software during its testing
6	FRP in the software is modeled by Non Homogeneous Poisson Process (NHPP) while the initial fault content is assumed to be a Poisson random variable
7	From failure observation point of view, all the faults are mutually independent
8	FRR may not remains constant but it may changes during execution of any test cases due to varying nature of bug reporting or OSS population user growth
9	There may exist multiple change points during different releases of the software

4 Model Development

Suppose $\{N(n), n \geq 0\}$ be a discrete time NHPP with $m(n)$ as the MVF which describes software failure phenomenon. It may be shown that:

$$\Pr(N(n) = k) = \frac{(m(n))^k}{k!}\exp((-m(n)) \quad = 0, 1, 2.\ldots \tag{1}$$

Using the assumption (3) stated above, the difference equation for FRP may be given as:

$$m(n+1) - m(n) = \frac{F(n+1) - F(n)}{1 - F(n)}(a - m(n)). \tag{2}$$

Now on solving Eq. (2) with initial condition $m(0) = 0$, $m(n)$ for FRP may be expressed as:

$$m(n) = aF(n). \tag{3}$$

By selecting an appropriate $F(n)$, we can derive MVF for the discrete SRGM. Using Eq. (3), intensity function $\lambda(n)$ is given by:

$$\lambda(n) = a\, f(n). \tag{4}$$

where

$$f(n) = \Delta F(n) = F(n+1) - F(n), \lambda(n) = m(n+1) - m(n). \tag{5}$$

We can rewrite

$$\lambda(n) = \frac{f(n)}{1 - F(n)}\,(a - m\,(n)). \tag{6}$$

Introducing $s(n) = \frac{f(n)}{1-F(n)}$

Where $s(n)$ denotes hazard rate function or the rate at which failure occurs per leftover fault. Rewriting Eq. (2) as:

$$m(n+1) - m(n) = s(n)[a - m(n)]. \tag{7}$$

5 Modeling Framework for Multiple Releases of OSS with Change Point

On the basis of bug reports of previous release from the user end involving millions of spontaneous contributors or volunteers, OSS progressively improves during its operational phase [20]. In the meantime new version occupied with added functional requirements is made available by the developers for its clients, the modeling for which is done using the fact that during testing of newly integrated code a number of bugs reported after the release of just previous version are removed along with faults which were introduced in the new code. Multi release OSS development process continues with their bug reports and their debugging to bring software quality and enhancement.During testing, different phase of OSS development process as suggested by Jorgensen [21] relate to each other for multi release software as represented in Fig. 1.

This section proposes a generalized framework to model FRP for multiple releases of OSS using the concept of change point so as to represent change in the FRR for each release.

Using aforementioned assumptions along with the possibility of multiple change points (say k) viz. $n_{i1}, n_{i2}, \ldots n_{ik} (where\, 0 < n_{ic} \le \tau_i)$, FRP for ith release may be described as:

$$m_i(n+1) - m_i(n) = \begin{cases} \dfrac{f_{i1}(n)}{1 - F_{i1}(n)}(a_i - m_i(n));\ 0 \le n \le n_{i1} \\ \dfrac{f_{i2}(n)}{1 - F_{i2}(n)}(a_i - m_i(n));\quad n_{i1} < n \le n_{i2} \\ \quad\quad\cdot \\ \quad\quad\cdot \\ \quad\quad\cdot \\ \dfrac{f_{i(k+1)}(n)}{1 - F_{i(k+1)}(n)}(a_i - m_i(n));\quad n > n_{ik}. \end{cases} \tag{8}$$

where, $s_i(n) = \frac{f_{ij}(n)}{1 - F_{ij}(n)}$; j = 1, 2,…, k + 1 denotes hazard rate for ith release before and after change point. Solving above difference equations for mean value function

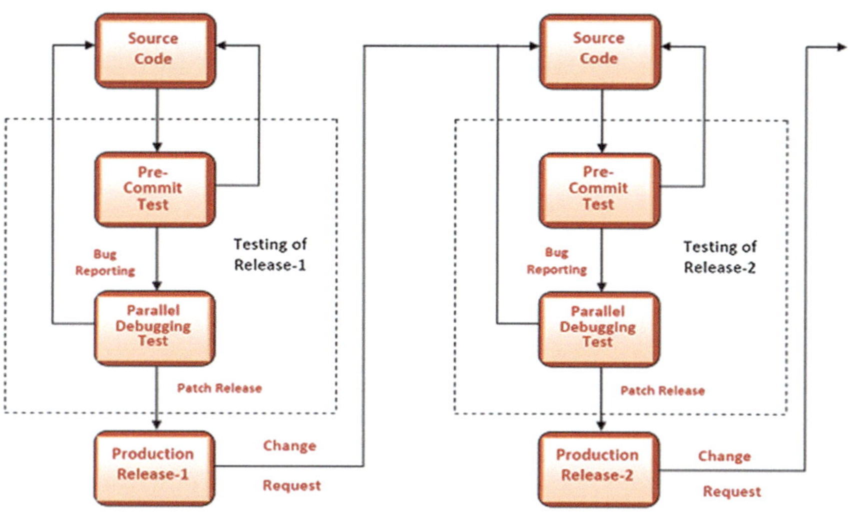

Fig. 1 Relation between different phases of testing process for multi release OSS

$m_i(n)$ with initial conditions $m_i(n=0)=0$ and $m_i(n=n_{ic})=m(n_{ic}))$; $c =$ 1, 2,…k, we get:

$$m_i(n) = \begin{cases} a_i^* F_{i1}(n) & ;0 \le n \le n_{i1} \\ a_i^*\left[1 - \dfrac{(1-F_{i1}(n_{i1}))}{(1-F_{i2}(n_{i1}))}.(1-F_{i2}(n))\right]; n_{i1} < n \le n_{i2} \\ a_i^*\left[1 - \dfrac{(1-F_{i1}(n_{i1}))}{(1-F_{i2}(n_{i1}))}\cdot\dfrac{(1-F_{i2}(n_{i2}))}{(1-F_{i3}(n_{i2}))}.(1-F_{i3}(n))\right]; n_{i2} < n \le n_{i3} \\ a_i^*\left[1 - \dfrac{(1-F_{i1}(n_{i1}))}{(1-F_{i2}(n_{i1}))}\cdot\dfrac{(1-F_{i2}(n_{i2}))}{(1-F_{i3}(n_{i2}))}\cdot\dfrac{(1-F_{i3}(n_{i3}))}{(1-F_{i4}(n_{i3}))}.(1-F_{i4}(n))\right]; n_{i3} < n \le n_{i4} \\ . \\ . \\ . \\ a_i^*\left[\begin{matrix} 1 - \dfrac{(1-F_{i1}(n_{i1}))}{(1-F_{i2}(n_{i1}))}\cdot\dfrac{(1-F_{i2}(n_{i2}))}{(1-F_{i3}(n_{i2}))}\cdot\dfrac{(1-F_{i3}(n_{i3}))}{(1-F_{i4}(n_{i3}))}\cdots \\ \cdots\cdots\dfrac{(1-F_{ik}(n_{ik}))}{(1-F_{i(k+1)}(n_{ik}))}.(1-F_{i(k+1)}(n)) \end{matrix}\right]; n > n_{ik}. \end{cases} \tag{9}$$

where

$$a_1^* = a_1,$$

$$a_{i+1}^* = a_i^*\left(\begin{matrix} \dfrac{(1-F_{i1}(n_{i1}))}{(1-F_{i2}(n_{i1}))}\cdot\dfrac{(1-F_{i2}(n_{i2}))}{(1-F_{i3}(n_{i2}))}\cdot\dfrac{(1-F_{i3}(n_{i3}))}{(1-F_{i4}(n_{i3}))}\cdots \\ \cdots\dfrac{(1-F_{ik}(n_{ik}))}{(1-F_{i(k+1)}(n_{ik}))}.(1-F_{i(k+1)}(\tau_i)) \end{matrix}\right); i = 1,2,3,\ldots. \tag{10}$$

Mean value function for failure count of discrete SRGMs corresponding to distinct debugging scenario may easily be derived on selecting suitable probability distribution function. To illustrate the accuracy of the model we have used discrete Weibull distribution function:

$$i.e.,\ F_i(n) = \left(1 - b_i^{n^{k_i}}\right);\quad n_{i-1} < n \le n_i. \tag{11}$$

The above functional form may be considered as one of the most appropriate distribution function to monitor user-driven process i.e., bug reports in OSS [22, 23, 24]. The highly flexible Weibull distribution function may changes its shape for different values of its shape parameter (here *k*) as shown in Fig. 2. This unique property is helpful to evaluate reliability across various applications and appraise the rate of bug reporting from the user-end in OSS [25]. In particular, when $k > 1$, it shows the increasing nature to represent huge bug reports as a consequence of user growth with time, $k = 1$ shows constant rate of failure and the corresponding steady bug reports and when $k < 1$, it depicts decreasing failure rate of undetected bugs and indicates product obsolescence due to reduction in its users and their bug reports.

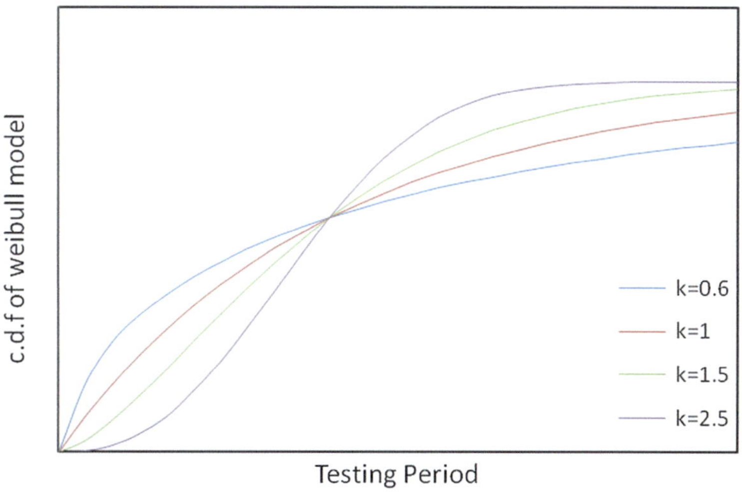

Fig. 2 Weibull distribution function for distinct values of shape parameter, *k*

6 Data Sets, Parameter Estimates

In this section, unknown parameters of the proposed SRGM are estimated taking into account the effect of multiple change points in different releases by using two software fault data sets of Mozilla and Apache OSS projects collected from bug tracking system of Bugzilla (https://bugzilla.mozilla.org/, (https://bugzilla.apache.org/)) [26] on three consecutive releases Firefox 3.0, 3.5, 3.6 and Apache 2.0.35, 2.0.36, 2.0.39.

In DS-1, faults count in three successive releases were observed during 53, 28 and 50 weeks. In the very first release there were 48 faults observed in 53 weeks and the second release was tested 28 weeks which reported 93 errors. In 50 weeks testing of third release 128 faults were found.

For DS-2, first release was tested for 32 days and there were found 72 defects whereas the second release was tested for 41 days and 47 fault counts were found. In third release 53 bugs were observed during testing of 53 days and. Here, we have used the actual release-update time for each version i.e., 32, 41 and 53 days from each release respectively.

To quantify goodness of fit of given model, mean square error (MSE) and regression coefficient (R-square) have been used. The parameter estimation for both the data sets (DS-1, DS-2) can be viewed from Tables 1 and 2 respectively. In addition, a comparative analysis has been done among the proposed model and an SRGM without change point [27] which may be seen through Tables 3 and 4. It can observed that the proposed model with change point gives better values of R^2 and MSE corresponding to each release for DS-1 and DS-2.

Table 1 Parameter estimates and goodness of fit criteria for DS-1

		Release 1 (with two change points)	Release 2 (with no change points)	Release 3 (with two change points)
Model parameters	a_i	49.585	55.628	34.561
	b_{i1}	0.941	0.987	0.967
	b_{i2}	0.963	–	0.929
	b_{i3}	0.983	–	0.999
	k_{i1}	0.809	1.423	0.843
	k_{i2}	1.026	–	0.809
	k_{i3}	1.44	–	1.71
Goodness of fit criteria	R^2	0.99	0.995	0.994
	MSE	1.422	1.253	0.624

Table 2 Parameter estimates and goodness of fit criteria for DS-2

		Release 1 (with one change points)	Release 2 (with two change points)	Release 3 (with no change points)
Model parameters	a_i	79.913	47.343	71.67
	b_{i1}	0.953	0.947	0.976
	b_{i2}	0.768	0.81	–
	b_{i3}	–	0.957	–
	k_{i1}	1.081	0.824	0.974
	k_{i2}	0.759	0.387	–
	k_{i3}	–	1.203	–
Goodness of fit criteria	R^2	0.994	0.996	0.996
	MSE	2.596	0.631	0.903

Table 3 Comparitive analysis for DS-1

Model	Comparison Criteria	Release 1	Release 2	Release 3
SRGM without Change Point [27]	R^2	0.973	0.995	0.987
	MSE	3.869	1.011	1.233
Proposed SRGM with change point	R^2	0.99	0.995	0.994
	MSE	1.422	1.253	0.624

Table 4 Comparitive analysis for DS-2

Model	Comparison criteria	Release 1	Release 2	Release 3
SRGM without Change Point [27]	R^2	0.994	0.958	0.996
	MSE	2.93	7.26	0.89
Proposed SRGM with change point	R^2	0.994	0.996	0.996
	MSE	2.596	0.631	0.903

7 Goodness of Fit Curves and Data Analysis

To locate change point, we have drawn curves corresponding to actual data sets and we look for the kinks on the curves.

The change-points of dataset DS-1 have been located around the twenty-ninth week and forty-second week of release-1. There is no significant change point seen on the data set of release-2 whereas eighth and twenty-third week (after second release) of third release are located as the change points. Figures 3, 4 and 5 represent graphically the real (or actual) and the estimated number of faults which are removed for three successive releases of the Mozilla corresponding to Firefox (DS-1) 3.0, 3.5, 3.6.

There is one change point corresponding to seventeenth week is observed for release-1 dataset of DS-2 and two change points corresponding to eighteenth and twenty-eighth week (after first release) have been located on release-2 data set whereas there is no change point is found on the data set of release-3. Goodness of fit of the three software releases of Apache (DS-2) 2.0.35, 2.0.36, 2.0.39 are given in Figs. 6, 7 and 8 respectively.

Fig. 3 Goodness of fit for release 1 of DS-1 with two change points

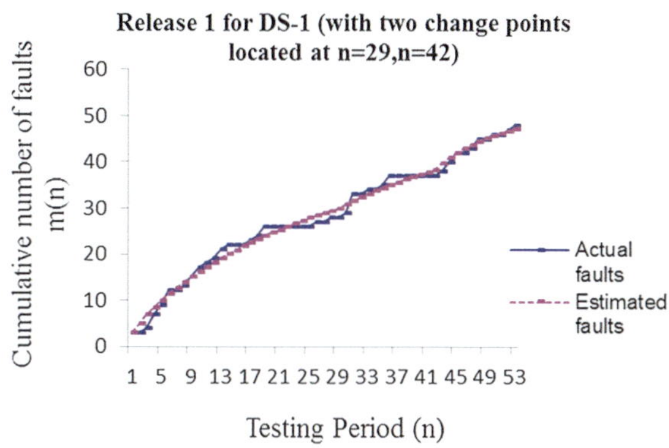

Fig. 4 Goodness of fit for release 2 of DS-1 with no change point

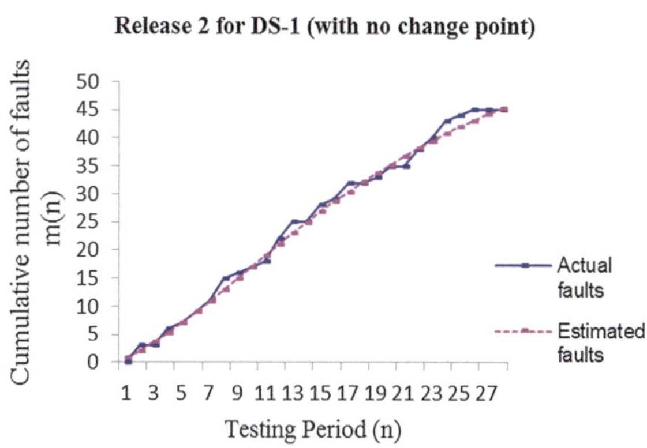

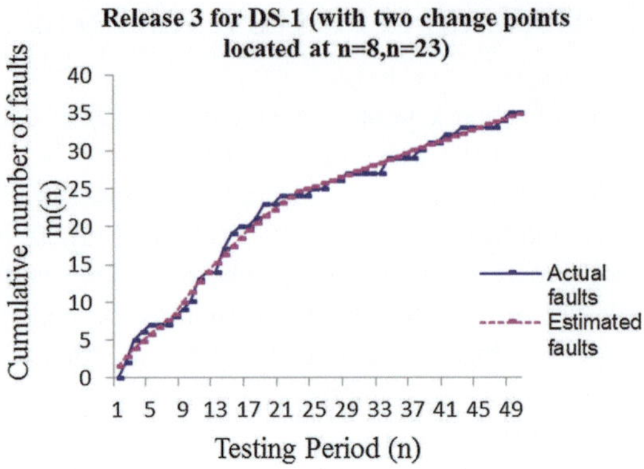

Fig. 5 Goodness of fit for release 3 of DS-1 with two change points

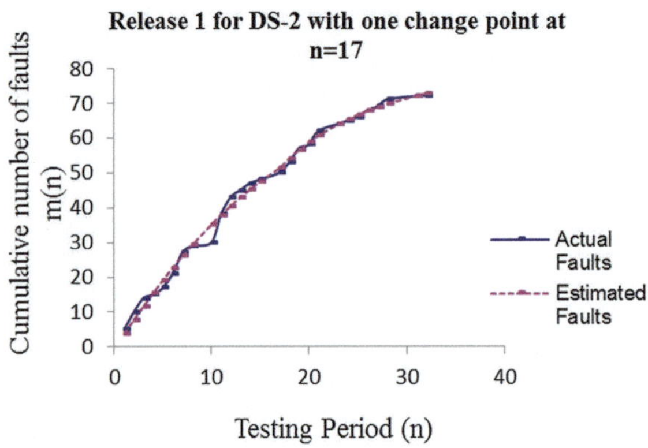

Fig. 6 Goodness of fit for release 1 of DS-2 with one change point

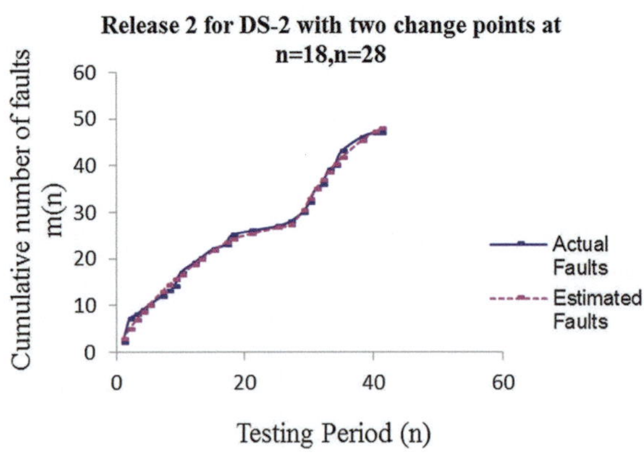

Fig. 7 Goodness of fit for release 2 of DS-2 with two change points

It may be observed from the Tables (located above) that the proposed model gives better values of R^2 and MSE corresponding to release 2 for DS-1 and corresponding to release 3 for DS-2. From the Figures, we may see that each release more or less represents similar pattern of bug frequencies and the Weibull

Fig. 8 Goodness of fit for release 3 of DS-2 with no change point

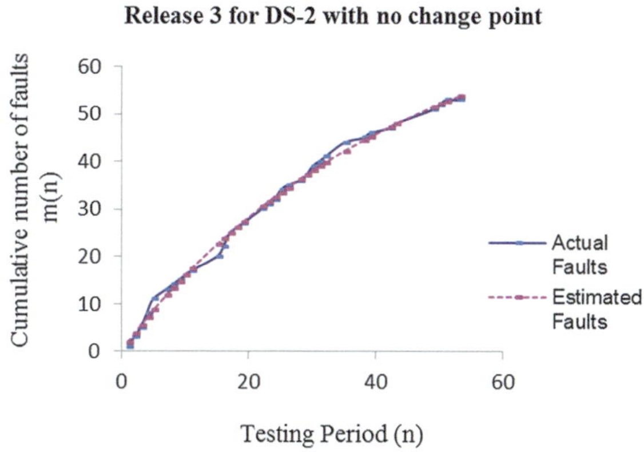

distribution function is found to be quite suitable to represent the variation incurred. Estimated values of model parameters are fairly close to the real values, thus the proposed SRGM fits the release data sets quite well.

8 Conclusion

In this paper, a generalized framework has been presented to model multiple releases of open source software (OSS) with the effect of change point. The proposed work has been illustrated in a numerical example using discrete Weibull distribution and taking into consideration multiple change points on two real life failure data sets cited in the text for three subsequent releases of OSS. The accuracy of the proposed model has been investigated using parameter estimates, goodness of fit curves and comparison criteria results and it may be concluded that proposed model fits the given data sets really well.

References

1. Raymond, E. S. (2001). The cathedral and the bazaar, musings on linux and open source by an accidental revolutionary, 2nd Ed. O'Reilly & Associates.
2. Tamura, Y., & Yamada, S. (2013). Reliability assessment based on hazard rate model for an embedded OSS porting phase. *Journal of Software Testing, Verification and Reliability, 23,* 77–88.
3. Aggarwal, A. G., Kapur, P. K., & Nijhawan, N. (2015). A discrete SRGM for multi release software system with faults of different severity. In Press. *International Journal of Operational Research.*
4. Aggarwal, A. G., Kapur, P. K., & Nijhawan, N. (2015, February). A discrete SRGM for multi-release software system with imperfect debugging and related release policy. In *Proceedings of IEEE 1st International Conference on Futuristic trend in Computational*

Analysis and Knowledge Management (ABLAZE-2015), held at Amity University, Uttar Pradesh, pp. 186–192.
5. Garmabaki, A. H. S., Aggarwal, A. G., & Kapur, P. K. (2011). Multi up-gradation software reliability growth model with faults of different severity. In *Proceedings of IEEE International Conference on Industrial Engineering and Engineering Management (IEEM)*, New York, pp. 1539–1543.
6. Hu, Q. P., Peng, R., Xie, M., Ng, S. H., & Levitin, G. (2011). Software reliability modelling and optimization for multi-release software development processes. In *Proceedings of IEEE International Conference on Industrial Engineering and Engineering Management*, pp. 1534–1538.
7. Kapur, P. K., Aggarwal, A. G., & Nijhawan, N. (2014). A discrete SRGM for multi-release software system. *International Journal of Industrial and Systems Engineering, 16,* 143–155.
8. Kapur, P. K., Aggarwal, A. G., & Nijhawan, N. (2014). A unified discrete modeling framework for detection and correction processes of multi-release software. In: O. Parkash (Ed.), mathematical modeling and applications, Chapter 9, pp. 129–149. (ISBN-13: 978-3-659-59470-0, ISBN-10: 3659594709). LAMBERT Academic Publishers.
9. Gratus, V. A., & Pratibha, P. (2013). Multi-release software: An approach for assessment of reliability metrics from field data. *Mining Intelligence and Knowledge Exploration, Lecture Notes in Computer Science, 8284,* 475–486.
10. Kim, J., Malaiya, Y. K., & Ray, I. (2007). Vulnerability discovery in multi-version software systems. In *9th Ninth IEEE International Symposium on High-Assurance Systems Engineering (HASE'05)*, pp. 141–148.
11. Li, X., Li, Y. F., Xie, M., & Ng, S. H. (2011). Reliability analysis and optimal version-updating for open source software. *International Journal of Information and Software Technology, 53,* 929–936.
12. Tamura, Y., & Yamada S. (2007). Software reliability assessment and optimal version-upgrade problem for open source software. In *Proceedings of IEEE International Conference on Systems, Man, and Cybernetics*, Montreal, Canada, pp. 1333–1338.
13. Tandon, A., Aggarwal, A. G., & Nijhawan, N. (2016). An NHPP SRGM with change point and multiple releases. *International Journal of Information Systems in the Service Sector (IJISSS-IGI Global)*, 8, 57–68.
14. Chang, Y. P. (2001). Estimation of parameters for non-homogenous poisson process software reliability with change-point model. *Communications in Statistics: Simulation and Computation, 30,* 623–635.
15. Inoue, S., & Yamada, S. (2011). A bivariate software reliability model with change-point and its applications. *American Journal of Operations Research, 1,* 1–7.
16. Aggarwal, A. G., Tandon, A., & Nijhawan, N. (2015, March). A change point based discrete SRGM for multi-release software system. In *Published in the Proceedings of International Conference on Evidence Based Management (ICEBM)*, held at BITS Pilani, Pilani, pp. 674–678.
17. Nijhawan, N., & Aggarwal, A. G. (2015). On development of change point based generalized SRGM for software with multiple releases. In: P*roceedings of IEEE 4th International Conference on Reliability, Infocom Technologies and Optimization (ICRITO) at Amity Institute of Information Technology*. Amity University, Uttar Pradesh pp. 1–6.
18. Yang, J., Liu, Y., Xie, M., & Zhao, M. (2016). Modeling and analysis of reliability of multi-release open source software incorporating both fault detection and correction processes. *The Journal of Systems and Software, 115,* 102–110.
19. Rafi, Sk. M. D., & Akthar, S. (2011). Discrete software reliability growth models with discrete test effort functions. *International Journal of Software Engineering & Applications*, *2,* 11–21.
20. Gyimothy, T., Ferenc, R., & Siket, I. (2005). Empirical validation of object-oriented metrics on open source software for fault prediction. *IEEE Transaction on Software Engineering, 31,* 897–910.

21. Jorgensen, N. (2001). Putting it all in the trunk: incremental software development in the free BSD open source project. *Information Systems Journal, 11,* 321–336.
22. Englehardt, J. D., & Li, R. C. (2011). The discrete Weibull distribution: An alternative for correlated counts with confirmation for Microbial counts in water. *Risk Analysis, 31,* 370–381.
23. Nakagawa, T., & Osaki, S. (1975). The discrete weibull distribution. *IEEE Transactions on Reliability, 24,* 300–301.
24. Nekoukhou, V., & Bidram, H. (2015). The ExponentiatedDiscrete Weibull Distribution. *Statistics & Operations Research (SORT), 39,* 127–146.
25. Rahmani, C., Siy, H. & Azadmanesh, A. (2009). An experimental analysis of open source software reliability. In *F2DA Workshop on 28th IEEE Symposium on Reliable Distributed Systems, Niagara Falls.*
26. Bugzilla, https://bugzilla.mozilla.org/, https://bugzilla.apache.org/.
27. Nijhawan, N., Aggarwal, A. G. & Dhaka, V. (2017). A discrete modeling framework for multi-release open source software systems. *Accepted for Publication in International Journal of Innovation and Technology Management (IJITM).* World Scientific.

Improvisation of Reusability Oriented Software Testing

Varun Gupta, Akshay Mehta and Chetna Gupta

Abstract The study involves the factors that are responsible for software testing and determining the extent of reusability on the basis of test outcomes. It deals with improving and promoting practices of reusability along with providing a method to improve such practices. A case study was conducted in some of the leading organizations related to reusability practices involved in developing a new software keeping in consideration the test cases generated. According to the results, the factors that emphasize the software testing process are majorly cost and time that play an efficient role in the development of software. It is also necessary to focus on test process definition, testing automation along with the testing tools.

Keywords Reusability · Software testing

1 Introduction

In the arena of Software Engineering, reusability is always a challenge. Software Reuse plays an important during the development of a new software as it reduces cost and enhances the quality of the designed software. According to Gupta et al. [1], reusability refers to reusing the existing assets in some form or other. These assets may include product or immediate products of the software development lifecycle like code modules, software component, test suites, designs documents, requirement artifacts and documentation etc. During the development of a new software, the organization focuses on the testing automation techniques as well keeping in concern the reusability aspect as well. Testing is of two types-Automated Testing and Manual Testing. According to Dustin et al. [2], automation testing

V. Gupta · A. Mehta (✉)
Amity University, Noida, India
e-mail: akshay.mehta3121@gmail.com

C. Gupta
JIIT, Noida, India

P. K. Kapur et al. (eds.), *System Performance and Management Analytics*,
Asset Analytics, https://doi.org/10.1007/978-981-10-7323-6_13

refers to the use of automation of software testing activities including the development of the software and execution of the test scripts along with verification of requirements for testing and use of testing tools.

Software Reusability can minimize the efforts and resources that are being used to develop an incremental software [3]. Likewise, the majority of software developing organization are opting reusability practices and encouraging it thereby increasing the productivity, decreasing time and cost and overcome software development crisis. Kits [4] estimated that reusability reduces more than 50% of the efforts during the development phase.

Different studies have shown that testing automation is a significant area of interest with the aim to improvise the degree of automation which can be implemented by developing advanced techniques for generating test inputs or by finding the suitable automated solutions for the same as proposed by Bertolino [5].

Although there are studies and report that have been conducted on the automation testing practices but there seems to be lack of studies that monitors the practices undergoing in the organization during the development of incremental software. According to Briand [6], "empirical studies help in comparing and improving the software testing practices and techniques." To provide enhanced knowledge about the practices being held at the organization, we've conducted a survey in five big organizations with the purpose to understand the factors which play the key role in such practices.

The organizations were selected from the companies who produce telecommunication software and practice testing automation. The data was collected by interviewing different organizational positions in the respective organizations. The data which is collected after being interviewed in these organization are analyzed and then observations were taken in concern. As no such research is done regarding the practices of testing automation in recent years, this will help us to get the ground reality of the practices opted in these organizations.

The paper is structured as follows: Literature Review in Sect. 2 & Research Process in Sect. 3. The analysis result in Sect. 4. To conclude the research, the discussion and conclusion is there in the Sect. 5.

2 Related Work

Sharma et al. [7] presented an Artificial Neural Network (ANN) framework based approach to determine the reusability of the software based on its test cases by which developer can take the decision to choose a component to reuse."

According to Kawal et al. [8] defined various reusability factors and dependencies along with the relationship among the reusable factors. The model proposed by them defines the probability of success and failure for the reusability of the software and assured that the proposed system gave the results that were accurate by 80%.

Cho et al. [9] proposed a set of metrics for measuring the extent of complexity, reusability and customizability. They proposed an approach of Component Reusability Level (CRL) to measure the extent of component's reuse level per application in the component based development. But this approach was based on the Line of Code (LoC) so the complexity of the software was consider from the developer point of view.

Boxall et al. [10] proposed that the understandability of the components affect their reusability. Their interconnectivity and their extent of communication through message passing is a major factor. It also stated that the size of the interface of the component, argument counter and argument repetition scale are the other factors that play a crucial part in the reuse of the component. However the proposed approach didn't focused much on other aspects of the interface which included complexity of the arguments and the return types.

Gill [11] discussed on the issues concerning the component reusability and its benefits in term of cost and time-saving. He provided some guidelines to enhance the level of software reusability in component-based development. While Mili et al. [12] focused on two aspects, usability and usefulness including portability, flexibility, understandability and confidence to assess the reusability. Further, in this research, we mainly focused on the organization's role in measuring the reusability of the software components. Further, we give a direction to confine several reusability measures together."

3 Research Process

This paper defines the later stages of the analysis results after a long empirical study that included a set of questions. First, we collected the quantitative data using the surveying method from 10 software development organizational units and analyzed the data statistically. This survey involved the team leaders of the software developing department of these organizations.

To understand the practices undergoing in different software developing organization and to explain the phenomenon of the questions, the explanatory grounded theory approach was selected. This involved the interview data and other collected data. In particular, the questions during the survey were:

RQ1: Are you applying reusability practices in your organization and up to what extent?
RQ2: What types of testing practices are being engaged in the organization to promote reusability?
RQ3: How often are the reusability practices exercised?

Secondly, authors study the quality and change-proneness of the components and how does the reusability practice gets affected. Specifically, we explored the following questions:

RQ4: How is the efficiency of the end product is rendered by exercising reusability?
RQ5: How is the quality of the product affected by these practices?
RQ6: Is there any significant change in the level of complexity by practicing reusability?

The research questions in first phase and second phase assumed a cross- sectional analysis, i.e. the data was collected at the end of the development and testing phase. Our third phase involved a set of questions which were based on the economic terms of the organizations. These included:

RQ7: How efficient is reusability practice in terms of Cost and Time?
RQ8: What are success and failure rate of the project exercising reusability and what are the disadvantages involved in it?

The first set of research questions addressed the frequency of the reusability practices. This mainly focused on the regularity of these practices and how often they are being practiced in the organization. It also addresses the extent of it along with the types of testing practices involved with it.

While the second set of the research questions focused more on the quality aspect of the software. As reusability is meant to increase the efficiency and reduce the complexity, sometimes it goes the other way round. So to analyze this aspect, such questions were included.

The third phase covered the most important aspect of the proprietary software is the economic aspect. This aspect is important from the investor point of view and is important from the elapsed time for development point of view. Reusability of test cases encourage such development as it enhances the efficiency and reduces the cost factor.

4 Result Analysis

The grounded theory method was opted for analyzing the survey data from these organizations. According to Strauss and Corbin [13], grounded theory method provides three data analysis steps. These can be stated as: Open Coding—Where the essential studies are extracted from the data; Axial Coding—Where the connections between the studies are laid down; and Selective Coding—Where the Core information about the category is identified and explained. Such categorization can reduce the number of units to work with."

It is often assumed that a component should be reusable from the software engineering point of view. It should exhibit portability and flexibility like properties as per new requirements. In other words, the components with complex interface leads to much more efforts from the developers in customizing them. Therefore, a reusable comes handy as it is very easy to customize it when needed. Also, some non-functional organizational attributes define the quality of the reusable component. These facts can be summarized in the following factors which effect the reusability of software components:"

- Component's Quality
- Adaptability of the Component
- Frequency of Reuse
- Space-time Cost.

These are some of the factors which do effect the practice of reusability. Based on the survey conducted in the organization, we came across such issues that renders its practice.

A. Component's Quality

The real-time factors that influences the practice of reusability is the quality of the component. It depends on the structure of the component and the technology being used to develop that component. There are two such variables being used for the same. These can be stated as: State variable and Control variable. The state variable are the one that can be termed as dynamic while on the other hand, the control variable is the one that remains constant. So while maintaining the quality of the component, the major emphasis is on the control variable and is given much attention as compared to the state variable.

So, according to the study conducted, technology places a major role in the development and initiating reusability purpose. So it can be stated through a relation between the technology and the complexity of the components.

$$\text{Technology Constraints} \propto \text{Complexity}$$

This equations states that higher the technology constraints, higher is the complexity of the component.

B. Adaptability of the Component

An adaptable component is one which has the tendency to bear to any kind of functional or non-functional changes. It should be robust enough to cope up those changes of its environment without going any external intervention.

There are several shortcomings of adaptability in terms of the complexity. The most adaptable component is also the most complex one. While on the other hand, the component with more compose- ability is less adaptive in the sense of reusability. So, it can be stated as a much higher adaptive component has a much higher rate of reuse.

C. Frequency of Reuse

The frequency of Reuse of the component depends on a lot of factors. This involves whether there is any bug or flaw associated with that component which is undergoing reuse. Second, while undergoing is there any technology constraint which will render its service. While performing the test process, which type of testing techniques are being opted. Whether they are Manual Testing Techniques or Automated Techniques.

The study reveals that the organizations opting reuse of the components makes sure of all these issues that are relent the reuse practice. So these conditions are taken care of by the architects.

D. Space—Time Cost

It is convenient to broaden a component based on the amount of time it requires or the relative amount of space it consumes along with cost dimensions. So for an optimal software, the cost and time plays a major role in its development.

While going through the empirical study, it states that the big giant which are investing their money for the production of the component often prefer the development of newly developed component. This is so because they feel that by reusing the already developed software component, whatever are the bugs and flaws therein gets inherited in the incremental model and thereby increases the complexity. And the survey also states that the complexity using the reusable components get doubled when it comes to cost and time dimensions. But significant measure and techniques are practices to eliminate that risk and complexity. And it was also stated by the head of the organization that "*The cost for setting up of automated practices for testing and making the component reuse is much higher, but when there are too many units involved with it, it decreases the cost quite significantly at the same point.*"

5 Conclusion and Future Work

The objective of the study was to observe and find out what are the factors that are responsible for test cases reusability and up to what extent are these practices being exercised. We interviewed the employees of different organizations and at different organizational posts. The cases which we came across were analyzed using the grounded theory approach.

Building software using reusable component bring a lot of advantages to the organization. For the development of large project, such practices provides integration between the newly developed and reusable components. Therefore, choosing the right component which integrates successfully and functions well save the time and cost of the organization. However, the main perceived benefit of the testing automation included the improvisation in quality going through a better test coverage and it also stated that more test cases can be generated in less time.

We came across the practices wherein the reusable component are being implemented in the organization. These included almost all the development that is taking place in and out of the organization and moreover they prefer to use the reusable component. It also stated that somehow the complexity increases when there are technology constraints but gets eliminated.

This study was done in order to cover the marketing practices and we came to know that main disadvantage of the implementing was the Cost Factor. The term cost includes implementation cost, maintenance cost and training cost. So we came across to the conclusion that the maintenance and implementation are linked to each other. If we tend to design a system with low maintenance cost, then there is a lot of implementation cost involved and vice versa.

The reusability of test cases is essential in making the investment worthwhile. There are often some problem related to time and funds which hinders the reusability practices along with the technology playing the major role. The aim of the research was to study the factors influencing the reusability of test cases which in our opinion get affected by both from organization and developer's perspective. Much commitment, proper training is required to be inculcated within the organization to make such practices up to the money.

Currently, many researches are going on to make the integrated and adaptable component more efficient for reuse during the compilation time. However, there must be practices involving the know-how of engineering along with the coding so that it can be applied when the code is running.

References

1. Gupta, N. K., & Rohil, M. K. (2015). *Software Quality Measurement for Reusability.*
2. Dustin, E., Rashka, J., & Paul, J. (1999). *Automated software testing: introduction, management, and performance*. Boston: Addison-Wesley.
3. Kamalraj, R., Geetha, B. G., & Singaravel, G. (2009, March) Reducing Efforts on Software Project Management using Software Package Reusability. In *Advance Computing Conference, 2009. IACC 2009. IEEE International* (pp. 1624–1627).
4. Hummel, O., & Atkinson, C. (2006). Using the web as a reuse repository. In *Reuse of Off-the-Shelf Components*, (pp. 298–311). Springer,
5. Ahmaro, I. Y. T., Yusoff, M. Z. M., & Abdallah, M. Y. M. (2015). *The Current Practices of Software Reusability Approaches in Malaysia*. Selangor, Malaysia: College of Information Technology, Department of Software Engineering, Universiti Tenaga Nasional (UNITEN).
6. Karhu, K., Repo, T., & Taipale, O. (2009). *Empirical Observations on Software Testing Automation.*
7. Sharma, A., Grover, P., & Kumar, R. (2009). Reusability assessment for software components. *ACM SIGSOFT Software Engineering Notes, 34*(2), 1–6.
8. Jeet, K., Rana, Y., & Xin, R. (2012). A bayesian network based approach for software reusability prediction. *ACM SIGSOFT Software Engineering Notes, 37*(4), 1–5.
9. Cho, E. S., Kim, M. S., & Kim, S. D. (2001). Component metrics to measure component quality. In *Software Engineering Conference, 2001. APSEC 2001. Eighth Asia-Pacific*, pp. 419–426.
10. Boxall M. A., & Araban, S. (2004). Interface metrics for reusability analysis of components. In *Software Engineering Conference, 2004. Proceedings. 2004 Australian*. IEEE, pp. 40–51.
11. Gill, N. S. (2003). Reusability issues in component-based development. *ACM SIGSOFT Software Engineering Notes, 28*(4), 4.
12. Mili, H., Mili, F., & Mili, A. (1995). Reusing software: Issues and research directions. *Software Engineering, IEEE Transactions on, 21*(6), 528–562.
13. Strauss, A., & Corbin, J. (1990). *Basics of Qualitative Research: Grounded Theory Procedures and Techniques*. Newbury Park, CA: SAGE Publications.

Water Treatment System Performance Evaluation Under Maintenance Policies

Anshumaan Saxena, Sourabh Kumar Singh, Amit Kumar and Mangey Ram

Abstract The main determination of this work is to analyze the performance of the water treatment plant (WTP) and tries to find that which of the subpart/subparts of water treatment plant affects it. The problem that generally occurs in the WTP is based upon the poor maintenance and material used during manufacturing for its subparts. These types of the problem could be prevented if safety measures and maintenance techniques are followed properly and regularly. For analyzing WTP, pump plays very important role in supplying water to different components; thus, other machines can also perform their function as well. Along with pumps, valves also need regular maintenance for better performance. Other components also have their significance. Except this, the WTP had various components which need maintenance and replacement over a different span of time period.

Keywords Water treatment plant (WTP) · Sensitivity analysis Safety and reliability analysis · Multistate system

1 Introduction

Reliability is the most significant/valuable thing which is essential for an industrial system/industrial plant. In the past, so many industrial systems were analyzed through reliability approach including thermal power plant [9], casting process [11, 17, 19], sugar mill [12, 14, 15], marine power plant [10], hydropower station Ye

A. Saxena (✉) · S. K. Singh
Department of Mechanical Engineering, Graphic Era Deemed to be University, Dehradun, Uttarakhand, India
e-mail: saxena.ams@gmail.com

A. Kumar
Department of Mathematics, Lovely Professional University, Phagwara, Punjab, India

M. Ram
Department of Mathematics, Computer Science and Engineering, Graphic Era Deemed to University, Dehradun, Uttarakhand, India

P. K. Kapur et al. (eds.), *System Performance and Management Analytics*, Asset Analytics, https://doi.org/10.1007/978-981-10-7323-6_14

et al. [21], etc., to discuss their different performance measures. Keeping these works in mind here authors tried to find the various reliability measure of a water treatment plant to analyze its performance.

WTP plays a very significant role in various places where water is used. In the past, so many researchers work on it [13, 20]. As the name WTP or water purification system stands for a system which used to remove undesirable chemicals [1] biological contaminants and suspended solids from the water taken from the rivers, lakes, or seas [4]. The main motive of WTP is to provide water which is good for domestic, industrial, and other purposes at reasonable costs [5]. To achieve this goal, a variety of treatment processes are utilized that employ various kinds of physical and chemical phenomena. Process which is used for treatment of water is selected normally on the overall quality of raw water [7] and the combination of all the processes follows the sequence of the standard process.

The main factors that must be taken into consideration while developing a treatment process chain include the following [20]:

- The quality of preprocessed water,
- The required quality of treated water,
- Plant size (capacity), site conditions,
- Availability of skilled labor, and
- The project cost.

The conversational water treatment plant includes a series of processes such as coagulation of small colloidal particles [8] flocculation of the small particles to form larger ones, followed by sedimentation and sand filtration. Some advanced techniques are also used in the treatment process like reverse osmosis (RO), nanofiltration (NF), ultrafiltration, electrodialysis, ozonation, and activated carbon absorption [1] for the removal of iron and manganese. Some mechanical components played a very important role like—pumps, valves, stirrer, and air compressors [16].

The series of processes are as follows, which are used in the treatment:

(a) **Coagulation**
This is the process by which the colloidal particles in the water get separated. So that they further form flocs by means of flocculation process. Chemicals used in this process are called as coagulants [2], e.g., aluminum sulfate, ferric chloride, and hydrated lime.

(b) **Flocculation**
In this process, the individual separates colloidal particles that collide with each other in order to form aggregates that can be removed easily by the sedimentation [2]. It involves a slow stirring of water that causes the small coagulated particles to form flocs. This stirring is created by some mechanical or hydraulic means of mixing.

(c) **Sedimentation**
In this process, the aggregates which have been formed by coagulation and flocculation are separated out from the water [18]. Remaining flocs get collected as sludge from the bottom of the sedimentation tank with the help of sludge return pump on regular basis.

(d) **Flotation**
It is an effective process in which removal of relatively light flocs is carried out. Flotation involves the formation of small air bubbles in water that has to be flocculated [6]. The air bubbles raise the flocs out on the surface of the water where they can easily collect and removed from the top of the flotation unit.

(e) **Sand Filtration**
In this process, water is filtered through layer of fine sand in a specially designed container. During this process, the small remaining floc particle is removed by sand grains. Rapid sand filtration and slow sand filtration are other two types of sand filtration.

(f) **Disinfection**
Water is disinfected before it enters the distribution system to ensure that any disease-causing bacteria, virus, and parasites are destroyed. Disinfection involves the addition of the required number of chemical agents called disinfectants to the water. The most commonly used disinfectant is chlorine gas. Other disinfectants are ozone, chlorine dioxide, and other chlorine compounds such as calcium hypochlorite (HTH), sodium hypochlorite, etc. [3]. Physical methods of disinfection of water include radiation with ultraviolet light and boiling. Disinfection by means of UV radiation is more popular and effective that is why here author considers UV disinfector.

(g) **Stabilization**
It refers to the chemical stability of the water. It involves the addition of chemicals to the water to adjust its chemical properties in order to prevent corrosion or scale formation, P^H correction by addition of acids or bases [13].

(h) **Sludge Treatment/Disposal**
Sludge from sedimentation tank can cause pollution in huge amount and having large pollution potential because it contains suspended material and chemical that have been already removed from the process. Therefore, it must be disposed and treated in a proper manner to prevent contamination of water sources.

2 Assumptions

The following assumptions are taken throughout the problem:

(i) No other failure (which was not considered through the paper) occurs during the process.
(ii) The average failure rates are taken to be constant.

(iii) Initially, the WTP is free from all types of defects.
(iv) The WTP may work with reduced capacity, i.e., in a degraded state.
(v) The WTP never stop working due to unavailability of a repairman.

3 Nomenclature

The following notations are taken throughout the problem:

t/s	Time scale variable/Laplace transforms variable
$P_0(t)$	The state probability at time t in which WTP working in good condition
$P_1(t)$	The state probability at time t in which WTP working in degraded condition by the failure of the valve
$P_2(t)$	The state probability at time t in which WTP working in degraded condition by the failure of RO filter and valve
$P_3(t)$	The state probability at time t in which WTP working in degraded condition by the failure pump
$P_4(t)$	The state probability at time t in which WTP working in degraded condition by the failure of coagulator
$P_5(t)$	The state probability at time t in which WTP working in degraded condition by the failure of UV disinfector
$P_6(t)$	The state probability at time t in which WTP working in degraded condition by the failure of UV disinfector and fluoridation
$P_7(t)$	The state probability at time t in which WTP working in degraded condition by the failure of power supply
$P_8(x,t)$	The state probability at time t in which WTP is in failed condition by the failure of the valve, RO filter, and aerator
$P_9(x,t)$	The state probability at time t in which WTP is in failed condition by the failure of pump and standby pump
$P_{10}(x,t)$	The state probability at time t in which WTP is in failed condition by the failure of coagulator and flocculator
$P_{11}(x,t)$	The state probability at time t in which WTP is in failed condition by the failure of UV Disinfector, fluoridator, and Ph. controller
$P_{12}(x,t)$	The state probability at time t in which WTP is in failed condition by the failure of power supply and standby power supply
$\lambda_{PS}/\lambda_{SPS}/\lambda_V/\lambda_{RF}/\lambda_A/\lambda_P/\lambda_{SP}/\lambda_C/\lambda_F/\lambda_{UVD}/\lambda_{FD}/\lambda_{PhC}$	Failure rate of power supply/standby power supply/valve/RO filter/aerator/pump/standby pump/coagulator/flocculator/UV disinfector/fluoridator/Ph. controller
$\phi_{PS}(x)/\phi_V(x)/\phi_{RF}(x)/\phi_P(x)/\phi_C(x)/\phi_{UVD}(x)/\phi_{FD}(x)$	Repair rates of power supply/valve/RO filter/pump/coagulator/UV disinfector/fluoridator
$\mu_1(x)/\mu_2(x)/\mu_3(x)/\mu_4(x)/\mu_5(x)$	Simultaneous repair rate of UV disinfector, fluoridator, and Ph. controller/power supply and standby power supply/valve, RO filter and aerator/pump and standby pump/coagulator and flocculator

4 State Transition Diagram

On the basis of repairs and failures, with the help of Markov process and configuration diagram as shown in Fig. 1, the following state transition diagram (Fig. 2) is developed.

5 Mathematical Formulation and Solution

With the aid of the above state transition diagram (Fig. 2), the following set of the intro-differential equation is generated to find various reliability characteristics of WTP.

$$\left(\frac{\partial}{\partial t}+\lambda_{\mathrm{V}}+\lambda_{\mathrm{C}}+\lambda_{\mathrm{UVD}}+\lambda_{\mathrm{PS}}+\lambda_{\mathrm{P}}\right)P_0(t)=\sum_{i,j}\phi_i(x)P_j(t)+\sum_{k,l}\int_0^{\infty}\mu_k(x)P_l(x,t)\,dx \quad (1)$$

$$\text{where}\quad \begin{aligned} i &= \mathrm{V,P,C,UVD,PS} \\ j &= 1,3,4,5,7, \\ k &= 1,2,3,4,5 \\ l &= 11,12,8,9,10,\ \text{respectively} \end{aligned}$$

$$\left(\frac{\partial}{\partial t}+\phi_{\mathrm{V}}(x)+\lambda_{\mathrm{RF}}\right)P_1(t)=\lambda_{\mathrm{V}}P_0(t)+\phi_{\mathrm{RF}}(x)P_2(t) \quad (2)$$

$$\left(\frac{\partial}{\partial t}+\phi_{\mathrm{RF}}(x)+\lambda_{\mathrm{A}}\right)P_2(t)=\lambda_{\mathrm{RF}}(x)P_1(t) \quad (3)$$

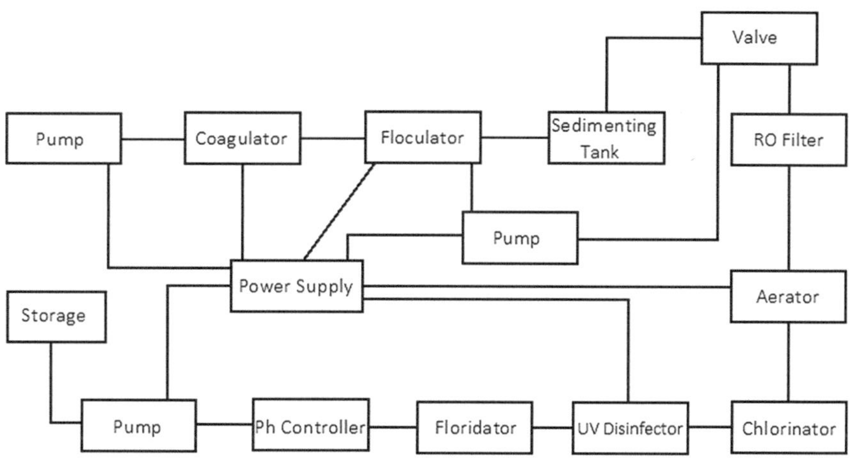

Fig. 1 Configuration diagram

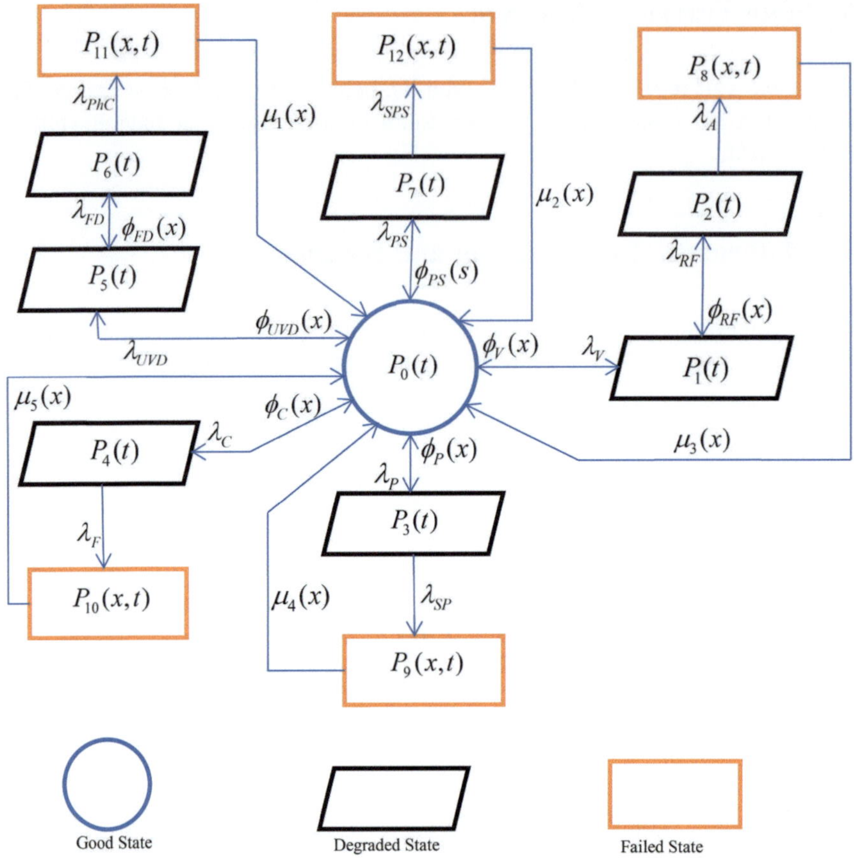

Fig. 2 State transition diagram

$$\left(\frac{\partial}{\partial t} + \phi_{\mathrm{P}}(x) + \lambda_{\mathrm{SP}}\right)P_3(t) = \lambda_{\mathrm{P}}P_0(t) \tag{4}$$

$$\left(\frac{\partial}{\partial t} + \phi_{\mathrm{C}}(x) + \lambda_{\mathrm{F}}\right)P_4(t) = \lambda_{\mathrm{C}}P_0(t) \tag{5}$$

$$\left(\frac{\partial}{\partial t} + \phi_{\mathrm{UVD}}(x) + \lambda_{\mathrm{FD}}\right)P_5(t) = \lambda_{\mathrm{UVD1}}P_0(t) + \phi_{\mathrm{FD}}(x)P_6(t) \tag{6}$$

$$\left(\frac{\partial}{\partial t} + \phi_{\mathrm{FD}}(x) + \lambda_{\mathrm{PhC}}\right)P_6(t) = \lambda_{\mathrm{FD}}P_5(t) \tag{7}$$

$$\left(\frac{\partial}{\partial t} + \phi_{\mathrm{PS}}(x) + \lambda_{\mathrm{SPS}}\right)P_7(t) = \lambda_{\mathrm{PS}}P_0(t) \tag{8}$$

$$\left(\frac{\partial}{\partial x}+\frac{\partial}{\partial t}+\mu_i(x)\right)P_j(t)=0, \quad \text{where} \quad \begin{array}{l} i=1,2,3,4,5 \\ j=11,12,8,9,10 \end{array} \tag{9}$$

Boundary conditions

$$P_k(0,t)=\lambda_l P_m(t) \quad \text{where} \quad \begin{array}{l} k=8,9,10,11,12 \\ l=\text{A, SP, F, PhC, SPS} \\ m=2,3,4,6,7 \end{array} \tag{10}$$

Initial condition

$$P_i(t)=\begin{cases} 1, & \text{when } i=0 \text{ and } t=0 \\ 0, & \text{otherwise} \end{cases} \tag{11}$$

The above set of intro-differential equation can be rewritten as (by the help of Laplace transformation)

$$(s+\lambda_{\text{V}}+\lambda_{\text{C}}+\lambda_{\text{UVD}}+\lambda_{\text{PS}}+\lambda_{\text{P}})\overline{P}_0(s)=\sum_{i,j}\phi_i(x)\overline{P}_j(s)+\sum_{k,l}\int_0^{\infty}\mu_k(x)\overline{P}_l(x,s)\,\mathrm{d}x \tag{12}$$

$$\text{where} \quad \begin{array}{l} i=\text{V, P, C, UVD, PS} \\ j=1,3,4,5,7, \\ k=1,2,3,4,5 \\ l=11,12,8,9,10, \text{ respectively} \end{array}$$

$$(s+\phi_{\text{V}}(x)+\lambda_{\text{RF}})\overline{P}_1(s)=\lambda_{\text{V}}\overline{P}_0(s)+\phi_{\text{RF}}(x)\overline{P}_2(s) \tag{13}$$

$$(s+\phi_{\text{RF}}(x)+\lambda_{\text{A}})\overline{P}_2(s)=\lambda_{\text{RF}}\overline{P}_1(s) \tag{14}$$

$$(s+\phi_{\text{P}}(x)+\lambda_{\text{SP}})\overline{P}_3(s)=\lambda_{\text{P}}\overline{P}_0(s) \tag{15}$$

$$(s+\phi_{\text{C}}(x)+\lambda_{\text{F}})\overline{P}_4(s)=\lambda_{\text{C}}\overline{P}_0(s) \tag{16}$$

$$(s+\phi_{\text{UVD}}(x)+\lambda_{\text{FD}})\overline{P}_5(s)=\lambda_{\text{UVD}}\overline{P}_0(s)+\phi_{\text{FD}}(x)\overline{P}_6(s) \tag{17}$$

$$(s+\phi_{\text{FD}}(x)+\lambda_{\text{PhC}})\overline{P}_6(s)=\lambda_{\text{FD}}\overline{P}_5(s) \tag{18}$$

$$(s+\phi_{\text{PS}}(x)+\lambda_{\text{SPS}})\overline{P}_7(s)=\lambda_{\text{PS}}\overline{P}_0(s) \tag{19}$$

$$\left(\frac{\partial}{\partial x}+s+\mu_i(x)\right)\overline{P}_j(s)=0, \quad \text{where} \quad \begin{array}{l} i=1,2,3,4,5 \\ j=11,12,8,9,10 \end{array} \tag{20}$$

$$\overline{P}_k(0,s) = \lambda_l \overline{P}_m(s) \quad \text{where} \quad \begin{array}{l} k = 8, 9, 10, 11, 12 \\ l = \text{A, SP, F, PhC, SPS} \\ m = 2, 3, 4, 6, 7 \end{array} \tag{21}$$

By solving Eqs. (1–21), the following transition state probabilities are obtained:

$$\overline{P}_0(s) = \frac{1}{\{H_3 - \phi_{\text{V}}(x)H_1 - H_1H_4 - H_5 - H_6 - \phi_{\text{UVD}}(x)H_2 - H_7 - H_8\}}$$

$$\overline{P}_1(s) = H_1P_0(s),\ \overline{P}_2(s) = \frac{\lambda_{\text{RF}}H_2}{(s + \phi_{\text{RF}}(x) + \lambda_{\text{A}})}P_0(s),\ \overline{P}_3(s) = \frac{\lambda_{\text{P}}}{(s + \phi_{\text{P}}(x) + \lambda_{\text{SP}})}P_0(s),$$

$$\overline{P}_4(s) = \frac{\lambda_C}{(s + \phi_{\text{C}}(x) + \lambda_{\text{F}})}P_0(s),\ \overline{P}_5(s) = H_2P_0(s),\ \overline{P}_6(s) = \frac{\lambda_{\text{FD}}H_2}{(s + \phi_{\text{FD}}(x) + \lambda_{\text{PhC}})}P_0(s),$$

$$\overline{P}_7(s) = \frac{\lambda_{\text{PS}}}{(s + \phi_{\text{PS}}(x) + \lambda_{\text{SPS}})}P_0(s),$$

where

$$H_1 = \frac{\lambda_{\text{V}}}{\left[s + \phi_{\text{V}}(x) + \lambda_{\text{RF}} - \frac{\phi_{\text{RF}}(x)\lambda_{\text{RF}}}{(s + \phi_{\text{RF}}(x) + \lambda_{\text{A}})}\right]},$$

$$H_2 = \frac{\lambda_{\text{UVD}}}{\left[s + \phi_{\text{UVD}}(x) + \lambda_{\text{FD}} - \frac{\phi_{\text{FD}}(x)\lambda_{\text{FD}}}{(s + \phi_{\text{FD}}(x) + \lambda_{\text{PhC}})}\right]},$$

$$H_3 = (s + \lambda_{\text{V}} + \lambda_{\text{C}} + \lambda_{\text{UVD}} + \lambda_{\text{PS}} + \lambda_{\text{P}}),$$

$$H_4 = \frac{\mu_3(x)\lambda_{\text{A}}\lambda_{\text{RF}}}{(s + \phi_{\text{RF}}(x) + \lambda_{\text{A}})(s + \mu_3(x))}$$

$$H_5 = \left[\phi_{\text{P}}(x) + \frac{\mu_4(x)\lambda_{\text{SP}}}{\{s + \mu_4(x)\}}\right]\left[\frac{\lambda_{\text{P}}}{\{s + \phi_{\text{P}}(x) + \lambda_{\text{SP}}\}}\right],$$

$$H_6 = \left[\phi_{\text{C}}(x) + \frac{\mu_5(x)\lambda_{\text{F}}}{\{s + \mu_5(x)\}}\right]\left[\frac{\lambda_{\text{C}}}{\{s + \phi_{\text{C}}(x) + \lambda_{\text{F}}\}}\right]$$

$$H_7 = \left[\frac{\mu_1(x)\lambda_{\text{PhC}}\lambda_{\text{FD}}H_2}{\{s + \phi_{\text{FD}}(x) + \lambda_{\text{PhC}}\}\{s + \mu_1(x)\}}\right],$$

$$H_8 = \left[\phi_{\text{PS}}(x) + \frac{\mu_2(x)\lambda_{\text{SPS}}}{\{s + \mu_2(x)\}}\right]\left[\frac{\lambda_{\text{PS}}}{\{s + \phi_{\text{PS}}(x) + \lambda_{\text{SPS}}\}}\right].$$

From the state transition diagram, one can observe that the working state (up-state) and failed state (downstate) probability of WTP are given as

$$\overline{P}_{\text{up}}(s) = \overline{P}_0(s) + \overline{P}_1(s) + \overline{P}_2(s) + \overline{P}_3(s) + \overline{P}_4(s) + \overline{P}_5(s) + \overline{P}_6(s) + \overline{P}_7(s) \quad (22)$$

$$\overline{P}_{\text{down}}(s) = \overline{P}_8(x,s) + \overline{P}_9(x,s) + \overline{P}_{10}(x,s) + \overline{P}_{11}(x,s) + \overline{P}_{12}(x,s) \quad (23)$$

6 Numerical Study and Certain Case for WTP

6.1 Reliability of WTP

Reliability is the measure of a system performance to analyze that how much one can trust on a system. It is a collective contribution by all the components of the system. If any of the component of the system not working in a good manner, then this affects the reliability of the system directly. So, in order to make a system highly, reliable one has to control the performance of each and every component of the system as much as possible. In order to calculate the reliability of the considered water treatment plant, we put all repairs equal to zero and various failure rates as:

$\lambda_{PS} = 0.22, \lambda_{SPS} = 0.05, \lambda_P = 0.20, \lambda_C = 0.08, \lambda_{SP} = 0.07, \lambda_V = 0.03, \lambda_{RF} = 0.07, \lambda_A = 0.15, \lambda_F = 0.04, \lambda_{UVD} = 0.03, \lambda_{FD} = 0.05, \lambda_{PhC} = 0.01$ in (22) and taking the inverse Laplace transform, the reliability of the system is given as

$$R(t) = \left\{ \begin{array}{l} 1.015906\, e^{(-0.56\, t)} + 0.938775\, e^{(-0.315\, t)} \sinh(0.245\, t) - 0.032352\, e^{(-0.05\, t)} - 0.051219\, e^{(-0.15\, t)} \\ + 0.307692\, e^{(-0.3\, t)} \sinh(0.26\, t) + 0.980392\, e^{(-0.305\, t)} \sinh(0.255\, t) + 0.068181\, e^{(-0.01\, t)} \end{array} \right\} \quad (24)$$

Now varying the time unit in the (24), we obtain Table 1 and corresponding Fig. 3, which shows the behavior of reliability of water treatment plant.

6.2 Meantime to Failure (MTTF) of WTP

This reliability measure of the system shows the average number of failures of the components of the system per unit time. With the help of MTTF, one can identify that which components failure occurs most frequently. It is calculated as

$$MTTF = \text{limt}_{S \to 0} R(s) \quad (25)$$

Table 1 Reliability versus time

Time (t)	Reliability $R(t)$
0	1.000000
1	0.988399
2	0.961256
3	0.926232
4	0.887666
5	0.847999
6	0.808589
7	0.770173
8	0.733136
9	0.697661
10	0.663815

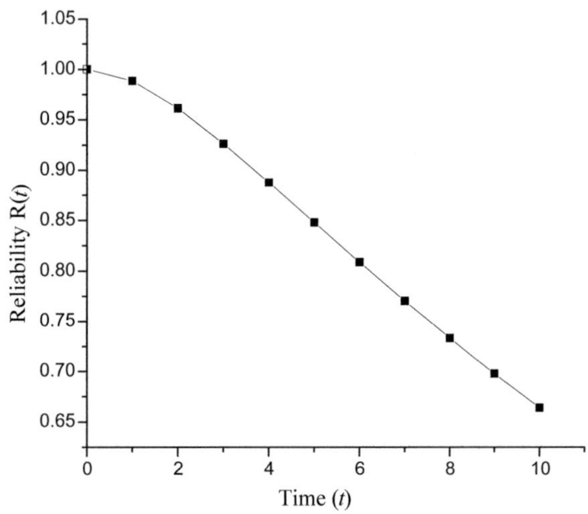

Fig. 3 Reliability of WTP versus time

The MTTF of the considered WTP has calculated as

$$\mathrm{MTTF} = \frac{1}{(\lambda_V + \lambda_C + \lambda_{UVD} + \lambda_{PS} + \lambda_P)}\left\{1 + \frac{\lambda_V}{\lambda_{RF}} + \frac{\lambda_{RF}\lambda_{UVD}}{\lambda_{FD}\lambda_A} + \frac{\lambda_P}{\lambda_{SP}} + \frac{\lambda_C}{\lambda_F} + \frac{\lambda_{UVD}}{\lambda_{FD}} + \frac{\lambda_{UVD}}{\lambda_{PHC}} + \frac{\lambda_{PS}}{\lambda_{SPS}}\right\} \tag{26}$$

Setting $\lambda_{SP} = 0.07, \lambda_V = 0.03, \lambda_{RF} = 0.07, \lambda_A = 0.15, \lambda_F = 0.04, \lambda_{UVD} = 0.03,$ $\lambda_{FD} = 0.05,$ $\lambda_{PhC} = 0.01$ $\lambda_{PS} = 0.22,$ and $\lambda_{SPS} = 0.05, \lambda_P = 0.20, \lambda_C = 0.08,$ and varying failure rates one by one from 0.01 to 0.09 with a step length 0.01 in (26), we get the MTTF of WTP as tabulated in Table 2a, b and Fig. 4a, b.

Table 2 MTTF of WTP versus failure rates

(a)

Variations in failures	MTTF						
	λ_{PS}	λ_{SPS}	λ_{P}	λ_{SP}	λ_{RF}	λ_{A}	λ_{C}
0.01	29.61632	57.43877	32.03088	56.62244	30.17346	33.01020	26.15451
0.02	29.34920	37.79591	31.56390	38.76530	27.56632	29.26020	26.13142
0.03	29.09652	31.24829	31.12087	32.81292	26.74489	28.01020	26.10924
0.04	28.85714	27.97448	30.70000	29.83673	26.36989	27.38520	26.08791
0.05	28.63003	26.01020	30.29965	28.05120	26.17346	27.01020	26.06738
0.06	28.41428	24.70068	29.91836	26.86054	26.06632	26.76020	26.04761
0.07	28.20905	23.76530	29.55481	26.01020	26.01020	26.58163	26.02857
0.08	28.01360	23.06377	29.20779	25.37244	25.98596	26.44770	26.01020
0.09	27.82724	22.51814	28.87619	24.87641	25.98299	26.34353	25.99248
0.10	27.64935	22.08163	28.55900	24.47959	25.99489	26.26020	25.97536

(b)

Variations in failures	MTTF				
	λ_{V}	λ_{F}	λ_{UVD}	λ_{FD}	λ_{PhC}
0.01	26.44444	36.72448	22.18342	32.29591	26.10120
0.02	26.22337	29.58163	24.13160	28.36734	23.33163
0.03	26.01012	27.20068	26.01020	27.05782	22.43877
0.04	25.80451	26.01020	27.82289	26.40306	21.99234
0.05	25.60591	25.29591	29.57307	26.01020	21.72448
0.06	25.41404	24.81972	31.26392	25.74829	21.54591
0.07	25.22857	24.47959	32.89841	25.56122	21.41836
0.08	25.04918	24.22448	34.47931	25.42091	21.32270
0.09	24.87557	24.02607	36.00961	25.31179	21.24829
0.10	24.70748	23.86734	37.49055	25.22448	21.18877

6.3 *Sensitivity Analysis for WTP*

Sensitivity analysis is a measure by which one can identify that which failure affects the system performance most. It is calculated by the partial derivatives of that measure with respect to input parameters. In this paper, it is calculated with respect to reliability and MTTF of the WTP.

6.3.1 Sensitivity of MTTF

For sensitivity analysis with respect to MTTF, one has differentiated () with respect to failure rates and then putting the values of various failure rates as $\lambda_{SPS} = 0.05, \lambda_{P} = 0.20, \lambda_{C} = 0.08, \lambda_{SP} = 0.07, \lambda_{V} = 0.03, \lambda_{RF} = 0.07, \lambda_{A} = 0.15,$

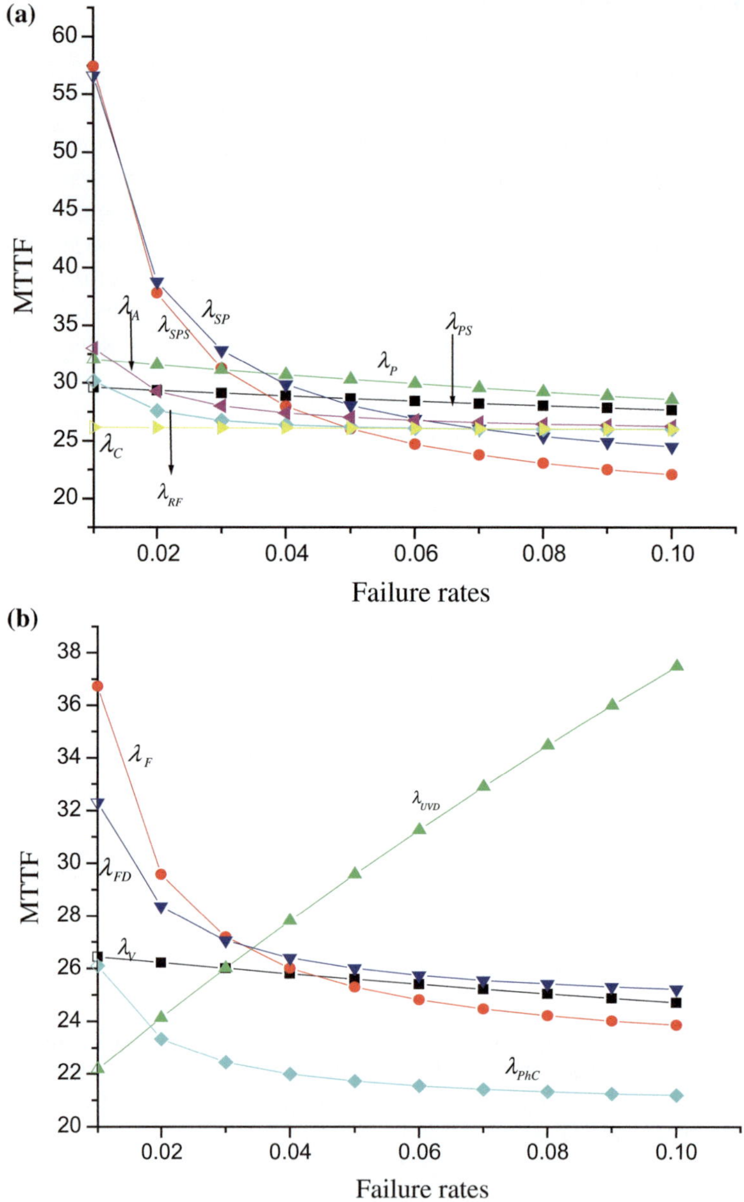

Fig. 4 **a** MTTF of WTP versus failure rates, and **b** MTTF of WTP versus failure rates

$\lambda_F = 0.04$, $\lambda_{UVD} = 0.03, \lambda_{FD} = 0.05$, and $\lambda_{PhC} = 0.01, \lambda_{PS} = 0.22$, we get the values of

$$\frac{\partial(\text{MTTF})}{\partial\lambda_{SPS}}, \frac{\partial(\text{MTTF})}{\partial\lambda_P}, \frac{\partial(\text{MTTF})}{\partial\lambda_C}, \frac{\partial(\text{MTTF})}{\partial\lambda_{SP}}, \frac{\partial(\text{MTTF})}{\partial\lambda_V}, \frac{\partial(\text{MTTF})}{\partial\lambda_{RF}}, \frac{\partial(\text{MTTF})}{\partial\lambda_A}, \frac{\partial(\text{MTTF})}{\partial\lambda_F}, \frac{\partial(\text{MTTF})}{\partial\lambda_{UVD}}, \frac{\partial(\text{MTTF})}{\partial\lambda_{FD}}, \frac{\partial(\text{MTTF})}{\partial\lambda_{PhC}}, \frac{\partial(\text{MTTF})}{\partial\lambda_{PS}}.$$

Varying the failure rates one by one, respectively, as 0.01, 0.02, 0.03, 0.04, 0.05, 0.06, 0.07, 0.08, 0.09 in the partial derivatives, one obtained Table 3a, b and Fig. 5a, b, respectively, for sensitivity analysis of WTP with respect to MTTF.

6.3.2 Sensitivity of Reliability

For sensitivity analysis with respect to reliability, one has differentiated reliability expression with respect to various failure rates and then putting the values of various failure rates as $\lambda_{SPS} = 0.05, \lambda_P = 0.20, \lambda_C = 0.08$, $\lambda_F = 0.04, \lambda_{UVD} = 0.03, \lambda_{FD} = 0.05$, $\lambda_{PhC} = 0.01, \lambda_{PS} = 0.22$, $\lambda_{SP} = 0.07, \lambda_V = 0.03, \lambda_{RF} = 0.07$, $\lambda_A = 0.15$, we get the values of $\frac{\partial R(t)}{\partial\lambda_{SPS}}, \frac{\partial R(t)}{\partial\lambda_P}, \frac{\partial R(t)}{\partial\lambda_C}, \frac{\partial R(t)}{\partial\lambda_{SP}}, \frac{\partial R(t)}{\partial\lambda_V}, \frac{\partial R(t)}{\partial\lambda_{RF}}, \frac{\partial R(t)}{\partial\lambda_A}, \frac{\partial R(t)}{\partial\lambda_F}, \frac{\partial R(t)}{\partial\lambda_{UVD}}, \frac{\partial R(t)}{\partial\lambda_{FD}}, \frac{\partial R(t)}{\partial\lambda_{PhC}}, \frac{\partial R(t)}{\partial\lambda_{PS}}$. Varying the failure rates one by one, respectively, as 0.01, 0.02, 0.03, 0.04, 0.05, 0.06, 0.07, 0.08, 0.09 in the partial derivatives, one obtained Table 4a, b and Fig. 6a, b, respectively, for sensitivity analysis of WTP with respect to MTTF.

7 Result Discussion and Conclusion

On the basis of the above calculation, the following results are obtained for WTP.

The nature of reliability of WTP is shown in Fig. 3. From this, it is observed that with respect to time the reliability of WTP decrease in a state line. The MTTF for WTP with respect to various failure rates is shown in Fig. 4a, b. It can be observed that the MTTF of the system is highest with respect to the failure rate of UV disinfector, it means that as time passes, failure of UV Disinfector occurs more frequently compared to other. The sensitivity analysis with respect to MTTF of WTP is shown in Fig. 5a, b. The graph shows that the MTTF of WTP is most sensitive with respect to the failure rate of flocculator. Hence, much more required on the failure rate of flocculator to control the MTTF of WTP. The sensitivity analysis with respect to the reliability of WTP is shown in Fig. 6a, b. The graph shows that the reliability of WTP is most sensitive with respect to the failure rate of UV Disinfector. It is equally sensitive with respect to the failure rate of fluoridator and Ph. Controller. Here, much more required on the failure rate of UV disinfector to enhance the reliability of WTP.

Table 3 Sensitivity of MTTF of WTP versus failure rates

(a)

Variations in failures	Sensitivity of MTTF						
	$\frac{\partial(\mathrm{MTTF})}{\partial\lambda_{PS}}$	$\frac{\partial(\mathrm{MTTF})}{\partial\lambda_{SPS}}$	$\frac{\partial(\mathrm{MTTF})}{\partial\lambda_{P}}$	$\frac{\partial(\mathrm{MTTF})}{\partial\lambda_{SP}}$	$\frac{\partial(\mathrm{MTTF})}{\partial\lambda_{RF}}$	$\frac{\partial(\mathrm{MTTF})}{\partial\lambda_{A}}$	$\frac{\partial(\mathrm{MTTF})}{\partial\lambda_{C}}$
0.01	−27.47521	−3928.571	−47.95992	−3571.428	−28.5714	−750.0000	−2.356161
0.02	−25.97001	−982.1428	−45.46893	−892.8571	−26.7857	−187.5000	−2.262857
0.03	−24.58520	−436.5079	−43.16708	−396.8253	−2.38095	−83.33333	−2.174987
0.04	−23.30827	−245.5357	−41.03571	−223.2142	−6.33928	−46.87500	−2.092138
0.05	−22.12829	−157.1428	−39.05838	−142.8571	−4.28571	−30.00000	−2.013934
0.06	−21.03571	−109.1269	−37.22060	−99.20634	−0.738095	−20.83333	−1.940035
0.07	−20.02209	−80.17492	−35.50954	−72.88629	−0.790087	−15.30612	−1.870129
0.08	−19.08001	−61.38392	−33.91381	−55.80357	−0.227678	−11.71875	−1.803935
0.09	−18.20228	−48.50088	−32.42328	−44.09171	0.529100	−9.259259	−1.741195
0.10	−17.38488	−39.28571	−31.02889	−35.71428	1.785714	−7.500000	−1.681671

(b)

Variations in failures	Sensitivity of MTTF				
	$\frac{\partial(\mathrm{MTTF})}{\partial\lambda_{V}}$	$\frac{\partial(\mathrm{MTTF})}{\partial\lambda_{F}}$	$\frac{\partial(\mathrm{MTTF})}{\partial\lambda_{UVD}}$	$\frac{\partial(\mathrm{MTTF})}{\partial\lambda_{FD}}$	$\frac{\partial(\mathrm{MTTF})}{\partial\lambda_{PhC}}$
0.01	−21.6198979	−1724.1379	191.15949	−758.62068	−517.2413
0.02	−20.8679593	−431.03448	184.51098	−189.65517	−129.3103
0.03	−20.1545778	−191.57088	178.20338	−84.291187	−57.47126
0.04	−19.4771617	−107.75862	172.21378	−47.413793	−32.32758
0.05	−18.8333333	−68.965517	166.52116	−30.344827	−20.68965
0.06	−18.2209083	−47.892720	161.10620	−21.072796	−14.36781

(continued)

Table 3 (continued)

(b)					
Variations in failures	Sensitivity of MTTF				
	$\frac{\partial(\mathrm{MTTF})}{\partial \lambda_{\mathrm{V}}}$	$\frac{\partial(\mathrm{MTTF})}{\partial \lambda_{\mathrm{F}}}$	$\frac{\partial(\mathrm{MTTF})}{\partial \lambda_{\mathrm{UVD}}}$	$\frac{\partial(\mathrm{MTTF})}{\partial \lambda_{\mathrm{FD}}}$	$\frac{\partial(\mathrm{MTTF})}{\partial \lambda_{\mathrm{PhC}}}$
0.07	−17.6378772	−35.186488	155.95114	−15.482054	−10.55594
0.08	−17.0823885	−26.939655	151.03960	−11.853448	−8.081896
0.09	−16.5527343	−21.285653	146.35649	−9.3656875	−6.385696
0.10	−16.0473372	−17.241379	141.88785	−7.5862068	−5.172413

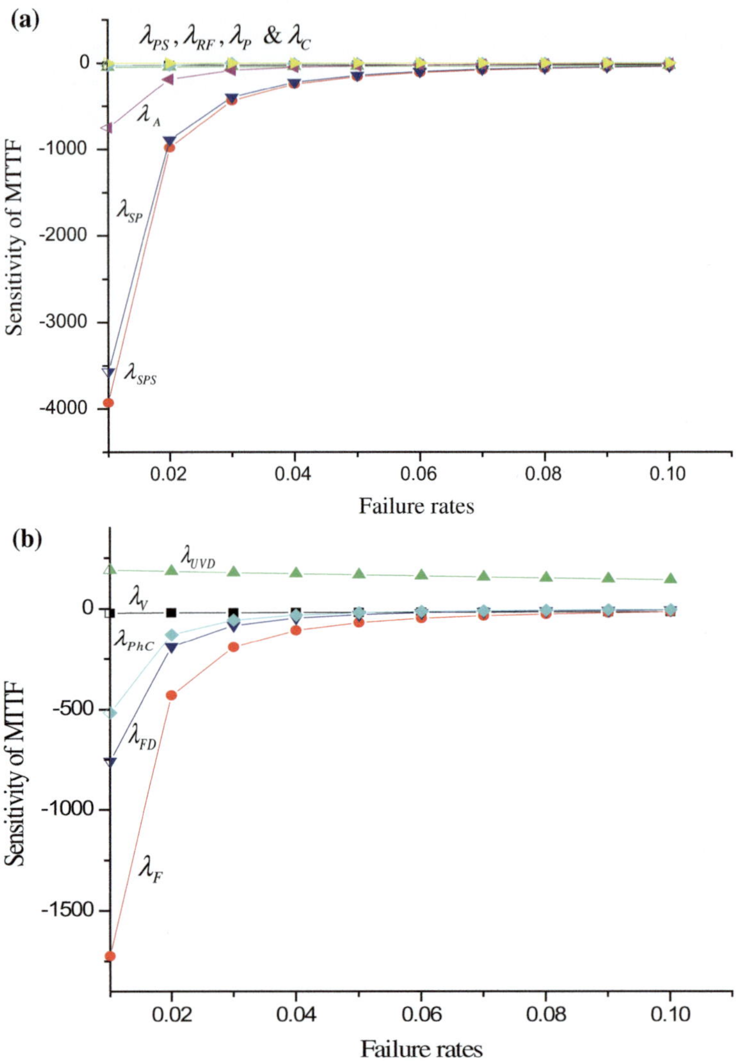

Fig. 5 **a** Sensitivity of MTTF of WTP versus failure rates, and **b** sensitivity of MTTF of WTP versus failure rates

Table 4 Sensitivity of reliability of WTP versus failure rates

(a)							
Time (t)	Sensitivity of reliability						
	$\frac{\partial R(t)}{\partial \lambda_{PS}}$	$\frac{\partial R(t)}{\partial \lambda_{SPS}}$	$\frac{\partial R(t)}{\partial \lambda_{P}}$	$\frac{\partial R(t)}{\partial \lambda_{SP}}$	$\frac{\partial R(t)}{\partial \lambda_{RF}}$	$\frac{\partial R(t)}{\partial \lambda_{A}}$	$\frac{\partial R(t)}{\partial \lambda_{C}}$
0	0	0	0	0	0	0	0
1	−0.017025	−0.088902	−0.025051	−0.079706	−0.000243	−0.000279	−0.012970
2	−0.047076	−0.291283	−0.073179	−0.257303	−0.001588	−0.001802	−0.033740
3	−0.074371	−0.543486	−0.122696	−0.472593	−0.004407	−0.004932	−0.049390
4	−0.094343	−0.810251	−0.165802	−0.693053	−0.008647	−0.009539	−0.056946
5	−0.106927	−1.072469	−0.200715	−0.901797	−0.014061	−0.015288	−0.057212
6	−0.113541	−1.320186	−0.227993	−1.090717	−0.020338	−0.021794	−0.052070
7	−0.115827	−1.548618	−0.248899	−1.256579	−0.027169	−0.028693	−0.043390
8	−0.115201	−1.755871	−0.264728	−1.398803	−0.034274	−0.035677	−0.032686
9	−0.112739	−1.941649	−0.276586	−1.518210	−0.414153	−0.042494	−0.021061
10	−0.109209	−2.106513	−0.285342	−1.616310	−0.048399	−0.048959	−0.009269

(b)					
Time (t)	Sensitivity of reliability				
	$\frac{\partial R(t)}{\partial \lambda_{V}}$	$\frac{\partial R(t)}{\partial \lambda_{F}}$	$\frac{\partial R(t)}{\partial \lambda_{UVD}}$	$\frac{\partial R(t)}{\partial \lambda_{FD}}$	$\frac{\partial R(t)}{\partial \lambda_{PhC}}$
0	0	0	0	0	0
1	−0.0250518	−0.0325532	0.0307928	−0.0003250	−0.0002146
2	−0.0731797	−0.1074569	0.1074196	−0.0021910	−0.0014897
3	−0.1226968	−0.2020856	0.2099520	−0.0061937	−0.0044002
4	−0.1658026	−0.3037792	0.3239434	−0.0123869	−0.0092027
5	−0.2007155	−0.4055559	0.4397917	−0.0205494	−0.0159757
6	−0.2279938	−0.5036645	0.5515878	−0.0303483	−0.0246996
7	−0.2488990	−0.5961864	0.6560155	−0.0414224	−0.0353020
8	−0.2647286	−0.6822379	0.7514764	−0.0534251	−0.0476840
9	−0.2765868	−0.7615155	0.8374545	−0.0660437	−0.0617345
10	−0.2853420	−0.8340361	0.9140804	−0.0790071	−0.0773382

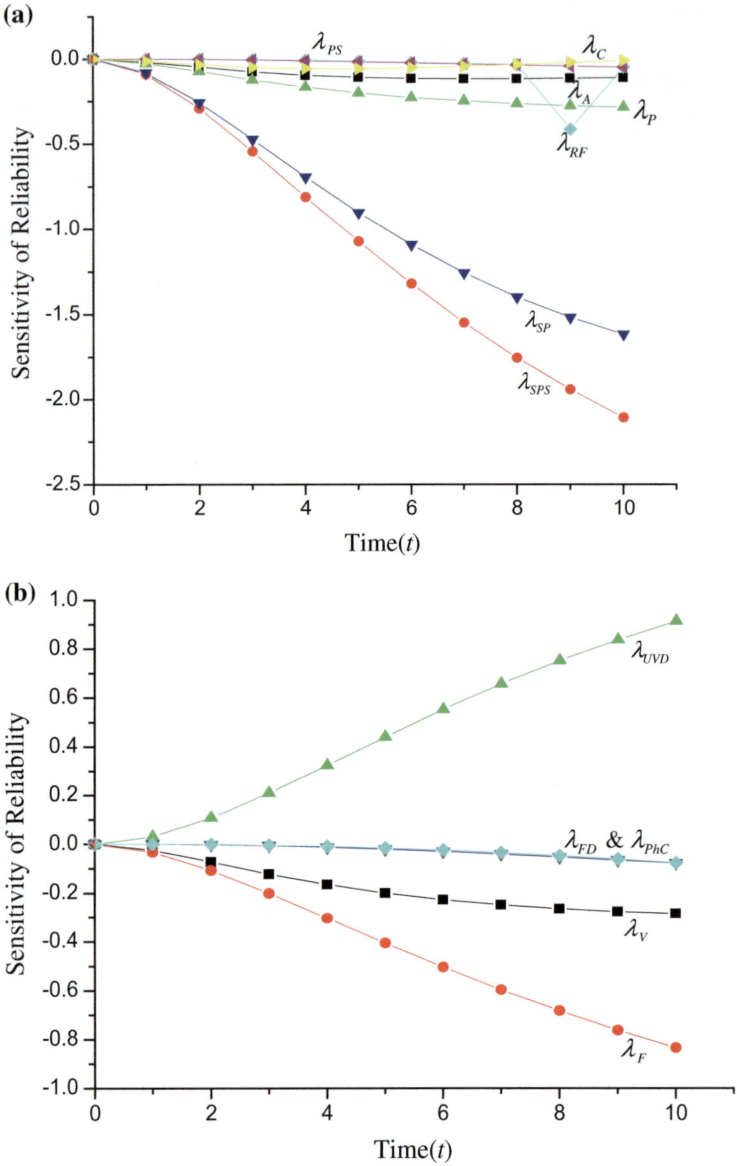

Fig. 6 **a** Sensitivity of reliability of WTP versus failure rates, and **b** sensitivity of reliability of WTP versus failure rates

References

1. Am Water Works Res, F., Langlais, B., Reckhow, D. A., & Brink, D. R. (1991). Ozone in water treatment: Application and engineering. CRC press.
2. Bao, Y., & Mays, L. W. (1990). Model for water distribution system reliability. *Journal of Hydraulic Engineering, 116*(9), 1119–1137.
3. Betancourt, W. Q., & Rose, J. B. (2004). Drinking water treatment processes for removal of *Cryptosporidium* and *Giardia*. *Veterinary Parasitology, 126*(1), 219–234.
4. Chang, E. E., Chiang, P. C., Huang, S. M., & Lin, Y. L. (2007). Development and implementation of performance evaluation system for a water treatment plant: Case study of Taipei water treatment plant. *Practice Periodical of Hazardous, Toxic, and Radioactive Waste Management, 11*(1), 36–47.
5. Eisenberg, D., Soller, J., Sakaji, R., & Olivieri, A. (2001). A methodology to evaluate water and wastewater treatment plant reliability. *Water Science and Technology, 43*(10), 91–99.
6. Faust, S. D., & Aly, O. M. (1998). Chemistry of water treatment. CRC Press.
7. Fujiwara, O., & Ganesharajah, T. (1993). Reliability assessment of water supply systems with storage and distribution networks. *Water Resources Research, 29*(8), 2917–2924.
8. Hashimoto, T., Stedinger, J. R., & Loucks, D. P. (1982). Reliability, resiliency, and vulnerability criteria for water resource system performance evaluation. *Water Resources Research, 18*(1), 14–20.
9. Kumar, A., & Ram, M. (2013). Reliability measures improvement and sensitivity analysis of a coal handling unit for thermal power plant. *International Journal of Engineering-Transactions C: Aspects, 26*(9), 1059.
10. Kumar, A., & Ram, M. (2015). Performance of marine power plant given generator, main and distribution switchboard failures. *Journal of Marine Science and Application, 14*(4), 450–458.
11. Kumar, A., Varshney, A., & Ram, M. (2015). Sensitivity analysis for casting process under stochastic modelling. *International Journal of Industrial Engineering Computations, 6*(3), 419–432.
12. Kumar, D., Singh, J., & Pandey, P. C. (1990). Design and cost analysis of a refining system in the sugar industry. *Microelectronics Reliability, 30*(6), 1025–1028.
13. Loucks, D. P. (1979). Water resources systems. *Reviews of Geophysics, 17*(6), 1335–1351.
14. Mathew, J., & Rajendran, C. (1993). Scheduling of maintenance activities in a sugar industry using simulation. *Computers in Industry, 21*(3), 331–334.
15. Sachdeva, A., Kumar, D., & Kumar, P. (2008). Reliability analysis of pulping system using Petri nets. *International Journal of Quality & Reliability Management, 25*(8), 860–877.
16. Su, Y. C., Mays, L. W., Duan, N., & Lansey, K. E. (1987). Reliability-based optimization model for water distribution systems. *Journal of Hydraulic Engineering, 113*(12), 1539–1556.
17. Tzong, R. Y., & Lee, S. L. (1992). Solidification of arbitrarily shaped casting in mold-casting system. *International Journal of Heat and Mass Transfer, 35*(11), 2795–2803.
18. Vasquez, J. A., Maier, H. R., Lence, B. J., Tolson, B. A., & Foschi, R. O. (2000). Achieving water quality system reliability using genetic algorithms. *Journal of Environmental Engineering, 26*(10), 954–962.
19. Venkatesan, A., Gopinath, V. M., & Rajadurai, A. (2005). Simulation of casting solidification and its grain structure prediction using FEM. *Journal of Materials Processing Technology, 168*(1), 10–15.
20. Wagner, J. M., Shamir, U., & Marks, D. H. (1988). Water distribution reliability: Analytical methods. *Journal of Water Resources Planning and Management, 114*(3), 253–275.
21. Ye, L., Wang, S., Bing, F., Malik, O. P., & Zeng, Y. (2001). Control/maintenance strategy fault tolerant mode and reliability analysis for hydro power stations. *IEEE Transactions on Power Systems, 16*(3), 340–345.

Prediction of El-Nino Year and Performance Analysis on the Calculated Correlation Coefficients

Malsa Nitima, Gautam Jyoti and Bairagee Nisha

Abstract El-Nino is a meteorological/oceanographic phenomenon that occurs at irregular intervals of time (every few years) at low latitudes. El-Nino can be related to an annual weak warm ocean current that runs southward along the coast of Peru and Ecuador about Christmastime. It is characterized by unusually large warming that occurs every few years and changes the local and regional ecology. El-Nino has been linked to climate change anomalies like global warming, etc. The data for this work has been taken from the websites mainly for India (Becker in Impacts of El-Nino and La Niña on the hurricane season, 2014 [1]; Hansen et al. in GISS surface temperature analysis (GISTEMP) NASA goddard institute for space studies, 2017 [2]; Cook in Pacific marine environmental laboratory national oceanic and atmospheric administration, 1999 [3]; Climate Prediction Center—Monitoring & Data [4]; Romm in Climate Deniers' favorite temperature dataset just confirmed global warming, 2016 [5]; World Bank Group, 2017 [6]; National Center for Atmospheric Research Staff (Eds) in The climate data guide: global temperature data sets: overview & comparison table, 2014 [7]; Global Climate Change Data, 1750–2015 [8]). Data have been preprocessed using imputation, F-measure, and maximum likelihood missing value methods. Finally, the prediction has been made about the time of occurrence of the next El-Nino year by using a multiple linear regression algorithm. A comparative analysis has been done on the three approaches used. The work also calculates Karl Pearson's correlation coefficient between global warming and temperature change, temperature change and El-Nino, and finally global warming and El-Nino. Performance analysis has been done on the correlation coefficient calculated.

M. Nitima (✉) · G. Jyoti · B. Nisha
JSS Academy of Technical Education, Noida, India
e-mail: nitima.malsa@gmail.com

G. Jyoti
e-mail: jyotig@jssaten.ac.in; jyotijssaten@gmail.com

B. Nisha
e-mail: nisha1.bairagi@gmail.com

P. K. Kapur et al. (eds.), *System Performance and Management Analytics*, Asset Analytics, https://doi.org/10.1007/978-981-10-7323-6_15

Keywords El-Nino · Multiple linear regression · Karl Pearson's correlation coefficient · Global warming

1 Introduction

The chapter introduces El-Nino and how it develops, its impacts on human being, and thereafter how it affects the global warming. The chapter also describes a stepwise experimental methodology which has been used to predict the next El-Nino year by using a multiple linear regression algorithm. The chapter also describes the calculated Karl Pearson's correlation coefficient between global warming and temperature change, temperature change and El-Nino, and finally global warming and El-Nino. The Performance analysis has also been described on the correlation coefficient calculated.

1.1 El-Nino and Its Behaviour

El-Nino is a versatile and unsurprisingly weather phenomenon that occurs when ocean temperatures in the Pacific Ocean near the equator differ from the normal. The winds weaken or break down in an El-Nino year. Rain and storm occur as a consequence of the warm water that is normally pushed toward the western Pacific washes back across, piling up on the east side of the Pacific from California to Chile. The pattern usually occurs very frequently (two to seven years). The 2015–2016 El-Nino year is called a "super" El-Nino year. This is the worst in 15 years. The other two previous super El-Nino years are 1982–1983 and 1997–1998 [9].

When El-Nino develops, the strong winds blow from east to west across the Pacific Ocean around the equator. The winds force warm surface ocean water from South America west toward Asia and Australia, and cold water to take its place along South America. The difference in temperature is created as a consequence of this across the Pacific Ocean, which helps out to blow the strong winds (the temperature difference and the strong winds are strongly correlated). The accretion of warm water in the west adds heat to the air, causing it to rise and create unstable weather; this is the main reason why the western Pacific region is warm and rainy. In general, cool and dry air is found on the eastern side of the Pacific.

1.2 El-Nino and Its Effects: Global Warming

In the study, it has been observed that when El-Nino develops, it disturbs sea life in the Pacific and causes weather changes all over the world. It has also been observed that the El-Nino badly affects human life due to the abnormal atmospheric and oceanic conditions and also responsible for climate changes. Numerical models of El-Nino are used for weather forecasting. The forecasts can be presented in terms of five possibilities: (1) near normal conditions, (2) a weak El-Nino with a slightly wetter than normal growing season, (3) a full-blown El-Nino with flooding, (4) cooler than normal waters offshore, with higher than normal chance of drought, and (5) Global warming [10].

According to the study [11] shows the close relationship between ENSO and global temperature, hence global warming can be attributed to human activity. The global warming [12] is defined as the most recent and endless rise in earth surface temperature according to the environmental protection agency (EPA). The greenhouse gases are released as a result of climate pattern change. Carbon dioxide (CO_2), nitrous oxide (N_2O), methane (CH_4), water vapor (H_2O), and other gases including hydrofluorocarbons (HFCs), sulfur hexafluoride (SF_6), and perfluorocarbons (PFCs) are the constituents of the greenhouse gases (GHGs).

In our work, an experimental methodology has been used. Through prevalent study of the literature, a number of factors that are considered to have influence on the effects of global warming and El-Nino are identified. Input variables are identified through these factors. Real data from various sites have been collected online and then filtered out using manual techniques and preprocessed the data by using different methods. Then the data will transform into a standard format. After that, feature and parameter selection are identified.

Then the analysis is done on the identified parameters and implementation will be performed on the data by applying algorithms. Firstly, the data preprocessing is applied to remove the missing values from the data and apply 3 different methods and compare the results of them for finding the accuracy among them. Secondly, apply the Karl Pearson's correlation coefficient algorithm for finding the relationship between El-Nino and Global Warming. After that the prediction algorithm, i.e., multiple linear regression is applied and predicting the upcoming years of El-Nino. Then, the implementation results will produce and analyzed. The methodology used is illustrated stepwise with the help of diagram as shown in Fig. 1.

2 Literature Survey

Prediction is the forecasting of upcoming phases of ENSO. Many factors that are responsible for El-Nino are studied in this paper. During the El-Nino conditions, the SW Pacific Ocean extreme cyclone wave hazard is significantly larger, and the incidence of two extreme El-Nino events (1982–83 and 1997–98) in the satellite observation led to predict the highest extreme-wave atmosphere in the east of the region. This means that any prospect change in climate influences on the frequency and strength of such strong El-Nino events results in tremendous climate transform in the area [13].

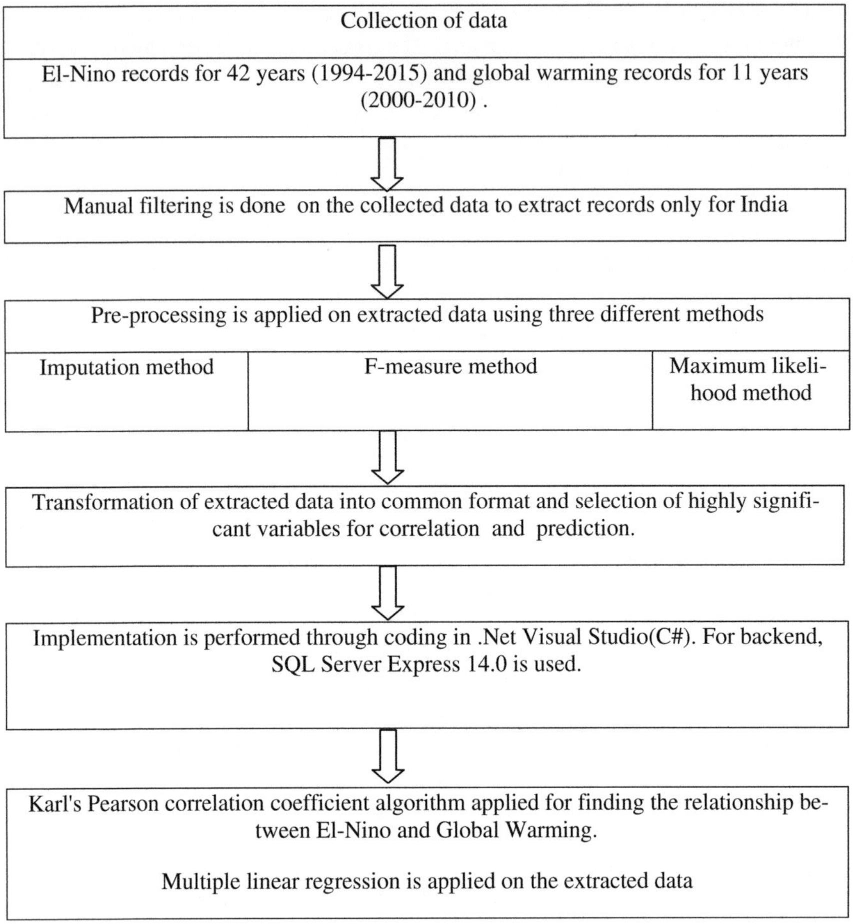

Fig. 1 Process flow diagram of the methodology used to predict the next El-Nino year

Data mining concepts and techniques, suggested by Han and Kamber, allows the users to analyze data from different sizes and magnitudes, classify it and recapitulate the associations which are identified during the mining progression [14].

Kaur et al. described the concept that is used in this paper which they used in education sector [2].

El-Nino the new and upcoming field in ecological sector and can be applied in different areas like sports, agriculture, transportation, Environmental risk, global monitoring, city planning, medical, education, etc. Yu et al. give the new model to predict the length of the day using extreme learning machine, and prediction results are analyzed and compared with those obtained by other machine learning-based prediction methods, including BPNN, generalization regression neural networks

(GRNN), and adaptive network-based fuzzy inference systems (ANFIS). It is shown that while achieving similar prediction accuracy, the developed method uses the much less training time than other methods [3].

To quantify the extreme significant wave height from tropical cyclones across the Southwest Pacific Ocean, Stephens and Ramsay describe the first use of a stochastic cyclone model (SCM). To quantify the effects of the El-Nino-Southern Oscillation (ENSO) on severe significant wave heights, and also the effects of projected climate change on cyclone intensity and frequency of happening, the SCM was used. During El-Nino conditions, cyclone formation and propagation are expected to occur. However, these cyclones are more likely to be powerful, mainly in extreme El-Nino event duration, leading to a higher long-term extreme-wave climate in the eastern SW Pacific, in spite of the relatively low cyclone observation rate there [4].

Forecasting should be alienated from its behavior and impacts, once the occurrence of the event has been assured. Whenever a forecast is made, someone is responding to it. Therefore, it is necessary to report societies and economic sectors affected by El-Nino. [5].

Richman and Leslie's work examines changes in mean values of maximum daily temperatures or each summer month, in southeastern Australia. A 10-site dataset, for 1958–2013, was collected and resample to quantify temporal changes and uncertainty in decades monthly maximum temperatures. The result confirms the regional environment of the warming [1].

Nerudová and Solilová discussed the three methods of missing data: regression method, imputation, and multiple imputation. They also discussed the impact of the three methods on the CCCTB determination. On the basis of the results, the most appropriate method will be selected that leads to the least distortion. The results obtained from the three methods are compared [6].

To handle longitudinal data with missing values, Li and Yi discover a new method, i.e., pairwise likelihood method. Performance is measured under different situations for the proposed method and particularly, efficiency and robustness are examined. Then, longitudinal survey data are analyzed with the proposed method that has taken from the Waterloo Smoking Prevention Project [7].

The data have been collected from different websites and an experimental methodology is used to generate the database. Then, different algorithms for removing missing values have been used, i.e., imputation method, F-measures, and maximum likelihood. Correlation between El-Nino and global warming has been determined through temperature change by using Karl Pearson's correlation coefficient and finally, the prediction has been made about the time of occurrence of the next upcoming El-Nino year by using a multiple linear regression algorithm.

3 Methodology

An experimental methodology has been used for determining the correlation between El-Nino and global warming. It also predicts the upcoming El-Nino year. The methodology is described in the following steps and illustrated with the help of the diagram as shown in Fig. 1.

Step 1. Some of the factors have been identified which are considered to be the causes for global warming and El-Nino.
Step 2. Real world data have been collected from various websites [8, 15–21]. For this, a record of 42 years (1994–2015) for El-Nino and a record of 11 years (2000–2010) for global warming have been taken.
Step 3. Collected data have been filtered out using manual techniques, mainly for India. El-Nino and global warming-related variables along with their domain values are defined in Tables 1 and 2, respectively.
Step 4. Preprocessing has been done on the extracted records by using three different methods named as imputation method, F-measure, and maximum likelihood method. The El-Nino data have many missing values. So, to fill the missing data, three different methods: imputation, F-measure, and maximum likelihood methods are used.

Table 1 El-Nino related variables

Variable name	Description	Domain values
SST	Sea surface temperature	1974–2015
SLP	Surface level pressure	1974–2015
OLW	Outgoing long wave	1974–2015
Z WIND	Zonal wind	1974–2015
MER WIND	Meridal wind	1974–2015
TEMP	Temperature	1974–2015
AIR HUMI	Air humidity	1974–2015
LONG	Longitude	1974–2015
LATI	Latitude	1974–2015
SOI	Southern oscillation index	1974–2015
WIND	Flow of wind	1974–2015

Table 2 Global warming related variables

Variable name	Description	Domain values
Greenhouse gases (CO_2)	Total fossil fuel, gases, liquids, solids, cement production, gas flaring, and per capita	2000–2010
Variations in earth's orbit	Earth's climate change	2000–2010
Deforestation	Population, standard of living	2000–2010
Burning fossil fuels	Gases, solids, and liquids	2000–2010

3.1 Imputation Method

- Missing values have been filled by the average of the column in which the missing value appeared.
- **Average**. The arithmetic mean is calculated by adding sum of given numbers and then dividing by the count of the numbers.

$$\text{average} = (a_1 + a_2 + a_3 \ldots\ldots\ldots\ldots\ldots\ldots + a_n/n) \tag{1}$$

Here, a_1, a_2, a_3, a_n are representing factors a_1 = longitude; a_2 = latitude; a_3 = temperature; a_4 = SLP; a_5 = SOI; a_6 = wind; a_7 = SST; a_8 = OLW; a_9 = air humidity; a_{10} = zonal wind; and a_{11} = meridal wind, and n is the total count of these factors.

3.2 F-Measures Method

- Missing values have been filled through harmonic mean of the column in which column the missing value appeared. The formula used for harmonic mean is

$$\frac{1}{H_y} = \frac{1}{n}\sum \frac{1}{Y_j}. \tag{2}$$

3.3 Maximum Likelihood Method

- Missing values have been filled by the most likely occurred values of the particular column.
- Sort the data column-wise, and fill the missing data by most common values in the dataset.

Step 5. Then the data has been transformed into a common format. After that, highly significant variables were identified using the select attribute facility of MYSQL, which were used in prediction of upcoming El-Nino year and also considered in finding out the correlation between El-Nino and global warming. The highly significant variables along with their domain values are listed below in Table 3.

Table 3 Highly significant variables for prediction and correlation

Variable name	Description	Domain values
SST	Sea surface temperature	1974–2015
SLP	Surface level pressure	1974–2015
TEMP	Temperature	1974–2015
SOI	Southern oscillation index	1974–2015
WIND	Flow of wind	1974–2015
GHG	Green house gas (CO_2)	2000–2010

Step 6. Calculation of the correlation coefficient between global warming and temperature change, temperature change and El-Nino, and finally global warming and El-Nino is done through Karl Pearson's correlation coefficient. It also predicts the time of occurrence of the next El-Nino year by using a multiple linear regression algorithm.

3.4 *Karl Pearson's Correlation Coefficient*

To establish the relationship between global warming and El-Nino by showing transitive relation between them.

$a \rightarrow$ global warming,
$b \rightarrow$ temperature change,
c $\rightarrow$ El-Nino,
a $\rightarrow$ b,
b $\rightarrow$ c, therefore a $\rightarrow$ c.

Evaluate the correlation between two or more attributes for numerical attributes. A and B, by computing the correlation coefficient (also known as Pearson's product moment coefficient).

$$\sum\nolimits_{i=1}^{N} \left(a_i b_{i\_} \overline{\mathrm{NAB}} / N\sigma_A \sigma_B\right) \tag{3}$$

3.5 *Multiple Linear Regression Algorithm*

Multiple linear regression analysis technique has been applied to predict the upcoming El-Nino year. The general linear regression model, with normal error terms, simply in terms of *X* variables is shown in

$$Y = X\beta + \varepsilon; \tag{4}$$

$$\beta_0 n + \beta_1 \sum\nolimits_{i=1}^{n} xi1 + \ldots\ldots\ldots + \beta_p \sum\nolimits_{i=1}^{n} x_{ip} + \varepsilon_i = \sum\nolimits_{i=1}^{n} y_i \tag{5}$$

Here, y represent the years in the dataset, i.e., y_1 = 1974, y_2 = 1975… y_{42} = 2015.

x denotes the factors, i.e., x_1 = longitude; x_2 = latitude; x_3 = temperature; x_4 = SLP; x_5 = SOI; x_6 = wind; x_7 = SST; x_8 = OLW; x_9 = air humidity; x_{10} = zonal wind; and x_{11} = meridal wind, and n is the total number of years, i.e., 42.

Step 7. Implementation is performed through coding in.Net Visual Studio (C#). For backend, SQL server express 14.0 is used.

To build GUI applications, Windows forms designer is used. The user interface is linked with code using an event-driven programming model. Either C# or VB. NET code is generated for the application.

Connectivity of database is done by SQL Server Express 2014 to create a table in SQL server and data is inserted by import the values from Excel files.

Data Explorer is used to manage databases on Microsoft SQL Server instances. For creating queries and writing stored procedures, Data Explorer is used either with T-SQL or SQL CLR. Debugging as well as IntelliSense support is also available.

Manage database connections on an accessible computer can be done through Server Explorer. It can also be used to browse running Windows Services, performance counters, Windows Event Log, and message queues and use them as a data source.

The coding for the algorithms was in c# that is a backend process of the project. It is simple to create the form in .net c# without being coding for the form by tool. Projects have sections for properties, references, and a Program.cs file.

4 Comparative Analysis

The algorithm is analyzed with three different missing values methods: imputation method, F-measures, and maximum likelihood. All the statistical results are provided in Table 4.

From the literature review [22] the imputation method for filling out the missing value gives the best result as compared to F-measures and maximum likelihood. The results obtained through imputation method provide valid statistical inferences that properly reflect the uncertainty due to missing values. Therefore, after 16 years (i.e., 2033) El-Nino may occur as predicted.

Table 4 Statistical results of MLR on different missing values algorithms

Missing values methods applied in MLR	Predicted year
Imputation method	16.604
F-measures	27.0072
Maximum likelihood	31.07228

Table 5 Calculated correlation coefficient

Calculated correlation coefficient		
Global warming and temperature change	Temperature change and El-Nino	Global warming and El-Nino
−0.93175	**0.91974**	**−0.8139**

5 Performance Analysis

Performance analysis has been done on the correlation coefficient calculated between global warming and temperature change, temperature change and El-Nino, and finally global warming and El-Nino. Calculated correlation coefficients of the three are mentioned in Table 5.

Table 5 depicts the relationship between global warming and temperature change, which is the negative one (**−0.93175**). This implies that as the global warming decreases, the temperature increases. Next, it shows the relationship between temperature change and El-Nino, which is a positive one (**0.91974**). This implies that as the temperature increases, the El-Nino effect increases.

So, finally, as the global warming decreases, the El-Nino effect increases. This can be verified from the calculated correlation coefficient between global warming and El-Nino (**−0.8139**).

6 Conclusion

In this paper, multiple linear regressions were used for prediction of upcoming El-Nino year on the dataset of 42 years. The dataset contained missing values so preprocessing steps were applied in the dataset to remove or fill out the missing values. Three methods used for removing missing values are imputation method, F-Measure, and maximum likelihood. Among all missing value methods, imputation method is considered to give valid statistical inferences that properly reflect the uncertainty due to missing values. So, the next El-Nino may occur after 16 years from the current year as predicted.

Through the calculated values of Karl Pearson's correlation, it has been observed that as the global warming decreases, the El-Nino effect increases.

7 Future Scope

The Prediction can be made on the basis of La-Nina also. Since it is known that ENSO affects the drought and flood situations, it is going to affect the world as a whole, e.g., in terms of agriculture, stock markets, and economy as a whole.

References

1. Richman, M. B., & Leslie, L. M. (2013). Classification of changes in extreme heat over South eastern Australia. *Elsevier Procedia Computer Science, 20,* 148–155.
2. Kaur, P., Singh, M., & Singh Josan, G. (2015). Classification and prediction based data mining algorithms to predict slow learners in education sector. *Elsevier Procedia Computer science, 57,* 500–508.
3. Lei, Y., Zhao, D., & Cai, H. (2015). Prediction of length-of-day using extreme learning machine. *Elsevier, Geodesy and Geodynamics, 6*(2), 151–159.
4. Stephens, S. A., & Ramsay, D. L. (2014). Extreme cyclone wave climate in the Southwest Pacific Ocean: Influence of the El Niño southern oscillation and projected climate change. *Elsevier Procedia Computer Science, 123,* 13–26.
5. Glantz, M. H. (2015). Shades of chaos: Lessons learned about lessons learned about forecasting El Niño and Its Impacts. *Springer International Journal Disaster Risk Science, 6,* 94–103.
6. Nerudová, D., & Solilová, V. (2014). Missing data and its impact on the CCCTB determination. *Elsevier Procedia Economics and Finance, 12,* 462–471.
7. Li, H., & Yi, G. Y. (2013). A pairwise likelihood approach for longitudinal data with missing observations in both response and covariates. *Elsevier Computational Statistics and Data Analysis, 68,* 66–81.
8. Becker, E. (2014). Impacts of El Niño and La Niña on the hurricane season. Available at https://www.climate.gov/news-features/blogs/enso/impacts-el-ni%C3%B1o-and-la-nB1a-hurricane-season.
9. Cho, R. (2016). Climate, general earth institute El Niño and global warming—What's the connection? Available at http://blogs.ei.columbia.edu/2016/02/02/el-nino-and-global-warming-whats-the-connection.
10. El Nino and climate prediction Edit-Design Center edc(at)atmos.washington.edu. Available at https://atmos.washington.edu/gcg/RTN/rtnt.html.
11. Pattimer. (2015). Global warming and the El Niño Southern oscillation. Available at https://www.skepticalscience.com/el-nino-southern-oscillation.htm.
12. Shah, A. (2015). Global issues climate change and global warming introduction. Available at http://www.globalissues.org/article/233/climate-change-and-global-warming-introduction#WhatistheGreenhouseEffect.
13. Dane, S., & Thool, R. C. (2013). Imputation method for missing value estimation of mixed-attribute data sets. *International Journal of Advanced Research in Computer Science and Software Engineering, 3*(5), 729–734.
14. Han, J., Kamber, M., & Pie, J. (2011). *Data mining concepts and techniques.* Morgan Kaufmann series: Elsevier.
15. Hansen, J., Ruedy, R., Sato, M., & Lo, K. (2017). Global surface temperature change. In: *GISS surface temperature analysis (GISTEMP) NASA goddard institute for space studies.* Available at https://data.giss.nasa.gov/gistemp.

16. Cook, D. (1999). Pacific marine environmental laboratory national oceanic and atmospheric administration. US Department of Commerce. Available at https://archive.ics.uci.edu/ml/machine-learning-databases/el_nino-mld/.
17. Climate Prediction Center—Monitoring & Data. Available at http://www.cpc.ncep.noaa.gov/data/indices/.
18. Romm, J. (2016). Climate deniers' favorite temperature dataset just confirmed global warming. Available at http://thinkprogress.org/climate/2016/03/02/3755715/satellites-hottest-february-global-warming/.
19. World Bank Group. (2017). Available at http://data.worldbank.org/climate-change/.
20. National Center for Atmospheric Research Staff (Eds). (2014). The climate data guide: global temperature data sets: Overview & comparison table. Available at https://climatedataguide.ucar.edu/climate-data/global-temperature-data-sets-overview-comparison-table.
21. Global Climate Change Data. (1750–2015). Available at http://www.google.co.in/search=dataset+on+global+warming=dataset+on+deforestation.
22. Weichenthal, S., Ryswyk, K. V., Goldstein, A., Bagg, S., Shekkarizfard, M., & Hatzopoulou, M. (2016). A land use regression model for ambient ultrafine particles in Montreal, Canada: A comparison of linear regression and a machine learning approach. *Elsevier Environmental Research, 146*, 65–72.

Performance of Static Spatial Topologies in Fine-Grained QEA on a P-PEAKS Problem Instance

Nija Mani, Gur Saran and Ashish Mani

Abstract Population-based meta-heuristics can admit population models and neighborhood topologies, which have a significant influence on their performance. Quantum-inspired evolutionary algorithms (QEA) often use coarse-grained population model and have been successful in solving difficult search and optimization problems. However, it was recently shown that the performance of QEA can be improved by changing its population model and neighborhood topologies. This paper investigates the effect of static spatial topologies on the performance of QEA with fine-grained population model on well-known benchmark problem generator known as P-PEAKS.

Keywords Population structure · Evolutionary algorithm · Cellular NEWS

1 Introduction

Quantum-inspired evolutionary algorithms (QEAs) are a class of meta-heuristics, which are designed by drawing inspiration from quantum mechanical principles into the framework of evolutionary algorithms (EA). They are used to improve search and optimization potential of evolutionary algorithms [1] by providing a better balance between exploration and exploitation [2]. They have been successful in solving wide range of benchmark as well as real-world problems, in discrete as well as continuous domain variables ranging from dynamic multicast routing with network coding [3], placing distributed generators in distribution system [4], solve large size quadratic knapsack problems [5], ordering problems [6], sum-of-squares-based fuzzy controller design [7], 0–1 knapsack problem [8]. QEAs

N. Mani (✉) · G. Saran
Dayalbagh Educational Institute (Deemed University), Dayalbagh, Agra, India
e-mail: mani.nija@gmail.com

A. Mani
Amity University Uttar Pradesh, Noida, Uttar Pradesh, India

P. K. Kapur et al. (eds.), *System Performance and Management Analytics*,
Asset Analytics, https://doi.org/10.1007/978-981-10-7323-6_16

are popular for their Q-bit representation, which is probabilistic in nature. A number of modifications have been proposed to improve the performance of canonical QEA by changing the variation operator [9] and population structure as in case of versatile quantum-inspired evolutionary algorithm (vQEA) [10]. In a recent study [11], it has been noticed that there is an improvement in the performance of QEA by changing the population model of QEA [12, 13] from coarse grained to panmictic [10] on some problems. Further, it has been shown in [14] that a QEA with fine-grained population model performs better than coarse-grained and panmictic models. Fine-grained population model admits neighborhood topologies, i.e., the way an individual in the population is interacting or influencing the evolution of other individuals. There are two types of topologies viz. random and spatial. The effect of random static topologies on the performance of fine-grained QEA was investigated in [15]. This paper continues the investigation into the effect of static spatial topologies on the performance of fine-grained QEA (FQEA) [16, 17].

The individuals in fine-grained population model are spread on a grid of nodes. This model admits many different kinds of fine-grained topologies and grid shapes, which can be classified into two main categories, viz., spatial and random topologies. The spatial topologies can be designed by considering information from fitness landscape, phenotype, and genotype solution space as well as all the three of them, and their combinations in form of ratios can be used to arrange the population in a spatial structure [18]. Random topologies are much easier to design as the location of the individuals on the grid is assigned randomly [15].

A well-known benchmark problem generator called P-PEAKS has been used to investigate the effect of static spatial topologies on the performance of FQEA. A comparative study has been performed between five different sorting methods of seven grid shapes of fine-grained QEA with static spatial topologies (SSFQEA), two ring topologies, and fine-grained QEA with static random topology. The results have been validated by performing nonparametric tests for multiple comparisons using Friedman test and post hoc procedure [19] and Wilcoxon signed ranked test has been used for pair-wise comparison [20]. The paper is further organized as follows: population topologies have been explained in Sect. 2. Section 3 presents testing of QEA with static spatial topologies. Section 4 concludes the paper.

2 Population Topology

Spatial structures have been used in particle swarm optimization (PSO) to construct meaningful neighborhoods [18] and in principle, appear to facilitate better performance than random population topologies. The canonical QEA described in [12] divides the population into groups and uses attractors in its rotation gate, which is a variation operator. The selection of attractor decides whether coarse-grained population is being used or panmictic model [14, 21]. Similarly, the selection of attractor is performed by considering neighborhood topologies in case of SSFQEA in the following manner:

i. Von Neumann topology—In this topology, the individuals in the population are arranged on two-dimensional toroidal grid. The neighborhood of an individual has four other individuals, i.e., its immediate north, east, west, and south neighbors (NEWS). Grid size is given by $G_x \times G_y$, where G_x is number of rows and G_y is number of columns. Further, it has been applied on three different shapes of toroidal grid, viz., square, rectangle, and narrows as shown in Fig. 1. All individuals are placed in a unique position on a toroidal grid and are linked to four others in their respective neighborhood in a cubic lattice type arrangement as shown in Figs. 1 and 2.

ii. Ring topology: In this topology, the individuals in the population are arranged on a ring with each individual occupying a unique position. The neighborhood has been constructed in two ways viz. directed and undirected. In directed ring topology, there are only two members in the group and the individual that acts as the attractor for an individual cannot have that individual as its own attractor, so attractor's selection is in one direction only as shown in Fig. 3. The undirected ring topology has three members and each individual is attached to the one before it and the one after it in the immediate neighborhood and so the attractor's selection can be in either direction.

Five types of spatial arrangement have been constructed by sorting the population on best fitness value of solution vectors (fitness landscape) called "IBF",

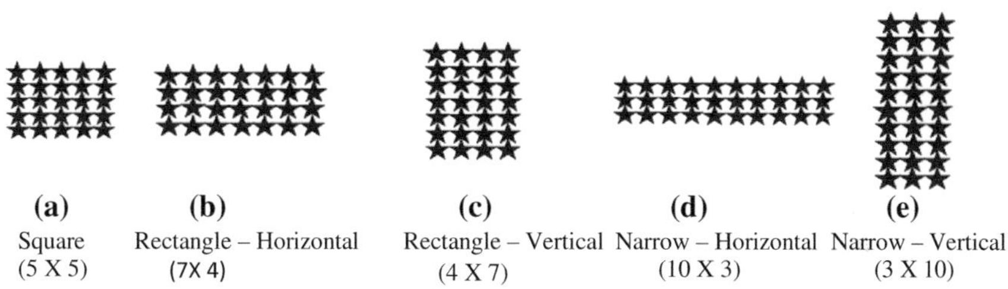

Fig. 1 Cellular grid structures

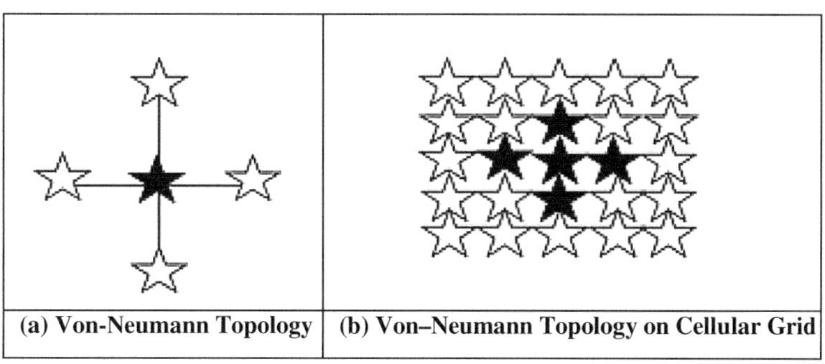

Fig. 2 Von Neumann topology

Fig. 3 Ring topology

binary solution vector (phenotype space) called "IBB", Euclidian distance of α values of Q-bit vector from origin (genotype space) called "QB", ratio of fitness to binary value of solution vector (combination of fitness landscape and phenotype space) called "IBFBPR", and ratio of fitness to Euclidian distance of α values of Q-bit vector from origin (combination of fitness landscape and genotype space) called "IBFQR".

The Spatial topologies can be implemented with ring and von Neumann only as G-Best [17] topology is not affected by spatial ordering of individual members in population because it is fully connected. This work describes the ring and von Neumann topologies by spatially distributing the individuals in the population and empirically compares their relative performance. The neighborhood list evaluates only once during the initialization stage and maintains the structure throughout the execution of the algorithm in a static spatial population topology and it sorts the population according to the above listed five spatial information in every generation.

The static spatial topologies are dynamic in nature as the sorting of population in each generation may create new neighbors for most of the individuals without changing the grid shape. Thus, static spatial topologies are static because grid shape is remaining static during the execution of the QEA. The algorithm with static spatial topology is as shows:

Step 1: t = 0; Population Size = NP, Topology = Type, Neighborhood Size = NS, Grid Size = G_x X G_y, Sorting = Type_Spatial_information;

Step 2: initialization of $Q_1(t) \ldots Q_{NP}(t)$ & Computation Neighborhood_list();

Step 3: obtain $P_1(t) \ldots P_{NP}(t)$ by measurement operation on $Q_1(t) \ldots Q_{NP}(t)$ respectively;

Step 4: if repairing required then perform repair $P_i(t)$, i = 1 ..NP;

Step 5: evaluation of $P_1(t) \ldots P_{NP}(t)$ & storing in $OP_1(t) \ldots OP_{NP}(t)$;

Step 6: sorting the population according to spatial Information;

Step 7: storing the global, neighborhood and individual best solutions into GB(t), $NB_i(t)$ & $IB_i(t)$ respectively, i = 1 .. NP;

```
  while (!condition_terminate) {
Step 8: t = t + 1;
        for each individual_i i = 1 .. NP {
Step 9:  determination of Attractor A_i(t) = NB_i(t-1);
Step 10: application of Q-gate(s) on Q_i(t-1) to update to Q_i(t);
Step 11: obtaining P_i(t) by measuring the states of Q_i(t);
Step 12: if repairing required then perform repair P_i(t);
Step 13: evaluation of P_i(t) & storing result into OP_i(t);
Step 14: storing the best solutions among IB_i(t-1) and
         OP_i(t) into IB_i(t);
Step 15: sorting the population according to spatial
         Information;
Step 16: storing the neighborhood best solution among
         NB_i(t-1) and Best_Neighbor_i(t) into NB_i(t) ;
 Step 17: storing the global best solution GB(t-1)
          among IB_i(t) into GB(t);}
```

In step 1, population size is initialized to NP and accordingly, topology is assigned to von Neumann or ring. The size of the neighborhood is assigned to NS depending on the type of topology, i.e., five in case of von Neumann, two in case of directed ring topology, and three in case of undirected ring topology. Grid size is assigned as $G_x \times G_y$ in case of von Neumann topology, and sorting method is assigned as the spatial information according to which population would be sorted. In step 2, the random initialization of Q-bit register $Q(t)$ which contains Q-bit strings $Q_1(t)$ … $Q_{\mathrm{NP}}(t)$ is performed along with the computation of neighborhood list with inputs as grid size, neighborhood size, and topology. In step 3, $P_1(t)$ … $P_{\mathrm{NP}}(t)$, which represents binary solutions, are obtained by application of measurement operator on $Q_1(t)…Q_{\mathrm{NP}}(t)$, respectively. In step 4, repairing is performed if necessary on binary solutions $P_i(t)$. In step 5, binary solution is evaluated to obtain its fitness $OP_i(t)$, where $OP_i(t)$ represents the objective function value. In step 6, the population is sorted according to the spatial information, viz., fitness landscape information (IBF), phenotype space information (IBB), genotype space information (QB), ratio of fitness landscape information and phenotype space information (IBFBPR), and ratio of fitness landscape information and genotype space information (IBFQR). In step 7, in the binary solutions $OP_i(t)$, the global, neighborhood, and individual best solutions are then selected, and stored into $GB(t)$, $NB_i(t)$, $IB_i(t)$, respectively, and neighborhood best solution is determined from the individuals in the neighborhood list of every individual. In step 9, the attractor $\mathrm{A_i}(t)$ is used for the *ith* individual as the $\mathrm{NB_i}(t)$. In step 10, using Q-Gates update $Q_{\mathrm{i}}(t-1)$ to $Q_{\mathrm{i}}(t)$, which is quantum rotation gate. In step 11, by measuring the states of $Q_{\mathrm{i}}(t)$ the binary solutions in $P_i(t)$ are formed as in step 3. In step 12, if the repair is needed, then it is performed as in step 4 and in step 13, each binary solution is observed and measured for the fitness as in step 5. In step 14, population is sorted as in step 6. In step 15, 16, and 17, the global, neighborhood, and individual best solutions are selected and stored into $GB(t)$, $NB_i(t)$, and $IB_i(t)$, respectively, based on a comparison between previous and current best solutions.

3 Testing, Results, and Analysis

A well-known benchmark problem generator called P-PEAKS has been used to investigate the effect of static spatial topologies on the performance of FQEA. In case of P-PEAKS problem, *P* number of *N*-bit strings are either randomly or heuristically generated, which are used as the position of P-PEAKS in the search space created by *N*-bits. The epistasis of fitness landscape is governed by *P*, i.e., higher the value of *P*, more is the ruggedness of the fitness landscape. The fitness value of remaining strings is computed by their hamming distance with the closest peak, divided by *N* (as shown in Eq. 1). P-PEAKS problem has a maximum fitness value of 1.0 [16].

$$f_{\text{P-PEAKS}}(\vec{x}) = \frac{1}{N}\max\{N - \text{Ham\_Dist}_{1 \le i \le p}(\vec{x} - \text{Peak}_i)\} \tag{1}$$

P-PEAKS problem generator provides for a large fairness for comparative studies between different instances of the algorithms. The P-PEAKS problem that has been used in this study, has *N* as 1000 and *P* as 20.

The value of parameter settings of QEA used for testing is given in Table 1. It has been deliberately kept same for all the instances of QEA so that as far as possible a fair comparative study can be statistically performed between them. The size of population is kept near 50 and the small variation in different instances is due to the grid size. One measurement has been performed on each *Q*-bit in one iteration of any instance. Fine-grained population model permits local migration only so it has been performed in every generation for all the instances. This work has used seven different toroidal grid sizes (3×16, 4×12, 6×8, 7×7, 16×3, 12×4, 8×6) with von Neumann topology having neighborhood size of five individuals to study their effect on the performance of QEA. Maximum number of iterations has been used as the stopping criterion, which is 3000 in this work.

Table 1 QEA parameters

Parameters	Value
θ_1 to θ_8	0, 0, 0.01π, 0, -0.01π, 0, 0, 0, respectively
Population size	50
Number of measurements	1
Local migration period (iterations)	1
Neighborhood topology	Von Neumann and ring
Neighborhood size	5 & 2, 3, respectively
Stopping criterion (iterations)	3000

Meta-heuristics are stochastic algorithms so statistical testing is necessary for finding their effectiveness. The descriptive statistics have been computed by executing a total of 30 independent runs for each instance and subsequently recording Best, Worst, Average, Median, and standard deviation (std) of the objective function value along with mean number of function evaluations to reach known best solution. Further, statistical analyses have been conducted using nonparametric tests as the conditions for parametric tests like independence, normality, and homoscedasticity are not guaranteed [19].

Single-problem analysis has been applied to results obtained over 30 runs of the algorithms over a given problem. In case of inferential statistics, hypothesis testing is employed to draw inferences about one or more populations from given test results. Two hypotheses, the null hypothesis H_0 and the alternative hypothesis H_1, are defined for statistical analyses. The null hypothesis is a statement of no difference between performance of two or more algorithms, whereas the alternative hypothesis represents the presence of a difference in the performance of two or more algorithms. When applying a statistical procedure to reject a hypothesis, a level of significance (=5%) is used to determine whether the hypothesis may be rejected.

There are several pair-wise and multiple comparison procedures available, however, in this work, we have used Wilcoxon's signed rank test for pair-wise comparison [20]. It has been implemented using spreadsheet given in companion CD of [20]. The multiple comparisons statistical test has been performed using Friedman test and associated post hoc procedures given as control test suite and multiple test suites [19]. The control test suite has been used for comparing the performance of one instance of QEA with the other instances, whereas multiple test suite has been used when multiple instances have to be compared without having preference for any particular instance.

The result of testing of SSFQEA with five different spatial distributions, i.e., sorting methods (IBB, IBF, IBFBPR, IBFQR, and QB) and grid size of 3 × 16, is given in Table 2. Further, the result of the corresponding FQEA with static random topology (SRT) has also been included from [15] for comparative study.

In order to compare the performance of the five instance of QEA with static spatial topologies, Friedman test (multiple comparison test suite [19]) has been

Table 2 Comparative study of SSFQEA with sorting methods and FQEA with static random topology (SRT) [15] on grid size 3 × 16

Grid shape	Sorting method	Best	Worst	Average	Median	Std	Avg. NFE
3 × 16	SRT	1.0000	1.0000	1.0000	1.0000	0.0000	89,699
	IBB	1.0000	1.0000	1.0000	1.0000	0.0000	85,197
	IBF	**1.0000**	**1.0000**	**1.0000**	**1.0000**	**0.0000**	**69,544**
	IBFBPR	1.0000	1.0000	1.0000	1.0000	0.0000	81,064
	IBFQR	1.0000	1.0000	1.0000	1.0000	0.0000	77,982
	QB	1.0000	1.0000	1.0000	1.0000	0.0000	83,762

Average NFE i.e. the row for which Average NFE is least is made bold or highlighted

applied on the SSFQEA, i.e., IBB, IBF, IBFBPR, IBFQR, and QB with performance metric being number of function evaluations to convergence to global optimum instead of objective function value as all the five instances of SSFQEA is converging to known global optimum in all the runs. The null hypothesis, H_0, is a statement of no difference between speed of convergence (measured by number of function evaluations to reach optimum) in all the five instances of SSFQEA, whereas the alternative hypothesis, H_1, represents the presence of difference in the performance of two or more algorithms. The ranking of IBF is best among all the five instances of SSFQEA and IBB has worst rank. Friedman statistic considering reduction performance (distributed according to chi-square with four degrees-of-freedom): 27.97. *P*-value computed by Friedman test: 1.26E-5. The null hypothesis, H_0, was rejected at significance level of 5% as *P*-value was less than 0.05 (multiple comparison test suite [19]).

SSFQEA with IBF sorting appears faster than all other algorithms as shown in Table 2 (Av. NFE.). Therefore, we want to investigate whether IBF is statistically superior to all other methods, so IBF is treated as the control method and control test suite [19] is applied. The null hypothesis, H_0, is a statement of no difference between speed of convergence (measured by number of function evaluations) between IBF and all other four instances of QEA with static spatial topologies, whereas the alternative hypothesis, H_1, represents the presence of difference in the performance of IBF and one or more of other four instances of SSFQEA. Friedman statistic (distributed according to chi-square with four degrees-of-freedom 27.97. *P*-value computed by Friedman test: 1.26E-5. The null hypothesis, H_0, was rejected at significance level of 5% as *P*-value was less than 0.05 (control test suite [19]). In pair-wise comparison between IBF and other four instances of SSFQEA, the null hypothesis, H_0, is a statement of no difference between speed of convergence (measured by number of function evaluations) between IBF and one of the other four instances of SSFQEA, whereas the alternative hypothesis, H_1, represents the presence of difference in the performance of IBF and one of the other four instances of SSFQEA. It was found that as per Hommel's and Holm's procedure (which rejects those hypotheses that have a p-value ≤ 0.05 at significance level of 5%), the null hypothesis could be rejected when comparison is performed with IBB, QB, IBFBPR, and IBFQR. In case of adjusted *P*-value, Hommel's, and Holm's procedure, the null hypothesis could be rejected when comparison is performed with QB, IBFBPR and IBB and IBFQR. Thus, it can be safely concluded that IBF outperforms IBB, QB, IBFBPR, and IBFQR.

In order to confirm the findings in Table 2 that SSFQEA with IBF performs better than FQEA with SRT, nonparametric Wilcoxon's signed rank test [20] was performed on number of function evaluations. The number of function evaluations of SSFQEA with IBF is μ_1 and the number of function evaluation of SRT $\mu_{2,}$ then the null hypothesis is H_0: $\mu_1 \geq \mu_2$ and the alternate hypothesis is H_1: $\mu_1 < \mu_2$. The result shows that null hypothesis has been rejected as Wilcoxon's signed rank test statistic is zero and less than the critical value of 152 at significance level of 5% which indicates that SSFQEA with IBF requires less number of function evaluations as compared to FQEA with SRT [15].

The result of testing of SSFQEA with five different spatial distributions, i.e., sorting methods (IBB, IBF, IBFBPR, IBFQR, and QB) and grid size of 4 × 12, is given in Table 3. Further, the result of the corresponding FQEA with static random topology (SRT) has also been included from [15] for comparative study.

In order to compare the performance of the five instance of QEA with static spatial topologies, Friedman test (multiple comparison test suite [19]) has been applied on the SSFQEA, i.e., IBB, IBF, IBFBPR, IBFQR, and QB with performance metric being number of function evaluations to convergence to global optimum instead of objective function value as all the five instances of SSFQEA is converging to known global optimum in all the runs. The null hypothesis, H_0, is a statement of no difference between speed of convergence (measured by number of function evaluations to reach optimum) in all the five instances of SSFQEA, whereas the alternative hypothesis, H_1, represents the presence of difference in the performance of two or more algorithms. The ranking of IBF is best among all the five instances of SSFQEA and IBB has worst rank. Friedman statistic (distributed according to chi-square with four degrees-of-freedom): 18.64. P-value computed by Friedman test: 9.25E-4. The null hypothesis, H_0, was rejected at significance level of 5% as P-value was less than 0.05 (multiple comparison test suite [19]).

SSFQEA with IBF sorting appears faster than all other algorithms as shown in Table 3 (Av. NFE.). Therefore, we want to investigate whether IBF is statistically superior to all other methods, so IBF is treated as the control method and control test suite [19] is applied. The null hypothesis, H_0, is a statement of no difference between speed of convergence (measured by number of function evaluations) between IBF and all other four instances of QEA with static spatial topologies, whereas the alternative hypothesis, H_1, represents the presence of difference in the performance of IBF and one or more of other four instances of SSFQEA. Friedman statistic (distributed according to chi-square with four degrees-of-freedom): 18.64. P-value computed by Friedman test: 9.25E-4. The null hypothesis, H_0, was rejected at significance level of 5% as P-value was less than 0.05 (control test suite [19]). In pair-wise comparison between IBF and other four instances of SSFQEA, the null hypothesis, H_0, is a statement of no difference between speed of convergence (measured by number of function evaluations) between IBF and one of the other

Table 3 Comparative study of SSFQEA with sorting methods and FQEA with static random topology (SRT) [15] on grid size 4 × 12

Grid shape	Sorting method	Best	Worst	Average	Median	Std	Avg. NFE
4 × 12	SRT	1.0000	1.0000	1.0000	1.0000	0.0000	83,923
	IBB	1.0000	1.0000	1.0000	1.0000	0.0000	77,010
	IBF	**1.0000**	**1.0000**	**1.0000**	**1.0000**	**0.0000**	**67,046**
	IBFBPR	1.0000	1.0000	1.0000	1.0000	0.0000	72,722
	IBFQR	1.0000	1.0000	1.0000	1.0000	0.0000	76,973
	QB	1.0000	1.0000	1.0000	1.0000	0.0000	74,310

Average NFE i.e. the row for which Average NFE is least is made bold or highlighted

four instances of SSFQEA, whereas the alternative hypothesis, H_1, represents the presence of difference in the performance of IBF and one of the other four instances of SSFQEA. It was found that as per Hommel's and Holm's procedure (which rejects those hypotheses that have a P-value ≤ 0.05 at significance level of 5%), the null hypothesis could be rejected when comparison is performed with IBB, QB, IBFBPR, and IBFQR. In case of adjusted P-value, Hommel's and Holm's procedure, the null hypothesis could be rejected when comparison is performed with QB, IBFBPR and IBB and IBFQR. Thus, it can be safely concluded that IBF outperforms IBB, QB, IBFBPR, and IBFQR.

In order to confirm the findings in Table 3, that SSFQEA with IBF performs better than FQEA with SRT, nonparametric Wilcoxon's signed rank test [20] was performed on number of function evaluations. The number of function evaluations of SSFQEA with IBF is μ_1 and the number of function evaluation of SRT μ_2, then the null hypothesis is H_0: $\mu_1 \geq \mu_2$ and the alternate hypothesis is H_1: $\mu_1 < \mu_2$. The result shows that null hypothesis has been rejected as Wilcoxon's signed rank test statistic is zero and less than the critical value of 152 at significance level of 5% which indicates that SSFQEA with IBF requires less number of function evaluations as compared to FQEA with SRT [15].

The result of testing of SSFQEA with five different spatial distributions, i.e., sorting methods (IBB, IBF, IBFBPR, IBFQR, and QB) and grid size of 6 × 8, is given in Table 4. Further, the result of the corresponding FQEA with static random topology (SRT) has also been included from [15] for comparative study.

In order to compare the performance of the five instance of QEA with static spatial topologies, Friedman test (multiple comparison test suite [19]) has been applied on the SSFQEA, i.e., IBB, IBF, IBFBPR, IBFQR, and QB with performance metric being number of function evaluations to convergence to global optimum instead of objective function value as all the five instances of SSFQEA is converging to known global optimum in all the runs. The null hypothesis, H_0, is a statement of no difference between speed of convergence (measured by number of function evaluations to reach optimum) in all the five instances of SSFQEA, whereas the alternative hypothesis, H_1, represents the presence of difference in the performance of two or more algorithms. The ranking of IBF is best among all the

Table 4 Comparative study of SSFQEA with sorting methods and FQEA with static random topology (SRT) [15] on grid size 6 × 8

Grid shape	Sorting method	Best	Worst	Average	Median	Std	Avg. NFE
6 × 8	SRT	1.0000	1.0000	1.0000	1.0000	0.0000	84,586
	IBB	1.0000	1.0000	1.0000	1.0000	0.0000	71,656
	IBF	**1.0000**	**1.0000**	**1.0000**	**1.0000**	**0.0000**	**65,408**
	IBFBPR	1.0000	1.0000	1.0000	1.0000	0.0000	71,826
	IBFQR	1.0000	1.0000	1.0000	1.0000	0.0000	69,218
	QB	1.0000	1.0000	1.0000	1.0000	0.0000	69,293

Average NFE i.e. the row for which Average NFE is least is made bold or highlighted

five instances of SSFQEA and IBB has worst rank. Friedman statistic (distributed according to chi-square with four degrees-of-freedom): 12.03. P-value computed by Friedman test: 0.01715. The null hypothesis, H_0, was rejected at significance level of 5% as P-value was less than 0.05 (multiple comparison test suite [19]).

SSFQEA with IBF sorting appears faster than all other algorithms as shown in Table 4 (Av. NFE.). Therefore, we want to investigate whether IBF is statistically superior to all other methods, so IBF is treated as the control method and control test suite [19] is applied. The null hypothesis, H_0, is a statement of no difference between speed of convergence (measured by number of function evaluations) between IBF and all other four instances of QEA with static spatial topologies, whereas the alternative hypothesis, H_1, represents the presence of difference in the performance of IBF and one or more of other four instances of SSFQEA. Friedman statistic (distributed according to chi-square with four degrees-of-freedom): 12.03. P-value computed by Friedman test: 0.01715. The null hypothesis, H_0, was rejected at significance level of 5% as P-value was less than 0.05 (control test suite [19]). In pair-wise comparison between IBF and other four instances of SSFQEA, the null hypothesis, H_0, is a statement of no difference between speed of convergence (measured by number of function evaluations) between IBF and one of the other four instances of SSFQEA, whereas the alternative hypothesis, H_1, represents the presence of difference in the performance of IBF and one of the other four instances of SSFQEA. It was found that as per Hommel's and Holm's procedure (which rejects those hypotheses that have a P-value ≤ 0.025 at significance level of 5%), the null hypothesis could be rejected when comparison is performed with IBB, IBFBPR, and IBFQR but it could not be rejected for QB. In case of adjusted P-value, Hommel's and Holm's procedure, the null hypothesis could be rejected when comparison is performed with IBFBPR and IBB but it could not be rejected for QB and IBFQR. However, average NFE is less for IBF as compared to QB and IBFQR. Thus, it can be safely concluded that IBF outperforms IBB and IBFBPR and is at least as good as QB and IBFQR.

In order to confirm the findings in Table 4, that SSFQEA with IBF performs better than FQEA with SRT, nonparametric Wilcoxon's signed rank test [20] was performed on number of function evaluations. The number of function evaluations of SSFQEA with IBF is μ_1 and the number of function evaluation of SRT μ_2, then the null hypothesis is H_0: $\mu_1 \geq \mu_2$ and the alternate hypothesis is H_1: $\mu_1 < \mu_2$. The result shows that null hypothesis has been rejected as Wilcoxon's signed rank test statistic is zero and less than the critical value of 152 at significance level of 5% which indicates that SSFQEA with IBF requires less number of function evaluations as compared to FQEA with SRT [15].

The result of testing of SSFQEA with five different spatial distributions, i.e., sorting methods (IBB, IBF, IBFBPR, IBFQR, and QB) and grid size of 7×7, is given in Table 5. Further, the result of the corresponding FQEA with static random topology (SRT) has also been included from [15] for comparative study.

In order to compare the performance of the five instance of QEA with static spatial topologies, Friedman test (multiple comparison test suite [19]) has been applied on the SSFQEA, i.e., IBB, IBF, IBFBPR, IBFQR, and QB with

Table 5 Comparative study of SSFQEA with sorting methods and FQEA with static random topology (SRT) [15] on grid size 7 × 7

Grid shape	Sorting method	Best	Worst	Average	Median	Std	Avg. NFE
7 × 7	SRT	1.0000	1.0000	1.0000	1.0000	0.0000	80,288
	IBB	1.0000	1.0000	1.0000	1.0000	0.0000	70,722
	IBF	**1.0000**	**1.0000**	**1.0000**	**1.0000**	**0.0000**	**65,382**
	IBFBPR	1.0000	1.0000	1.0000	1.0000	0.0000	69,110
	IBFQR	1.0000	1.0000	1.0000	1.0000	0.0000	69,392
	QB	1.0000	1.0000	1.0000	1.0000	0.0000	68,378

Average NFE i.e. the row for which Average NFE is least is made bold or highlighted

performance metric being number of function evaluations to convergence to global optimum instead of objective function value as all the five instances of SSFQEA is converging to known global optimum in all the runs. The null hypothesis, H_0, is a statement of no difference between speed of convergence (measured by number of function evaluations to reach optimum) in all the five instances of SSFQEA, whereas the alternative hypothesis, H_1, represents the presence of difference in the performance of two or more algorithms. The ranking of IBF is best among all the five instances of SSFQEA and IBB has worst rank. Friedman statistic (distributed according to chi-square with four degrees-of-freedom): 12.65. *P*-value computed by Friedman test: 0.01314. The null hypothesis, H_0, was rejected at significance level of 5% as *P*-value was less than 0.05 (multiple comparison test suite [19]).

SSFQEA with IBF sorting appears faster than all other algorithms as shown in Table 5 (Av. NFE.). Therefore, we want to investigate whether IBF is statistically superior to all other methods, so IBF is treated as the control method and control test suite [19] is applied. The null hypothesis, H_0, is a statement of no difference between speed of convergence (measured by number of function evaluations) between IBF and all other four instances of QEA with static spatial topologies, whereas the alternative hypothesis, H_1, represents the presence of difference in the performance of IBF and one or more of other four instances of SSFQEA. Friedman statistic (distributed according to chi-square with four degrees-of-freedom): 12.65. *P*-value computed by Friedman test: 0.01314. The null hypothesis, H_0, was rejected at significance level of 5% as *P*-value was less than 0.05 (control test suite [19]). In pair-wise comparison between IBF and other four instances of SSFQEA, the null hypothesis, H_0, is a statement of no difference between speed of convergence (measured by number of function evaluations) between IBF and one of the other four instances of SSFQEA, whereas the alternative hypothesis, H_1, represents the presence of difference in the performance of IBF and one of the other four instances of SSFQEA. It was found that as per Hommel's and Holm's procedure (which rejects those hypotheses that have a *P*-value ≤ 0.05 at significance level of 5%), the null hypothesis could be rejected when comparison is performed with IBB, QB, IBFBPR, and IBFQR. In case of adjusted *P*-value, Hommel's and Holm's procedure, the null hypothesis could be rejected when comparison is performed with QB,

IBFBPR and IBB and IBFQR. Thus, it can be safely concluded that IBF outperforms IBB, QB, IBFBPR, and IBFQR.

In order to confirm the findings in Table 5, that SSFQEA with IBF performs better than FQEA with SRT, nonparametric Wilcoxon's signed rank test [20] was performed on number of function evaluations. The number of function evaluations of SSFQEA with IBF is μ_1 and the number of function evaluation of SRT μ_2, then the null hypothesis is H_0: $\mu_1 \geq \mu_2$ and the alternate hypothesis is H_1: $\mu_1 < \mu_2$. The result shows that null hypothesis has been rejected as Wilcoxon's signed rank test statistic is zero and less than the critical value of 152 at significance level of 5% which indicates that SSFQEA with IBF requires less number of function evaluations as compared to FQEA with SRT [15].

The result of testing of SSFQEA with five different spatial distributions, i.e., sorting methods (IBB, IBF, IBFBPR, IBFQR, and QB) and grid size of 16 × 3, is given in Table 6. Further, the result of the corresponding FQEA with static random topology (SRT) has also been included from [15] for comparative study.

In order to compare the performance of the five instance of QEA with static spatial topologies, Friedman test (multiple comparison test suite [19]) has been applied on the SSFQEA, i.e., IBB, IBF, IBFBPR, IBFQR, and QB with performance metric being number of function evaluations to convergence to global optimum instead of objective function value as all the five instances of SSFQEA is converging to known global optimum in all the runs. The null hypothesis, H_0, is a statement of no difference between speed of convergence (measured by number of function evaluations to reach optimum) in all the five instances of SSFQEA, whereas the alternative hypothesis, H_1, represents the presence of difference in the performance of two or more algorithms. The ranking of IBF is best among all the five instances of SSFQEA and IBB has worst rank. Friedman statistic (distributed according to chi-square with four degrees-of-freedom): 14.95. *P*-value computed by Friedman test: 1.71E-8. The null hypothesis, H_0, was rejected at significance level of 5% as *P*-value was less than 0.05 (multiple comparison test suite [19]).

SSFQEA with IBF sorting appears faster than all other algorithms as shown in Table 6 (Av. NFE.). Therefore, we want to investigate whether IBF is statistically superior to all other methods, so IBF is treated as the control method and control

Table 6 Comparative study of SSFQEA with sorting methods and FQEA with static random topology (SRT) [15] on grid size 16 × 3

Grid shape	Sorting method	Best	Worst	Average	Median	Std	Avg. NFE
16 × 3	SRT	1.0000	1.0000	1.0000	1.0000	0.0000	94,216
	IBB	1.0000	1.0000	1.0000	1.0000	0.0000	70,144
	IBF	**1.0000**	**1.0000**	**1.0000**	**1.0000**	**0.0000**	**60,794**
	IBFBPR	1.0000	1.0000	1.0000	1.0000	0.0000	69,968
	IBFQR	1.0000	1.0000	1.0000	1.0000	0.0000	71,070
	QB	1.0000	1.0000	1.0000	1.0000	0.0000	68,181

Average NFE i.e. the row for which Average NFE is least is made bold or highlighted

test suite [19] is applied. The null hypothesis, H_0, is a statement of no difference between speed of convergence (measured by number of function evaluations) between IBF and all other four instances of QEA with static spatial topologies, whereas the alternative hypothesis, H_1, represents the presence of difference in the performance of IBF and one or more of other four instances of SSFQEA. Friedman statistic (distributed according to chi-square with four degrees-of-freedom): 14.95. P-value computed by Friedman test: 1.71E-8. The null hypothesis, H_0, was rejected at significance level of 5% as P-value was less than 0.05 (control test suite [19]). In pair-wise comparison between IBF and other four instances of SSFQEA, the null hypothesis, H_0, is a statement of no difference between speed of convergence (measured by number of function evaluations) between IBF and one of the other four instances of SSFQEA, whereas the alternative hypothesis, H_1, represents the presence of difference in the performance of IBF and one of the other four instances of SSFQEA. It was found that as per Hommel's and Holm's procedure (which rejects those hypotheses that have a P-value ≤ 0.05 at significance level of 5%), the null hypothesis could be rejected when comparison is performed with IBB, QB, IBFBPR, and IBFQR. In case of adjusted P-value, Hommel's and Holm's procedure, the null hypothesis could be rejected when comparison is performed with QB, IBFBPR and IBB and IBFQR. Thus, it can be safely concluded that IBF outperforms IBB, QB, IBFBPR, and IBFQR.

In order to confirm the findings in Table 6, that SSFQEA with IBF performs better than FQEA with SRT, nonparametric Wilcoxon's signed rank test [20] was performed on number of function evaluations. The number of function evaluations of SSFQEA with IBF is μ_1 and the number of function evaluation of SRT μ_2, then the null hypothesis is H_0: $\mu_1 \geq \mu_2$ and the alternate hypothesis is H_1: $\mu_1 < \mu_2$. The result shows that null hypothesis has been rejected as Wilcoxon's signed rank test statistic is zero and less than the critical value of 152 at significance level of 5% which indicates that SSFQEA with IBF requires less number of function evaluations as compared to FQEA with SRT [15].

The result of testing of SSFQEA with five different spatial distributions, i.e., sorting methods (IBB, IBF, IBFBPR, IBFQR, and QB) and grid size of 12 × 4, is

Table 7 Comparative study of SSFQEA with sorting methods and FQEA with static random topology (SRT) [15] on grid size 12 × 4

Grid shape	Sorting method	Best	Worst	Average	Median	Std	Avg. NFE
12 × 4	SRT	1.0000	1.0000	1.0000	1.0000	0.0000	83,709
	IBB	1.0000	1.0000	1.0000	1.0000	0.0000	67,178
	IBF	**1.0000**	**1.0000**	**1.0000**	**1.0000**	**0.0000**	**61,779**
	IBFBPR	1.0000	1.0000	1.0000	1.0000	0.0000	69,283
	IBFQR	1.0000	1.0000	1.0000	1.0000	0.0000	67,347
	QB	1.0000	1.0000	1.0000	1.0000	0.0000	66,414

Average NFE i.e. the row for which Average NFE is least is made bold or highlighted

given in Table 7. Further, the result of the corresponding FQEA with static random topology (SRT) has also been included from [15] for comparative study.

In order to compare the performance of the five instance of QEA with static spatial topologies, Friedman test (multiple comparison test suite [19]) has been applied on the SSFQEA, i.e., IBB, IBF, IBFBPR, IBFQR, and QB with performance metric being number of function evaluations to convergence to global optimum instead of objective function value as all the five instances of SSFQEA is converging to known global optimum in all the runs. The null hypothesis, H_0, is a statement of no difference between speed of convergence (measured by number of function evaluations to reach optimum) in all the five instances of SSFQEA, whereas the alternative hypothesis, H_1, represents the presence of difference in the performance of two or more algorithms. The ranking of IBF is best among all the five instances of SSFQEA and IBFBPR has worst rank. Friedman statistic (distributed according to chi-square with four degrees-of-freedom): 40.21. P-value computed by Friedman test: 3.91E-8. The null hypothesis, H_0, was rejected at significance level of 5% as P-value was less than 0.05 (multiple comparison test suite [19]).

SSFQEA with IBF sorting appears faster than all other algorithms as shown in Table 7 (Av. NFE.). Therefore, we want to investigate whether IBF is statistically superior to all other methods, so IBF is treated as the control method and control test suite [19] is applied. The null hypothesis, H_0, is a statement of no difference between speed of convergence (measured by number of function evaluations) between IBF and all other four instances of QEA with static spatial topologies, whereas the alternative hypothesis, H_1, represents the presence of difference in the performance of IBF and one or more of other four instances of SSFQEA. Friedman statistic (distributed according to chi-square with four degrees-of-freedom): 40.21. P-value computed by Friedman test: 3.91E-8. The null hypothesis, H_0, was rejected at significance level of 5% as P-value was less than 0.05 (control test suite [19]). In pair-wise comparison between IBF and other four instances of SSFQEA, the null hypothesis, H_0, is a statement of no difference between speed of convergence (measured by number of function evaluations) between IBF and one of the other four instances of SSFQEA, whereas the alternative hypothesis, H_1, represents the presence of difference in the performance of IBF and one of the other four instances of SSFQEA. It was found that as per Hommel's and Holm's procedure (which rejects those hypotheses that have a P-value ≤ 0.05 at significance level of 5%), the null hypothesis could be rejected when comparison is performed with IBB, QB, IBFBPR, and IBFQR. In case of adjusted P-value, Hommel's and Holm's procedure, the null hypothesis could be rejected when comparison is performed with QB, IBFBPR and IBB and IBFQR. Thus, it can be safely concluded that IBF outperforms IBB, QB, IBFBPR, and IBFQR.

In order to confirm the findings in Table 7, that SSFQEA with IBF performs better than FQEA with SRT, nonparametric Wilcoxon's signed rank test [20] was performed on number of function evaluations. The number of function evaluations of SSFQEA with IBF is μ_1 and the number of function evaluation of SRT μ_2, then the null hypothesis is H_0: $\mu_1 \geq \mu_2$ and the alternate hypothesis is H_1: $\mu_1 < \mu_2$.

Table 8 Comparative study of SSFQEA with sorting methods and FQEA with static random topology (SRT) [15] on grid size 8 × 6

Grid shape	Sorting method	Best	Worst	Average	Median	Std	Avg. NFE
8 × 6	SRT	1.0000	1.0000	1.0000	1.0000	0.0000	79,440
	IBB	1.0000	1.0000	1.0000	1.0000	0.0000	66,989
	IBF	**1.0000**	**1.0000**	**1.0000**	**1.0000**	**0.0000**	**63,038**
	IBFBPR	1.0000	1.0000	1.0000	1.0000	0.0000	67,173
	IBFQR	1.0000	1.0000	1.0000	1.0000	0.0000	67,450
	QB	1.0000	1.0000	1.0000	1.0000	0.0000	67,187

Average NFE i.e. the row for which Average NFE is least is made bold or highlighted

The result shows that null hypothesis has been rejected as Wilcoxon's signed rank test statistic is zero and less than the critical value of 152 at significance level of 5% which indicates that SSFQEA with IBF requires less number of function evaluations as compared to FQEA with SRT [15].

The result of testing of SSFQEA with five different spatial distributions, i.e., sorting methods (IBB, IBF, IBFBPR, IBFQR, and QB) and grid size of 8 × 6, is given in Table 8. Further, the result of the corresponding FQEA with static random topology (SRT) has also been included from [15] for comparative study.

In order to compare the performance of the five instance of QEA with static spatial topologies, Friedman test (multiple comparison test suite [19]) has been applied on the SSFQEA, i.e., IBB, IBF, IBFBPR, IBFQR, and QB with performance metric being number of function evaluations to convergence to global optimum instead of objective function value as all the five instances of SSFQEA is converging to known global optimum in all the runs. The null hypothesis, H_0, is a statement of no difference between speed of convergence (measured by number of function evaluations to reach optimum) in all the five instances of SSFQEA, whereas the alternative hypothesis, H_1, represents the presence of difference in the performance of two or more algorithms. The ranking of IBF is best among all the five instances of SSFQEA and IBB has worst rank. Friedman statistic (distributed according to chi-square with four degrees-of-freedom): 17.27. P-value computed by Friedman test: 0.0017. The null hypothesis, H_0, was rejected at significance level of 5% as P-value was less than 0.05 (multiple comparison test suite [19]).

SSFQEA with IBF sorting appears faster than all other algorithms as shown in Table 8 (Av. NFE.). Therefore, we want to investigate whether IBF is statistically superior to all other methods, so IBF is treated as the control method and control test suite [19] is applied. The null hypothesis, H_0, is a statement of no difference between speed of convergence (measured by number of function evaluations) between IBF and all other four instances of QEA with static spatial topologies, whereas the alternative hypothesis, H_1, represents the presence of difference in the performance of IBF and one or more of other four instances of SSFQEA. Friedman statistic (distributed according to chi-square with four degrees-of-freedom): 17.27. P-value computed by Friedman test: 0.0017. The null hypothesis, H_0, was rejected at significance level of 5% as P-value was less than 0.05 (control test suite [19]). In pair-wise comparison between IBF and other four instances of SSFQEA, the null hypothesis, H_0, is a statement of no difference between speed of convergence (measured by number of function evaluations) between IBF and one of the other four instances of SSFQEA, whereas the alternative hypothesis, H_1, represents the presence of difference in the performance of IBF and one of the other four instances of SSFQEA. It was found that as per Hommel's and Holm's procedure (which rejects those hypotheses that have a P-value ≤ 0.05 at significance level of 5%), the null hypothesis could be rejected when comparison is performed with IBB, QB, IBFBPR, and IBFQR. In case of adjusted P-value, Hommel's and Holm's procedure, the null hypothesis could be rejected when comparison is performed with QB, IBFBPR and IBB and IBFQR. Thus, it can be safely concluded that IBF outperforms IBB, QB, IBFBPR, and IBFQR.

In order to confirm the findings in Table 8, that SSFQEA with IBF performs better than FQEA with SRT, nonparametric Wilcoxon's signed rank test [20] was performed on number of function evaluations. The number of function evaluations of SSFQEA with IBF is μ_1 and the number of function evaluation of SRT μ_2, then the null hypothesis is H_0: $\mu_1 \geq \mu_2$ and the alternate hypothesis is H_1: $\mu_1 < \mu_2$. The result shows that null hypothesis has been rejected as Wilcoxon's signed rank test statistic is zero and less than the critical value of 152 at significance level of 5% which indicates that SSFQEA with IBF requires less number of function evaluations as compared to FQEA with SRT [15].

Further, in order to compare the performance of the seven grid sizes (3×16, 4×12, 6×8, 7×7, 16×3, 12×4, 8×6) of QEA with IBF sorting, Friedman test (multiple comparison test suite [19]) has been applied with performance metric being number of function evaluations to convergence to global optimum. The null hypothesis, H_0, is a statement of no difference between speed of convergence (measured by number of function evaluations to reach optimum) in all the instances of SSFQEA with seven grid size and IBF sorting, whereas the alternative hypothesis, H_1, represents the presence of difference in the performance of two or more instances. The ranking of 16×3 is best among all the seven instances of SSFQEA and 3×16 has worst rank. Friedman statistic (distributed according to chi-square with four degrees-of-freedom): 113.18. P-value computed by Friedman test: 5.43E-11. The null hypothesis, H_0, was rejected at significance level of 5% as P-value was less than 0.05 (multiple comparison test suite [19]).

SSFQEA with 16×3 and IBF sorting appears faster than all other algorithms as shown in Tables 2, 3, 4, 5, 6, 7 and 8 (Av. NFE.). Therefore, we want to investigate whether 16×3 is statistically superior to all other methods, so, 16×3 is treated as the control method and control test suite [19] is applied. The null hypothesis, H_0, is a statement of no difference between speed of convergence (measured by number of function evaluations) between 16×3 and all other seven instances of QEA with static spatial topologies, whereas the alternative hypothesis, H_1, represents the presence of difference in the performance of 16×3 and one or more of other six instances of SSFQEA. Friedman statistic (distributed according to chi-square with four degrees-of-freedom): 113.18. P-value computed by Friedman test: 5.43E-11.

The null hypothesis, H_0, was rejected at significance level of 5% as P-value was less than 0.05 (control test suite [19]). In pair-wise comparison between 16 × 3 and other six instances of SSFQEA, the null hypothesis, H_0, is a statement of no difference between speed of convergence (measured by number of function evaluations) between 16 × 3 and one of the other six instances of SSFQEA, whereas the alternative hypothesis, H_1, represents the presence of difference in the performance of 16 × 3 and one of the other six instances of SSFQEA. It was found that as per Hommel's and Holm's procedure (which rejects those hypotheses that have a P-value ≤ 0.05 at significance level of 5%), the null hypothesis could be rejected when comparison is performed with 3 × 16, 4 × 12, 6 × 8, 7 × 7, 8 × 6, and 12 × 4. In case of adjusted P-value, Hommel's and Holm's procedure, the null hypothesis could be rejected when comparison is performed with 3 × 16, 4 × 12, 6 × 8, 7 × 7, and 8 × 6, however, it could not reject null hypothesis for 12 × 4. Thus, it can be safely concluded that 16 × 3 outperforms with 6 × 8, 7 × 7, 16 × 3, 4 × 12, and 8 × 6, and is at least as good as 12 × 4 and it has lesser average NFE as compared to 12 × 4. Therefore, SSFQEA with 16 × 3 grid size and IBF sorting method is the best performing method of fine-grained static spatial topology.

The result of testing of QEA implementation with directed ring topology and five different spatial distributions, i.e., sorting methods (IBB, IBF, IBFBPR, IBFQR, and QB), is given in Table 9. Further, the result of the corresponding FQEA with static random topology (SRT) has also been included from [15] for comparative study.

In order to compare the performance of the five instance of QEA with directed ring topology and static spatial topologies, Friedman test (multiple comparison test suite [19]) has been applied on the QEA with directed ring topology, i.e., IBB, IBF, IBFBPR, IBFQR, and QB with performance metric being number of function evaluations to convergence to global optimum instead of objective function value as all the five instances of the QEA is converging to known global optimum in all the runs. The null hypothesis, H_0, is a statement of no difference between speed of convergence (measured by number of function evaluations to reach optimum) in all the five instances of the QEA, whereas the alternative hypothesis, H_1, represents the presence of difference in the performance of two or more algorithms. The ranking

Table 9 Comparative study of QEA with directed ring topology on sorting methods and FQEA with static random topology (SRT) [15]

Grid shape	Sorting method	Best	Worst	Average	Median	Std	Avg. NFE
1 Neighbor	SRT	1.0000	1.0000	1.0000	1.0000	0.0000	117,815
	IBB	1.0000	1.0000	1.0000	1.0000	0.0000	87,345
	IBF	**1.0000**	**1.0000**	**1.0000**	**1.0000**	**0.0000**	**71,922**
	IBFBPR	1.0000	1.0000	1.0000	1.0000	0.0000	86,835
	IBFQR	1.0000	1.0000	1.0000	1.0000	0.0000	81,028
	QB	1.0000	1.0000	1.0000	1.0000	0.0000	89,172

Average NFE i.e. the row for which Average NFE is least is made bold or highlighted

of IBF is best among all the five instances of the QEA and IBB has worst rank. Friedman statistic (distributed according to chi-square with four degrees-of-freedom): 46.67. P-value computed by Friedman test: 1.81E-9. The null hypothesis, H_0, was rejected at significance level of 5% as P-value was less than 0.05 (multiple comparison test suite [19]).

QEA (directed ring topology) with IBF sorting appears faster than all other algorithms as shown in Table 9 (Av. NFE.). Therefore, we want to investigate whether IBF is statistically superior to all other methods, so IBF is treated as the control method and control test suite [19] is applied. The null hypothesis, H_0, is a statement of no difference between speed of convergence (measured by number of function evaluations) between IBF and all other four instances of QEA with static spatial topologies, whereas the alternative hypothesis, H_1, represents the presence of difference in the performance of IBF and one or more of other four instances of QEA. Friedman statistic (distributed according to chi-square with four degrees-of-freedom): 46.67. P-value computed by Friedman test: 1.81E-9. The null hypothesis, H_0, was rejected at significance level of 5% as P-value was less than 0.05 (control test suite [19]). In pair-wise comparison between IBF and other four instances of QEA, the null hypothesis, H_0, is a statement of no difference between speed of convergence (measured by number of function evaluations) between IBF and one of the other four instances of SSFQEA, whereas the alternative hypothesis, H_1, represents the presence of difference in the performance of IBF and one of the other four instances of QEA. It was found that as per Hommel's and Holm's procedure (which rejects those hypotheses that have a P-value ≤ 0.05 at significance level of 5%), the null hypothesis could be rejected when comparison is performed with IBB, QB, IBFBPR, and IBFQR. In case of adjusted P-value, Hommel's and Holm's procedure, the null hypothesis could be rejected when comparison is performed with QB, IBFBPR and IBB and IBFQR. Thus, it can be safely concluded that IBF outperforms IBB, QB, IBFBPR, and IBFQR.

In order to confirm the findings in Table 9, that QEA with IBF performs better than FQEA with SRT, nonparametric Wilcoxon's signed rank test [20] was performed on number of function evaluations. The number of function evaluations of QEA with IBF is μ_1 and the number of function evaluation of SRT μ_2, then the null

Table 10 Comparative study of QEA with undirected ring topology on sorting methods and FQEA with static random topology (SRT) [15]

Grid shape	Sorting method	Best	Worst	Average	Median	Std	Avg. NFE
2 Neighbors	SRT	1.0000	1.0000	1.0000	1.0000	0.0000	101,693
	IBB	1.0000	1.0000	1.0000	1.0000	0.0000	90,380
	IBF	**1.0000**	**1.0000**	**1.0000**	**1.0000**	**0.0000**	**71,410**
	IBFBPR	1.0000	1.0000	1.0000	1.0000	0.0000	82,417
	IBFQR	1.0000	1.0000	1.0000	1.0000	0.0000	81,265
	QB	1.0000	1.0000	1.0000	1.0000	0.0000	87,373

Average NFE i.e. the row for which Average NFE is least is made bold or highlighted

hypothesis is H_0: $\mu_1 \geq \mu_2$ and the alternate hypothesis is H_1: $\mu_1 < \mu_2$. The result shows that null hypothesis has been rejected as Wilcoxon's signed rank test statistic is zero and less than the critical value of 152 at significance level of 5% which indicates that QEA with IBF requires less number of function evaluations as compared to FQEA with SRT [15].

The result of testing of QEA implementation with undirected ring topology and five different spatial distributions, i.e., sorting methods (IBB, IBF, IBFBPR, IBFQR, and QB), is given in Table 10. Further, the result of the corresponding FQEA with static random topology (SRT) has also been included from [15] for comparative study.

In order to compare the performance of the five instance of QEA with undirected ring topology and static spatial topologies, Friedman test (multiple comparison test suite [19]) has been applied on the QEA with directed ring topology, i.e., IBB, IBF, IBFBPR, IBFQR, and QB with performance metric being number of function evaluations to convergence to global optimum instead of objective function value as all the five instances of the QEA are converging to known global optimum in all the runs. The null hypothesis, H_0, is a statement of no difference between speed of convergence (measured by number of function evaluations to reach optimum) in all the five instances of the QEA, whereas the alternative hypothesis, H_1, represents the presence of difference in the performance of two or more algorithms. The ranking of IBF is best among all the five instances of the QEA and IBB has worst rank. Friedman statistic (distributed according to chi-square with four degrees-of-freedom): 44.48. *P*-value computed by Friedman test: 5.13E-9. The null hypothesis, H_0, was rejected at significance level of 5% as *P*-value was less than 0.05 (multiple comparison test suite [19]).

QEA (undirected ring topology) with IBF sorting appears faster than all other algorithms as shown in Table 10 (Av. NFE.). Therefore, we want to investigate whether IBF is statistically superior to all other methods, so IBF is treated as the control method and control test suite [19] is applied. The null hypothesis, H_0, is a statement of no difference between speed of convergence (measured by number of function evaluations) between IBF and all other four instances of QEA with static spatial topologies, whereas the alternative hypothesis, H_1, represents the presence of difference in the performance of IBF and one or more of other four instances of QEA. Friedman statistic (distributed according to chi-square with four degrees-of-freedom): 44.48. *P*-value computed by Friedman test: 5.13E-9. The null hypothesis, H_0, was rejected at significance level of 5% as *P*-value was less than 0.05 (control test suite [19]). In pair-wise comparison between IBF and other four instances of QEA, the null hypothesis, H_0, is a statement of no difference between speed of convergence (measured by number of function evaluations) between IBF and one of the other four instances of SSFQEA, whereas the alternative hypothesis, H_1, represents the presence of difference in the performance of IBF and one of the other four instances of QEA. It was found that as per Hommel's and Holm's procedure (which rejects those hypotheses that have a *P*-value $\leq$ 0.05 at significance level of 5%), the null hypothesis could be rejected when comparison is performed with IBB, QB, IBFBPR, and IBFQR. In case of adjusted *P*-value,

Hommel's and Holm's procedure, the null hypothesis could be rejected when comparison is performed with QB, IBFBPR and IBB and IBFQR. Thus, it can be safely concluded that IBF outperforms IBB, QB, IBFBPR, and IBFQR.

In order to confirm the findings in Table 10, that QEA with IBF performs better than FQEA with SRT, nonparametric Wilcoxon's signed rank test [20] was performed on number of function evaluations. The number of function evaluations of QEA with IBF is μ_1 and the number of function evaluation of SRT μ_2, then the null hypothesis is H_0: $\mu_1 \geq \mu_2$ and the alternate hypothesis is H_1: $\mu_1 < \mu_2$. The result shows that null hypothesis has been rejected as Wilcoxon's signed rank test statistic is zero and less than the critical value of 152 at significance level of 5% which indicates that QEA with IBF requires less number of function evaluations as compared to FQEA with SRT [15].

In order to compare the performance of directed and undirected ring topology with IBF, Wilcoxon's signed rank test [20] was performed on number of function evaluations for results tabulated in Tables 9 and 10. The number of function evaluations of undirected ring topology is μ_1 and the number of function evaluation of directed ring topology is μ_2, then the null hypothesis is H_0: $\mu_1 \geq \mu_2$ and the alternate hypothesis is H_1: $\mu_1 < \mu_2$. The null hypothesis could not be rejected as Wilcoxon's signed rank test statistic is 201.5 and more than the critical value of 152 at significance level of 5%. However, the average NFE for undirected topology is less as compared to average NFE of directed topology, so we consider that QEA with undirected topology is better than directed topology, though statistical evidence cannot reject the Null hypothesis.

In order to compare the performance of QEA with undirected ring topology and SSFQEA having grid size 16×3 with IBF sorting, Wilcoxon's signed rank test [20] was performed on number of function evaluations for results tabulated in Tables 10 and 2. The number of function evaluations of undirected ring topology is μ_1 and the number of function evaluation of directed ring topology is μ_2, then the null hypothesis is H_0: $\mu_1 \leq \mu_2$ and the alternate hypothesis is H_1: $\mu_1 > \mu_2$. The null hypothesis has been rejected as Wilcoxon's signed rank test statistic is 3 and less than the critical value of 152 at significance level of 5%. Thus, we can safely conclude that SSFQEA having grid size 16×3 with IBF sorting is better than QEA with ring topologies discussed in this paper.

4 Conclusions

The effect of Static Spatial Topologies on the performance of QEA has been investigated with the popular parameter values used with QEA [12, 13]. The P-PEAKS problem instance has been used for testing the QEA with the Static Spatial topologies. A total of nine different static spatial topologies with five different sorting methods based on fitness landscape, phenotype, and genotype space were investigated to arrive at the topologies used for designing FQEA. In static spatial topologies, von Neumann topologies have performed better than the ring topologies. The QEA with von Neumann (SSFQEA) static spatial topology with grid size 16 × 3 has performed better than all other grid sizes. The spatial arrangement of the population using fitness landscape sorting, i.e., IBF, has performed the best among all the other spatial arrangements. The performance of other methods of spatial arrangements has varied with the grid shape, however, they all have performed better than corresponding static random topologies. The QEA with ring topology having two neighbors has performed better than the one with a single neighbor directed topology. Further, it was slower as compared to the SSFQEAs with von Neumann topology. SSFQEA with 16 × 3 grid size has outperformed all the other 53 instances of QEAs as it was able to reach optima quickly in this problem instance. The SSFQEA with narrow horizontal grid shapes have in general performed better than the other grid shapes. It indicates that the exploration provided by the narrow grid shape along with fitness landscape-based spatial structure is helping QEA in locating the optimum quickly in this P-PEAKS problem instance [16].

References

1. Narayanan, A., & Moore, M. (1996). Quantum-inspired genetic algorithms In *Proceedings of IEEE International Conference on Evolutionary Computation*, pp. 61–66, 20–22 May 2016.
2. Michalewicz, Z., & Fogel, D. B. (2004). *How to solve it: Modern heuristics*. Springer.
3. Xing, H., Xu, L., Qu, R., & Qu, Z. (2016). A quantum inspired evolutionary algorithm for dynamic multicast routing with network coding. In *2016 16th International Symposium on Communications and Information Technologies (ISCIT)*, Qingdao (pp. 186–190).
4. Manikanta, G., Mani, A., Singh, H. P. & Chaturvedi, D. K. (2016). Placing distributed generators in distribution system using adaptive quantum inspired evolutionary algorithm. In *2016 Second International Conference on Research in Computational Intelligence and Communication Networks (ICRCICN)* (pp. 157–162), Kolkata.
5. Patvardhan, C., Bansal, S., Srivastav, A. (2016, February). Parallel improved quantum inspired evolutionary algorithm to solve large size quadratic knapsack problems. *Swarm Evol Comput, 26*, 175–190. ISSN 2210-6502.
6. da Silveira, L. R., Tanscheit, R., Vellasco, M. M. B. R. (2017, January). Quantum inspired evolutionary algorithm for ordering problems. *Expert Syst Appl, 67*, 71–83. ISSN- 0957-4174.
7. Yu, G. R., Huang, Y. C., & Cheng, C. Y. (2016). Sum-of-squares-based fuzzy controller design using quantum-inspired evolutionary algorithm. *International Journal of Systems Science, 47*(9), 2225–2236.
8. Pavithr, R. S., & Gursaran. (2016, August). Quantum inspired social evolution (QSE) algorithm for 0–1 knapsack problem. *Swarm Evol Comput, 29*, 33–46. ISSN 2210-6502.
9. Patvardhan, C., Narain, A., & Srivastava, A. (2007, December). Enhanced quantum evolutionary algorithm for difficult knapsack problems. In *Proceedings of International Conference on Pattern Recognition and Machine Intelligence*, Lecture Notes in Computer Science. Kolkata: Springer.
10. Platelt, M. D., Schliebs, S., & Kasabov, N. (2007). A versatile quantum inspired evolutionary algorithm. *Proceedings of IEEE CEC, 2007,* 423–430.
11. Mani, N., Gursaran, Sinha, A. K., & Mani, A. (2012). An evaluation of cellular population model for improving QiEA. In *Proceedings of GECCO-2012.*

12. Han, K. H., & Kim, J. H. (2002). Quantum–inspired evolutionary algorithm for a class of combinatorial optimization. *IEEE Transactions on Evolutionary Computation, 6*(6), 580–593.
13. Han, K. H., & Kim, J. H. (2004). Quantum-inspired evolutionary algorithms with a new termination criterion, H gate and two phase scheme. *IEEE Trans Evol Comput, 8*(2), 156–168.
14. Mani, N., Gursaran, Sinha, A. K., & Mani, A. (2014). Effect of population structures on quantum-inspired evolutionary algorithm. *Appl Comput Intell Soft Comput, 2014*(2014), 22 p. Article ID 976202.
15. Mani, N., Gursaran, & Mani, A. (2015). Performance of static random topologies in fine-grained QEA on P-PEAKS problem instances. In *2015 IEEE International Conference on Research in Computational Intelligence and Communication Networks (ICRCICN)* (pp. 163–168), Kolkata. https://doi.org/10.1109/icrcicn.2015.7434229.
16. Alba, E., & Dorronsoro, B. (2005). The exploration/exploitation tradeoff in dynamic cellular genetic algorithms. *IEEE Trans Evol Comput, 8*(2), 126–142.
17. Kennedy, J., & Mendes, R. (2002). Population structure and particle swarm performance. In *Proceeding of the 2002 Congress on Evolutionary Computation*, Honolulu, Hawali, 12–17 May 2002.
18. Lane, J., Engelbrecht, A., & Gain, J. (2008, September). Particle swarm optimization with spatially meaningful neighbours. Swarm Intell Symp. SIS 2008. IEEE pp. 1, 8, 21–23. https://doi.org/10.1109/sis.2008.4668281.
19. Derrac, J., Garcia, S., Molina, D., & Herrera, F. (2011). A practical tutorial on the use of nonparametric statistical tests as a methodology for comparing evolutionary and swarm intelligence algorithms. *Swarm Evol Comput, 1*(1), 3–18.
20. Aczel, A. D., & Sounderpandian, J. (2006). *Complete business statistics*. Boston, Mass: McGraw-Hill/Irwin.
21. Mani, N., Gursaran, & Mani, A. (2016). Design of cellular quantum-inspired evolutionary algorithms with random topologies. Quantum Inspired Comput Intell Res Appl, 111–146.

Android Malware Detection Using Code Graphs

Shikha Badhani and Sunil Kumar Muttoo

Abstract The amount of Android malware is increasing faster every year along with the growing popularity of Android platform. Hence, detection and analysis of Android malware have become a critical topic in the area of computer security. This paper proposes a novel method of detecting Android malware that uses the semantics of the code in the form of code graphs extracted from Android apps. These code graphs are then used for classifying Android apps as benign or malicious by using the Jaccard index of the code graphs as a similarity metric. We have also evaluated code graph of real-world Android apps by using the *k*-NN classifier with Jaccard distance as the distance metric for classification. The result of our experiment shows that code graph of Android apps can be used effectively to detect Android malware with the *k*-NN classifier, giving a high accuracy of 98%.

Keywords Android · Malware · Code graph · Classification

1 Introduction

The growing popularity of Android platform has contributed to the prevalence of continuously thriving and evolving malware on the platform. According to a recent report by Quick Heal, mobile ransomware has gone up by 33% in Q3, 2016 in comparison with the previous quarter [1].

In response to the increase of Android malware, the need for detecting Android malware efficiently has also increased remarkably. A lot of research has been carried out in the area of Android malware detection. One of the most prevalent approaches

S. Badhani (✉)
Department of Computer Science, Maitreyi College, University of Delhi, New Delhi, India
e-mail: sbadhani@maitreyi.du.ac.in

S. K. Muttoo
Department of Computer Science, University of Delhi, New Delhi, India
e-mail: skmuttoo@cs.du.ac.in

P. K. Kapur et al. (eds.), *System Performance and Management Analytics*, Asset Analytics, https://doi.org/10.1007/978-981-10-7323-6_17

for detecting such Android malware as used by virus scanners is signature-based detection in which a large database of malicious syntactic signatures (such as API calls, bytecode, permissions requested, etc.) is created and an app is classified as a malware if a match is found in the database. Approaches that are based on syntactic properties ignore the semantics of the code and thus can easily be circumvented by using code obfuscation [2]. Code obfuscation transforms a code by inserting new code or modifying existing code to make understanding and detection difficult and at the same time preserves the malicious behavior. Since the same malware is likely to exist in different physical forms but exhibits the same malicious behavior, it becomes important to consider the semantics of the code for efficient Android malware detection. This gave rise to semantic-based malware detection systems [3, 4] which are based on the formal model of the behavior of the code.

In this paper, we propose an Android malware detection algorithm that uses the semantics of the code. We extend the use of code graphs for Windows executables [5] for Android malware detection. Consider an original malicious Android app M that contains a sequence of API calls, S, and its obfuscated variant M′ that preserves the semantics of M and contains a set of API calls, S′. Our work is based on the assumption that S and S′ will have high similarity despite M and M' having different syntax. For extracting the sequence of API calls from an Android app, we disassemble the apk file and then create a control flow graph containing nodes which represent the APIs and directed edges from nodes which represent the call to all the possible APIs during sequential traversal of the bytecode. We only consider the package to which an API belongs for creating the nodes of the graph thus shrinking the control flow graph to code graph. This not only saves memory but also increases the time efficiency.

After extracting code graphs from malicious apps, the next step is to classify a test app as malicious or benign. Our first approach is to use the method as described in [5] by creating a database of code graphs extracted from various malicious apps and then comparing the test app with this database to determine similarity. High similarity implies that the app is malicious and low similarity implies that it is benign. This approach can be computationally inefficient if the malicious code graphs database is large. Hence, to overcome this drawback, we use machine learning for classifying graphs. For graph domain in which the features lack mathematical structure, *k*-nearest neighbor (*k*-NN) [6] is suitable, since it only requires a pattern dissimilarity measure for classification.

In this paper, we propose the following:

1. Extract code graphs from Android apps which specify the sequence of API calls. These code graphs form the semantic signature of the app which can be used for comparing graphs for Android malware detection.
2. Perform Android malware detection by using these code graphs. For graph classification, we have used the following algorithms:

   a. Created a database of malicious code graphs of various known Android malwares and then by using Jaccard index as a similarity metric, we were able to detect malwares which show high similarity with the known malwares.

b. Used k-NN classifier for automatically classifying code graphs as k-NN classifier only requires a pattern dissimilarity measure for classification. We have used Jaccard distance as the dissimilarity measure.

3. Present experimental results that demonstrate that code graphs can be effectively used for detecting Android malware.

2 Related Work

In this section, we discuss the work done in the area of Android malware detection and graph-based classification.

2.1 *Android Malware Detection*

Android malware detection relies on features that are either obtained statically or dynamically from Android apps. In static analysis [7, 8], features are extracted without executing the Android app such as API calls, permissions requested, bytecode, etc., while in dynamic analysis [9, 10], the runtime behavior of Android app is monitored to extract features based on system calls. After extracting features, next step is to classify the apps as benign or malicious, for which machine learning has been used. In machine learning, a machine acquires the generalization power by learning, i.e., a machine is fed with training data, coming from a certain source, whereon it tries to detect significant inherent patterns or structures in the source data in order to make predictions about new data coming from the same source [11].

Static analysis can be further divided into two approaches—Syntactic and Semantic. The syntactic approach is based on extracting syntax related features from an Android app such as permissions requested, intents, APIs, etc. For example, DroidMat [12] extracts features such as permissions, intents, and API calls from Android apps and then performs clustering on these apps followed by classification. Another Android malware detection system—Drebin [13] uses features such as network addresses, requested and used permissions, activities, services, content providers, receivers, intents, suspicious, and restricted API calls for performing classification. These syntax-based approaches suffer from a drawback that they do not consider the behavior of the code and hence are susceptible to evasion as the same malware is likely to exist in different physical form. In our approach, we have used the semantics of Android app for Android malware detection. The soundness and completeness of semantic-aware malware detectors are explored in [3].

2.2 *Graph-Based Classification*

Graph-based feature representations offer an efficient alternative to feature vectors if structure plays a vital role in describing the data. Many real-world data are represented not as vectors but as graphs such as biological sequences, chemical compounds, and natural language texts. Our work is based on extracting code graphs from Android apps and then performing graph-based classification on code graphs for Android malware detection. The most basic approach we have used to detect whether an app is malicious or not is to compare its semantic graph with a database of malicious graphs extracted from known malwares and then by using a similarity metric such as Jaccard index [14], a similarity score between the graph of the input app and each of the malicious graphs can be computed. A high similarity with any of the malicious graphs implies that the input app is malicious. Such an approach is used in [15], which is based on dynamic analysis. System call subsequence is extracted from malwares to compute malware similarity matrix by using Jaccard similarity. The next approach is to use machine learning. Most machine learning algorithms assume that the features of the training data are in vectorial form due to their rich mathematical structure and hence cannot be applied to graph-based feature representations. However, *k*-NN is one such classifier that has been applied to structural data such as graphs [16].

3 Methodology

To detect Android malware, our technique requires the extraction of the semantics of Android apps in the form of code graphs and then performing classification based on these code graphs. The process followed is shown in Fig. 1.

In the following sections, we discuss our methodology and provide relevant technical details of code graph generation and classification.

3.1 Code Graph Generation

Code graph is a directed graph representing semantic properties of an Android app. The concept behind code graph is that the sequence in which APIs are called in an Android app may not signify the goal of an Android app but it can be used to predict malicious apps by comparing their code graphs with the code graphs of known malware. We have implemented the generation of code graph for Android apps in Python by using a popular Android analysis open-source tool Androguard [17]. The code graph generation process is explained below.

At first, an Android app is decompiled to generate its Dalvik bytecode [18]. We consider only those instructions (Table 1) from the bytecode set which alters the

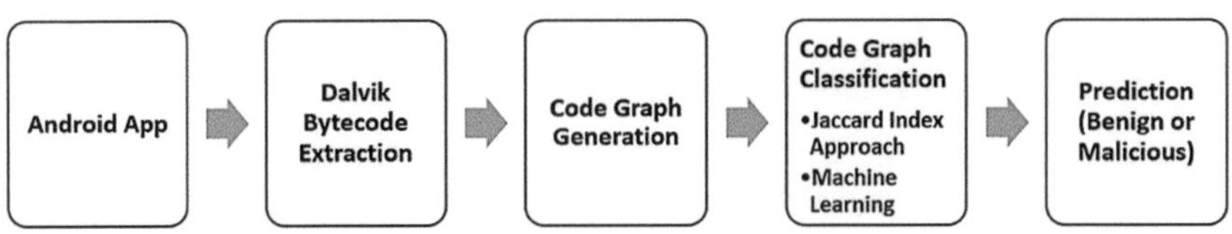

Fig. 1 Android malware detection

flow of the program. This Dalvik bytecode is then traversed to extract the used APIs. Instead of creating nodes for the APIs, we create nodes corresponding to the package to which an API belongs. This is done to reduce the complexity of the graph as there are a large number of APIs in Android.

In Algorithm 1, we give the code graph generation technique for an Android app.

```
Algorithm 1: Code Graph Generation of an Android App

Input: Android app (apk) A
Output: Code Graph G = (V,E,μ), a directed graph, where V is
the finite set of nodes representing packages in Android, E =
{ (ui,uj) | ui,uj ∈ V } is the set of directed edges where ui
denotes package to which an API call belongs and uj denotes
the package to which the next API call belongs which can be
called while traversing the bytecode.

V ← ∅; E ← ∅;
vi ← 0;
m ← FirstMethod (A);
while m is not the end of ListOfMethodsIn (A) do
  // The Address of First Instruction in Method m
  r ←  AddressOfFirstInstruction (m)
  while r is not the end of m do
    // The Instruction at address r in m
    It ← I[r];
    // The Parameter of the Instruction
    Pt ← P[r];
    if It is an invoke instruction then
       if It is an API call then
         // The Package Name of the API
         PN ← Package (Pt);
         if vi is equal to 0 then
            vi ← CreateNodeWithLabel (PN);
         else
            vj ← CreateNodeWithLabel (PN);
            E ←  E ∪ Edge (vi, vj);
            vi ←  vj;
         end if
       else if It is a Method call then
         vlist ← GetAllPossibleFirstAPICallPackages (r);
         for all API calls a ∈ vlist do
             vj ← CreateNodeWithLabel (a);
             E ← E ∪ Edge (vi, vj);
         end for
         vi ←CreateNodeWithLabel(
                 GetLastAPICallPackage(r));
    else if It is a goto instruction then
    // If the offset in Goto Instruction is 0 then it
    // implies it's a spin loop or a waiting loop hence
    // skip it
         if offset(It) is equal to 0 then
            continue;
         else
           vlist ← GetAllPossibleFirstAPICallPackages(r);
           for all API calls a ∈ vlist do
               vj ←  CreateNodeWithLabel(a);
               E ←  E ∪ Edge (vi, vj);
           end for
           vi ←  0
    else if It is a switch or an if instruction then
       vlist ←  GetAllPossibleFirstAPICallPackages(r);
       for all API calls a ∈ vlist do
            vj ←  CreateNodeWithLabel (a);
            E ←  E ∪ Edge (vi, vj);
       end for
    end if
    r ←  NextInstruction (m);
  end while
m ←  NextMethod (A);
end while
```

Table 1 Dalvik bytecode instructions

Instructions	Description
Goto	Unconditionally jump to the indicated instruction
Return	Returns from a method
Packed-switch/ sparse-switch	If a match is found with a value in the given register, then jump to a new instruction based on that value or go to the next instruction if no match is found
If	Branch to the given destination based on comparing values as specified
Invoke	Call the indicated method

Figure 2 shows the nodes and edges that are formed in a code graph using Algorithm 1 for an Android app having two methods *i* and *j*. The nodes are basically the packages to which an API belongs. Whenever an API_i is invoked, a node corresponding to the package P_i to which that API belongs is formed (if not already formed) in the code graph. Initially, API_1 is invoked, hence node P_1 is created. Next, when API_2 is invoked, node P_2 is created and a directed edge from P_1 to P_2 is inserted which signifies the flow that after API_1, API_2 is invoked. When Method j is called, there are two possibilities due to an if-else instruction in method *j*.

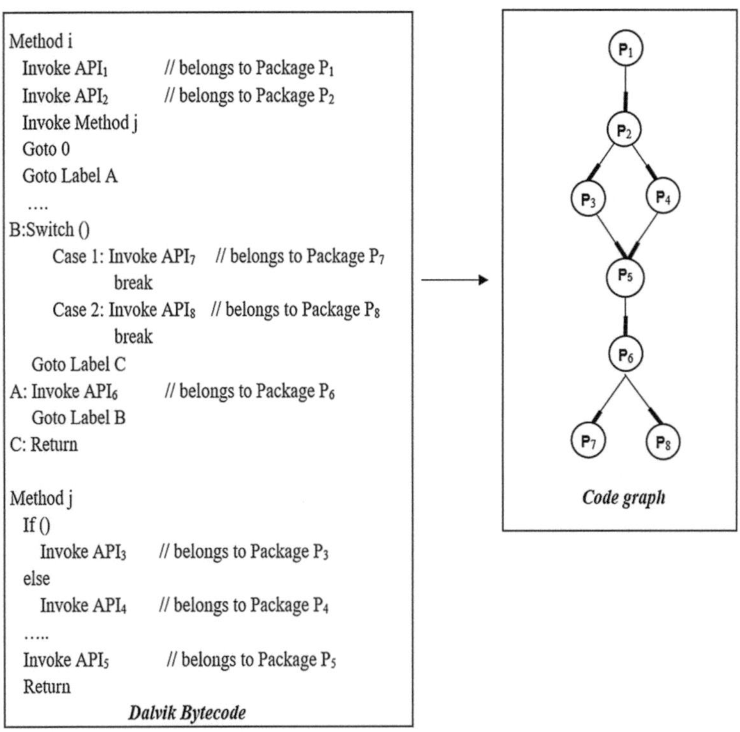

Fig. 2 Example of node and edge formation from Dalvik bytecode to code graph

Either API_3 will be invoked or API_4. Thus, in the code graph, two nodes P_3 and P_4 corresponding to API_3 and API_4 are created, and two directed edges ($P_2 \rightarrow P_3$ and $P_2 \rightarrow P_4$) are inserted. From API_3 and API_4, the next API that is invoked is API_5 (Node P_5 is created, and two directed edges $P_3 \rightarrow P_5$ and $P_4 \rightarrow P_5$ are inserted). After returning from method *j*, there is a Goto 0 instruction which implies a spin loop or a waiting loop, hence it is skipped. Next is Goto Label A where API_6 is invoked (Node P_6 is created and edge $P_5 \rightarrow P_6$ is inserted). Then, the control goes to Label B which contains a switch instruction with two cases corresponding to invoking API_7 and API_8. This results in two nodes creation (P_7 and P_8) and two directed edges ($P_6 \rightarrow P_7$ and $P_6 \rightarrow P_8$).

The code graph for an Android benign app Calculator.apk which simply performs basic arithmetic operations and for an Android malware FakeInstaller downloaded from AndroMalShare [19] are shown in Fig. 3. FakeInstaller usually appears as the installer for a legitimate application but secretly sends SMS messages to premium rate numbers, without the user's consent. Even in its code graph in Fig. 3, a node corresponding to Android.telephony [20] package is there which is used for malicious SMS operations.

In order to consider only the unique graphs, we have removed duplicate graphs from our code graphs database.

3.2 *Code Graph Classification*

For classifying graphs as malicious and benign, we have used two approaches which are discussed below:

Jaccard Index Approach. Our first approach is to create a database of malicious code graphs extracted from known malwares. Based on a similarity metric, we can compare the code graph of an input Android app with this database for classifying

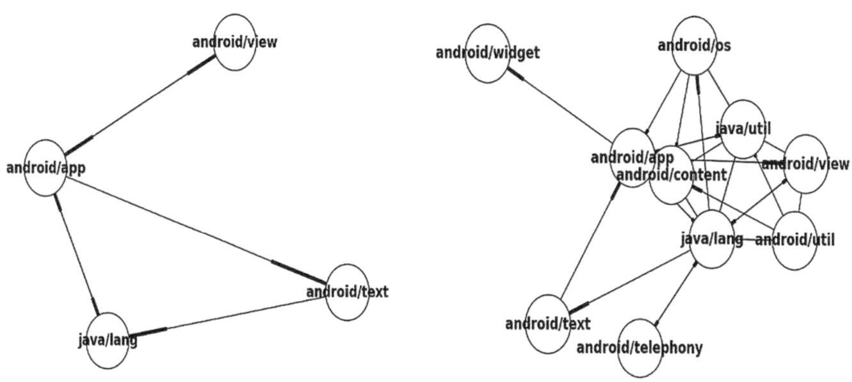

Code Graph of Calculator (Benign) Code Graph of FakeInstaller (Malware)

Fig. 3 Code graph examples

the app as malicious or not. Various similarity and dissimilarity metrics have been proposed in the literature for comparing two graphs. Graph Edit Distance (GED) [21, 22] is a dissimilarity metric that measures the minimum number of edit operations required to transform one graph to another. The disadvantage of GED is its exponential computational complexity in terms of the number of vertices. For malware detection, timely detection is one of the most important criteria, and hence, such high computational complexity approaches are avoided. We have used a fast method based on Jaccard index [14]. In the context of graphs, it is based on the rule that two graphs are similar if they have a lot of common vertices/edges. Hence, Jaccard index between two graphs A and B is defined as the intersection of edges or vertices divided by the union of edges or vertices. We have used the edges in our work since the code graphs are directed graphs and merely having common nodes will not be sufficient for comparison. We first compute the intersection and union graphs of A and B and then the Jaccard index J is computed as the number of edges in the intersection graph divided by the number of edges in the union graph:

$$J = |E(A \cap B)|/|E(A \cup B)|$$

The value of this similarity metric will lie between 0 and 1 because the union graph is always a super-graph of the intersection graph and hence the number of edges in the union graph will be greater than or equal to the number of edges in the intersection graph. When two code graphs are identical then the similarity will be 1 and when they have no common edges then the similarity will be 0. With experimental study, we consider an app as highly malicious if the similarity is greater than 0.8.

As an example, consider the code graphs of the benign app Calculator and malware FakeInstaller as shown in Fig. 3. There are eight directed edges in Calculator code graph (loops are not visible) and 34 in FakeInstaller code graph. Five edges are common in both the graphs. Thus, the Jaccard index of the two code graphs is calculated as = 5/(34 + 8 − 5) = 5/37 = 0.135. Such a low score implies they are not similar which indeed is true.

Machine Learning Approach. Our next approach is to use machine learning for graph-based representations to automatically classify the code graphs as benign or malicious. k-NN classifier [6] is one of the simplest machine learning models. This method classifies an unknown sample based on the class of the instances closest to it in the training space by measuring the distance between the training set and the unknown sample. It requires no knowledge about the data distribution which makes it apt and well suited for graph-based features. The classification process involves partitioning samples into a training set and a testing set. To train the classifier, a training set is used which contains the class of the samples also while testing aims to predict the class. The predicted test sample class is set equal to the true class among the k nearest training samples [23].

We have trained the k-NN classifier with the adjacency matrices of the code graphs as features. Adjacency matrix $A(G)$ of a directed graph G is a square matrix of size $N \times N$, where N is the number of nodes in graph G. If there is an edge from node i to node j, then we put 1 as the entry on row i, column j of the matrix $A(G)$ else 0. The k-NN classifier then classifies a test app based on its distance from k-nearest neighbors. The value of "k" is determined by using 10-fold cross-validation and the distance metric used for calculating the nearest neighbors is Jaccard distance which is a dissimilarity measure. Jaccard Distance is calculated as 1—Jaccard index.

4 Experimental Evaluation

4.1 *Dataset*

Our method is evaluated on a large data set of 1620 benign apps and 1620 malicious apps collected from Google Play [24] Store, and Andro-Tracker [25] and AndroMalShare [19], respectively. We performed the code graph generation and classification using this dataset.

4.2 *Feature Extraction and Code Graph*

We extracted the code graphs by using Algorithm 1 as explained in Sect. 3.1 by using a Python script. We have used the open-source static analysis Python tool Androguard [17] to extract the bytecode of an app. The extracted bytecode is then traversed to generate its code graph by using the code graph generation algorithm (Algorithm 1). NetworkX [26] Python package is used for generating the adjacency matrix of the code graph saved in sparse matrix form to reduce the space complexity for performing the graph classification. While generating the code graphs, the nodes correspond to the packages found in API level 23 [27]. Figure 4 summarizes the characteristics of code graphs generated. Around 65% of the benign code graphs and 84% of the malicious code graphs carry less than 20 nodes and 100 edges. Out of the 3240 code graphs generated, 2640 (81%) were generated in less than 50 s. These statistics are important as they are required to measure the capability of our technique in terms of runtime and space complexity that are dependent on the time required to generate code graphs, and the number of nodes and edges in them, respectively.

4.3 *Malware Detection*

We performed classification using these code graphs to evaluate their feasibility in detecting Android malware. Using the Jaccard index approach, we created a test set consisting of code graphs of 30 malicious and 30 benign Android apps and

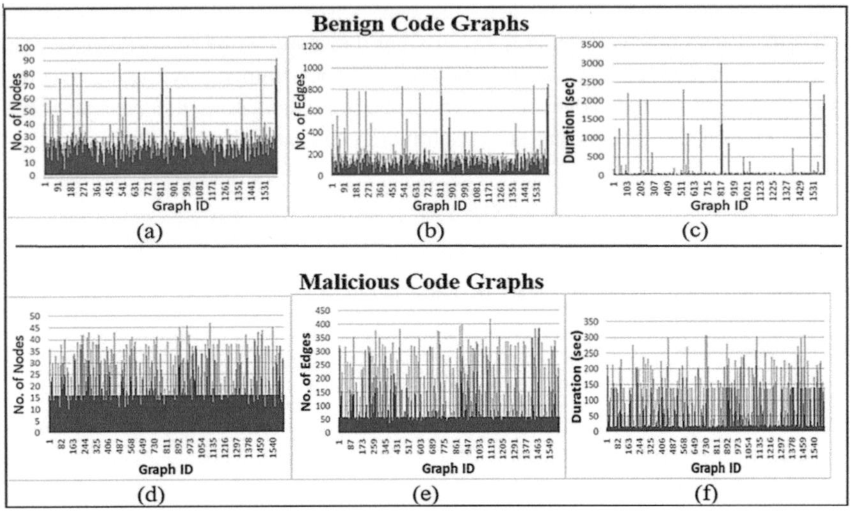

Fig. 4 Characteristics of code graphs: **a** & **d** number of nodes in benign and malicious code graphs, **b** & **e** number of edges in benign and malicious code graphs, and **c** & **f** duration in seconds of generating benign and malicious code graphs

compared them with the malicious data set of 1620 code graphs. We compute the Jaccard index of each test code graph with each of the malicious code graphs. The highest value amongst all Jaccard index values is the similarity score of that test app.

Out of 30 malicious test code graphs, we got a perfect similarity score of 1 for 29 apps and a high similarity score (>0.9) for 1 app. For all the 30 benign apps, a low similarity score (<0.5) was achieved which signifies dissimilarity with the malicious dataset. Thus, with such reliable results, code graphs can be used effectively for detecting malware.

A disadvantage of the above approach is that we need to compare the test app with the code graphs of all the known malicious code graphs for similarity calculation which can be computationally very inefficient if the malicious code graphs database is large. To automate this learning, we have performed *k*-NN classification on a corpus of 1620 benign apps and 1620 malicious apps. Out of the total 3240 apps, 70% of the apps were used for training the classifier and the remaining 30% were used to measure the efficiency in terms of accuracy. To measure the accuracy, a confusion matrix is constructed which uses the following metrics:

TP = Number of malicious apps correctly classified as malicious.
TN = Number of benign apps correctly classified.
FP = Number of benign apps incorrectly classified as malicious.
FN = Number of malicious apps not detected by the classifier.

Table 2 Confusion matrix

k-NN		Predicted	
		Benign	Malicious
Actual	Benign	455 (**TN**)	15 (**FP**)
	Malicious	4 (**FN**)	498 (**TP**)

$$\text{Accuracy is calculated as} = \frac{\text{TP} + \text{TN}}{\text{TP} + \text{FP} + \text{TN} + \text{FN}}$$

The confusion matrix is shown Table 2. *k*-NN classifier performed well and gave good results with 98% accuracy.

5 Conclusion

In this paper, we present a novel method of detecting Android malware by using code graphs. We extract code graphs from Android apps and then compare the code graph of 30 benign and 30 malicious Android apps with the database of malicious code graphs. A high similarity score based on Jaccard index implies it is malicious. In our experiment, we have achieved a 100% detection rate with a high similarity score (>0.9) for all the known malicious apps and a low similarity score (<0.5) for benign apps. To automate this learning of malicious behavior, we employ machine learning. We evaluated the code graphs of Android apps by using the *k*-NN classifier which gave a high accuracy of 98%. Thus, such high accuracy and reliable results show that code graphs can be effectively used for Android malware detection.

Our work is dependent on the Dalvik bytecode to generate code graphs. In general, native code cannot be handled by bytecode-level analysis because it is hidden from the bytecode perspective which is one of the limitations of such analysis. Moreover, the high complexity involved in similarity calculation among graphs hinders the use of such features for malware detection.

Apart from the *k*-NN classifier, we are exploring the application of different graph-based approaches on code graphs for classification of Android apps. Also, another future work is to investigate the effect of obfuscation on code graphs and how resilient they are to such transformations. We have progressed in these directions and the same would be presented in our next paper.

References

1. Quick Heal Threat Report Q1. (2016). http://dlupdate.quickheal.com/documents/others/quick_heal_quarterly_threat_report_q1_2016.pdf.
2. Rastogi, V., Chen, Y., & Jiang, X. (2013). Evaluating android anti-malware against transformation attacks. Northwest University, 329–334.
3. Preda, M. D., Christodorescu, M., Jha, S., & Debray, S. (2008). A semantics-based approach to malware detection. *ACM Transactions on Programming Languages and Systems, 30,* 1–54.
4. Zhang, M., Duan, Y., Yin, H., & Zhao, Z. (2014). Semantics-aware android malware classification using weighted contextual api dependency graphs. In *Proceedings of the 2014 ACM SIGSAC Conference on Computer and Communications Security* (pp. 1105–1116).
5. Lee, J., Jeong, K., & Lee, H. (2010). Detecting metamorphic malwares using code graphs. In *Proceedings of the 2010 ACM Symposium on Applied Computing*. SAC '10. 1970.
6. Aha, D. W., Kibler, D., & Albert, M. K. (1991). Instance-based learning algorithms. *Machine Learning, 6,* 37–66.
7. Enck, W., Ongtang, M., & McDaniel, P. (2009). On lightweight mobile phone application certification. In *Proceedings of the 16th ACM Conference on Computer and Communications Security*, CCS '09 (pp. 235–245).
8. Sanz, B., Santos, I., Laorden, C., Ugarte-Pedrero, X., Bringas, P.G., & Álvarez, G. (2013). PUMA: Permission usage to detect malware in android. *Advances in Intelligent Systems and Computing (AISC), 189*, 289–298.
9. Burguera, I., Zurutuza, U., & Nadjm-Tehrani, S. (2011). Crowdroid: Behavior-based malware detection system for android. In *Proceedings of the 1st ACM Workshop on Security and Privacy in Smartphones and Mobile Devices*. SPSM '11, Vol. 15.
10. Enck, W., Gilbert, P., Chun, B.-G., Cox, L. P., Jung, J., McDaniel, P., & Sheth, A. N. (2010). *TaintDroid: An information-flow tracking system for realtime privacy monitoring on smartphones* (Vol. 49, pp. 1–6). Osdi '10.
11. Shawe-Taylor, J., & Cristianini, N. (2004). *Kernel Methods for Pattern Analysis* (pp. 140–193).
12. Wu, D. J., Mao, C. H., Wei, T. E., Lee, H. M., & Wu, K. P. (2012). DroidMat: Android malware detection through manifest and API calls tracing. In *2012 Seventh Asia Joint Conference on Information Security (Asia JCIS). IEEE* (pp. 62–69). Asia JCIS 2012.
13. Arp, D., Spreitzenbarth, M., Malte, H., Gascon, H., & Rieck, K. (2014). Drebin: Effective and explainable detection of android malware in your pocket. In *Symposium on Network and Distributed System Security* (pp. 23–26).
14. Jaccard, P. (1901). Distribution de la flore alpine dans le Bassin des Drouces et dans quelques regions voisines. *Bulletin de la Société Vaudoise des Sciences Naturelles, 37,* 241–272.
15. Blokhin, K., Saxe, J., & Mentis, D. (2012). Malware similarity identification using call graph based system call subsequence features. In *2013 IEEE 33rd International Conference on Distributed Computing Systems Workshops (ICDCSW)* (pp. 6–10).
16. Schenker, A., Last, M., Bunke, H., & Kandel, A. (2003). Classification of web documents using a graph model. In *Proceedings Seventh International Conference on Document Analysis and Recognition.*
17. Androguard. https://github.com/androguard/androguard.
18. Dalvik bytecode. https://source.android.com/devices/tech/dalvik/dalvik-bytecode.html.
19. AndroMalShare. http://sanddroid.xjtu.edu.cn:8080/.
20. Android.Telephony package. https://developer.android.com/reference/android/telephony/package-summary.html.
21. Bunke, H., & Allermann, G. (1983). Inexact graph matching for structural pattern recognition. *Pattern Recognition Letters, 1,* 245–253.
22. Sanfeliu, A., Sanfeliu, A., & Fu, K. S. (1983). A distance measure between attributed relational graphs for pattern recognition. *IEEE Transactions on Systems, Man, and Cybernetics SMC, 13,* 353–362.

23. Liao, Y., & Vemuri, V. R. (2002). Use of k-nearest neighbor classifier for intrusion detection. *Computers & Security, 21,* 439–448.
24. Google Play. https://play.google.com.
25. Kang, H., Jang, J. W., Mohaisen, A., & Kim, H. K. (2015). Detecting and classifying android malware using static analysis along with creator information. International Journal of Distributed Sensor Networks, 11(6), 479174.
26. NetworkX. https://networkx.github.io/.
27. Android Package Index. https://developer.android.com/reference/packages.html.

Security in ZigBee Using Steganography for IoT Communications

Iqra Hussain, Mukesh Chandra Negi and Nitin Pandey

Abstract ZigBee is an IEEE 802.15.4-sourced arrangement for a suite of high-level communication protocols employed to generate personal area networks. ZigBee is a less price, low-complex and low power consumption wireless personal area network (WPAN) norm that targets at the extensive developments of instruments and devices with prolonged battery life that are employed in wireless controls or applications that are used for monitoring purposes. It has an extensive utilization in industries and operations that are conducted physically. Hence, ZigBee is mostly correlated with IoT and M2M. Therefore, security in these WPANs becomes a major interest and has gained a good amount of notice currently. The security methods used in these networks over a period of time usually include practices that are cryptographic in nature. Then again these practices recommended till date can have good scopes of improvements, as a result, to turn up with additional assured and protected data communication. The chapter proposes a technique to enhance security in ZigBee using steganography over the secret data being communicated between communicating parties. However in cryptographic practices, the message even if it stands strongly resilient, can stimulate doubts and, therefore, could be adequate enough for a third party, spying the systems that something that are of significant use have been exposed. Hence, to keep the security features consistent in these networks this chapter proposes a technique to protect data by means of Steganography over the data to be communicated, this allows two communicating parties to transmit covert communication through a shared route in a way with the purpose, no adversary can even discover as in the covert communication is being transferred. Hence, making use of cryptographic practices helps protecting the

I. Hussain (✉) · N. Pandey
Amity Institute of Information Technology, Amity University Uttar Pradesh, Noida, India
e-mail: iqrahussain4@gmail.com

N. Pandey
e-mail: npandeyg@gmail.com

M. C. Negi
Tech Mahindra Ltd., A7, Sector 64, Noida, India
e-mail: MN00330419@techmahindra.com

P. K. Kapur et al. (eds.), *System Performance and Management Analytics*, Asset Analytics, https://doi.org/10.1007/978-981-10-7323-6_18

insides of the messages only, on the other hand using steganographic practices can help in protecting the message contents and even the fact that a covert message has been transferred. The exclusive plan is to come up with a practice that is Steganographic in nature and ultimately has a resistance to any sort of steganalysis.

Keywords ZigBee · IEEE 802.15.4 · Steganography · Reserved bits Stego object · IoT

1 Introduction

1.1 ZigBee

ZigBee is an IEEE 802.15.4 sourced arrangement for a suite of high-level communication protocols employed to generate personal area networks. ZigBee is a less price, low-complex and low power consumption wireless personal area network (WPAN) norm that targets at the extensive developments of instruments and devices with prolonged battery life that are employed in wireless controls or applications that are used for monitoring purposes. ZigBee applications are used in home automation, industrial automation, health/medical care, smart grids, etc. We can say that ZigBee finds an extensive use in the world of Internet of Things and M2M. In advance, ZigBee has more than a few benefits for instance self-organizing, lower energy utilization, depleted price, lesser dimension of protocol stacks, and bigger addressing modes. ZigBee is the lone universal; standards-based wireless explanation that is capable of appropriately and reasonably be in charge of the extensive variety of appliances to enhance contentment, protection, and expediency for clients. This actually makes it an equipment of preference for world-chief service suppliers, installers and sellers who get a hold of the benefits of the Internet of things interested in the Smart Homes [1].

ZigBee constitutes two categories of hardware devices, the Full Functional Device (FFD) and the Reduced Function Devices (RFD). FFD devices are proficient for corresponding mutually with both FFD and RFD devices. Alternatively, RFD devices are capable of corresponding merely with the FFD devices. Further the ZigBee network devices consist of three classes of devices, the coordinator, the router, and the end devices. The ZigBee network protocol stack has been established on the basis of Open Systems Interconnection (OSI) [1].

1.2 ZigBee Architecture

As shown in Fig. 1, the ZigBee Architecture is alienated into two major divisions, the IEEE 802.15.4 part which incorporates the MAC layer and the Physical layer.

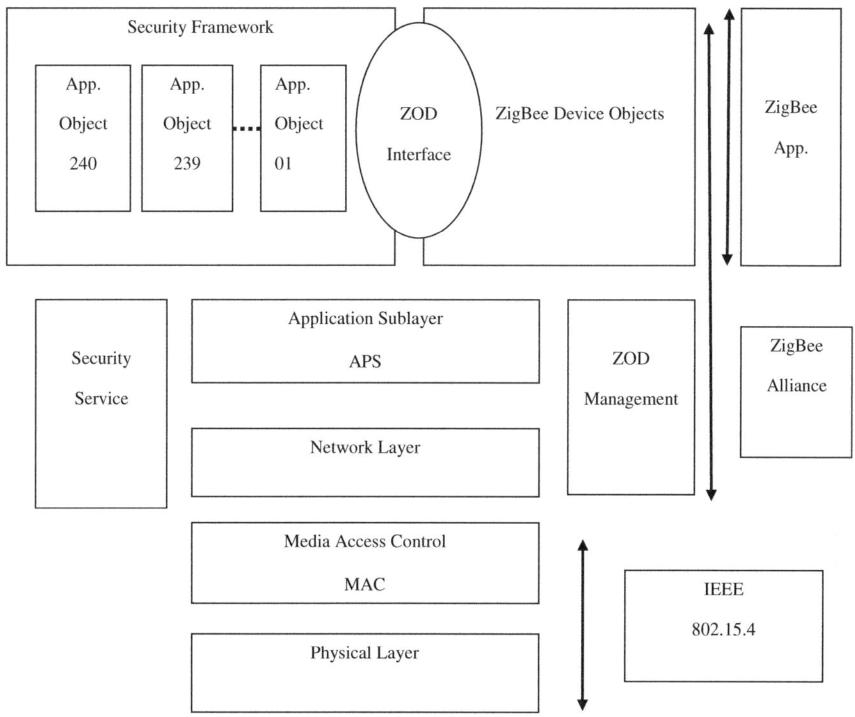

Fig. 1 ZigBee Architecture

The further division is the ZigBee layers, which incorporate the network layer, the ZigBee Device Objects (ZDO), the Application Support Sublayer (APS), and the Security management. Every layer executes separate operations. The modulation and the demodulation on the incoming signals and outgoing signals accordingly are performed by the physical layer. It broadcasts and collects information from a resource. The MAC layer accesses the network using Carrier-Sense Multiple Access with Collision Avoidance (CSMA/CA), to pass on frames for organization and impart a consistent transmission. The network layer positioned between the MAC layer and the APS offers tasks such as opening a network, organization of end devices, joining or exit a network, route detection, neighbor detection, etc. The Application Support Sublayer (APS) offers functions required for Application Objects (endpoints) and the ZDO to line with the network layer for data and management functions. The Application Objects\End Points describe input and output to the Application Support Sublayer. The ZigBee Device Objects carry out monitoring and management of Application Objects and execute by and large the device management jobs, i.e., identify the class of devices in a network, whether it is a coordinator, a router or an end device.

1.3 Steganography

Steganography refers to the practice of transferring messages concealed in irreproachable appearing communications on top of a public channel so that an adversary eavesdropping on the channel cannot even detect the existence of the concealed messages [2].

The intent of steganography comprises of embedding data such as text, movie, picture, etc., described as the encrypted message, in an alternative media [3]. The alternative media or support which incorporates the hidden data is specified as the cover object. And once the encrypted message is rooted in the cover message, the end result is called as the stego object. For example, we can encrypt a picture in an additional picture, and in this second picture, we cannot notice that the first picture is encrypted within the first picture. When we talk about steganography, we pass on to the similarity of Alice and Bob [3]. Alice and Bob are in jail and are watched by a warden, Wendy. If Alice desires to convey a message to Bob, this message has to go through Wendy. If Wendy understands the message which includes the curtail information, for instance the time of the escape. Wendy will not give that message to Bob in any case. For that reason, Alice is assumed to find a way out to encrypt facts in the message devoid of Wendy analyzing it. For instance, Alice will conceal a message in an alternate message and if every other letter is understood, the concealed message can be read; and if somehow Wendy reads this message, she will not be able to notice the concealed message. So in steganography, the above case states that the Steganographic technique should be kept hidden and if Wendy understands the Steganographic method, she will know how to interpret the message and all the participants who are involved in the communication should be able to understand this technique to hide the data and as well as later read the data [3].

On the further part, steganography permits participants to converse steganographically with no previous swap of secrets. Like by means of public-key encryption, the correspondent of a information even requires to be familiar with the receiver's public key or else take part in a key exchange protocol. At the same time as it is true that if there is no worldwide PKI, the use of public keys may provoke doubt, as seen in several circumstances it is the dispatcher of a message who is concerned in covering up his/her message and there is no requirement for him/her to broadcast any keys. A number of Steganographic methods set sights on to make use of specificities of communication rules to conceal message or information and employ the communication layer fields as the cover objects. Utilizing the Steganographic information in communication layer fields offers the conception of an encrypted channel in the network. And merely the devices that are aware of in which fields the data or information is encrypted can interpret or write down data. It can even be possible to unnoticeably swap records or information in the system if the system isn't aware about the Steganographic method used. References [4–6] show various potentials for hiding concealing the records or information, by means of order of the protocols to generate concealed channels or the Steganographic channels. The mainly employed techniques constitute of using the reserved field bits of the protocols.

The chapter proposes implementing Steganographic technique on the reserved field bits used to hide secret messages, in the layers of IEEE 802.14.5 to boost the overall ZigBee protocol security that is based on the similar IEEE 802.14.5 specification. The rest of this article is ordered as following: Segment II is going to put forward the associated design with taking in account the suggestion to conceal data into the MAC layers of 802.15.4 [7], Segment III is going to put forward the block representation scheme of correlating steganography with the data present inside the reserved bits of the MAC layer to enforce security, Segment IV will put forward the advantage of using this technique, Segment V will put forward the execution and Segment VI will put forward the termination of this document.

2 Related Work

This segment illustrates the suggestion of concealing records in the MAC layers of IEEE 802.14.5 protocol. This method comprises employing the reserved field bits to conceal data or records in them. IEEE 802.15.4 uses different forms of frames, depending on the type of data packet sent. The MAC layer of 802.14.5 makes use of 4 different categories of the data frames [3]:

1. Data Frame
2. Beacon Frame
3. Acknowledgement Frame
4. MAC Command Frame

1. **Data Frame**

The structure of MAC data frame is given in Fig. 2.

In the above given field, Frame Control field and the Address Information field impart potential for concealing the records or data explained below.

1.1 **Frame Control Field**

Structure of the Frame Control Field is given in Fig. 3

In the above given structure in Fig. 3, we can see that the 7–9 and 12–13 bits are reserved and can be used to encode a 3 bit and a 2 bit stego object respectively.

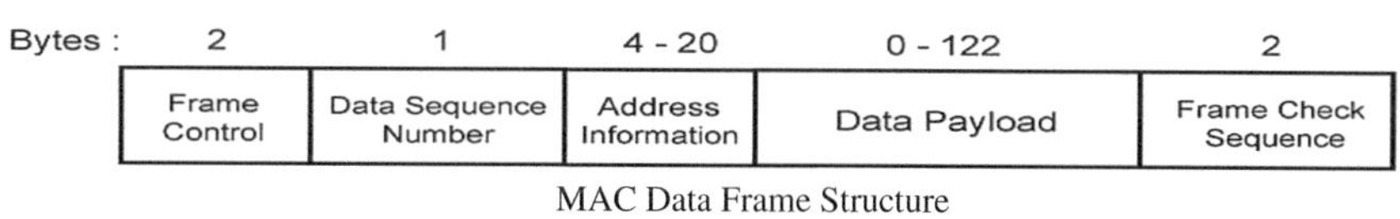

Fig. 2 MAC data frame structure

Bits:	0-2	3	4	5	6	7-9	10-11	12-13	14-15
	Frame Type	Security enabled	Frame Pending	Ack. request	Intra PAN	Reserved	Dest. addressing mode	Reserved	Source addressing mode

Frame Control Field Structure

Fig. 3 Frame control field structure

1.2 Address Information Field

The structure of the address information field is given in Fig. 4.

This Source Address in the above representation can be used for having small 16 bits or either an extensive 64 bits source address. This field can be used to hide data in a way, e.g., specifying some not existing source address and using this not existing source address, stego object with range till 64 bits could be hidden in it.

2. **Beacon Frame**

Structure of beacon frame is given in Fig. 5.

Possibility for hiding information is similar here as in Data Frame, i.e., frame control field and source address information field could be employed to conceal the steganographic object. The only distinction is that, source address information field here limits to 10 bytes only.

3. **Acknowledgement Frame**

Structure of the acknowledgement frame is given in Fig. 6.

Bytes:	0/2	0/2/8	0/2	0/2/8
	Destination PAN identifier	Destination Address	Source PAN identifier	Source Address

Address Information Field Arrangement

Fig. 4 Address information field arrangement

Bytes:	2	1	4/10	2 - 112	2
	Frame Control	Beacon Sequence Number	Source Address Information	MSDU	Frame Check Sequence

MAC Beacon Frame Arrangement

Fig. 5 MAC beacon frame arrangement

Bytes:	2	1	2
	Frame Control	Data Sequence Number	Frame Check Sequence

MAC Acknowledgement Frame Structure

Fig. 6 MAC acknowledgement frame structure

Bytes : 2	1	4 - 20	1	0 - 117	2
Frame Control	Data Sequence Number	Address Information	Command Type	Command Payload	Frame Check Sequence

MAC Command Frame Structure

Fig. 7 MAC command frame structure

The potential for hiding data or information here is in Frame Control field, similar to field of MAC data frame.

4. **MAC Command Frame**

Structure of the MAC command frame is given in Fig. 7.

Frame Control field and Address Information field offer similar capacity of concealing records in MAC command frame similar to the data frame.

3 Our Proposition

To come up with enhanced security in ZigBee protocol for IoT communication, our proposition will be associating the asymmetric steganographic technique with the data present in the reserved field bits for security constraints. As shown in Fig. 8, if two participants communicating with each other require sharing some sensitive information or a secret message in a network based on the ZigBee specification, the secret message can be hidden in the reserved field bits as discussed above in the related work [7]. But the fact remains that intruders can always try to make attempts to gain access to the sensitive information being transferred once they get an idea regarding the fact a hidden message is being transferred along, and therefore the communication needs to be kept private so that no other third party is allowed to have access to this sensitive information. For that reason a public key and a pair of private keys can be generated and associated with steganography to keep the communication more secure. The sender will hide the data in the reserved field bits and generate a public key to be associated with the stego object and then at the receiver part the stego object will be decrypted using the private key. Figure 8 shows a block representation of the Steganographic technique to be employed. Here, we make an assumption that a sensitive information or data is required to be shared between two participants and no other third party should have access to this data or information. This secret message or sensitive information can be put in the reserved field bits of the MAC layer and associating a public key and a pair of private keys that are known only to the sender and receiver. Once this secret message is encrypted using the public key, the formed stego object passes through the channel to the other participant. And so the other participant receives the stego object and decrypts this stego object using the private key known only to the participants.

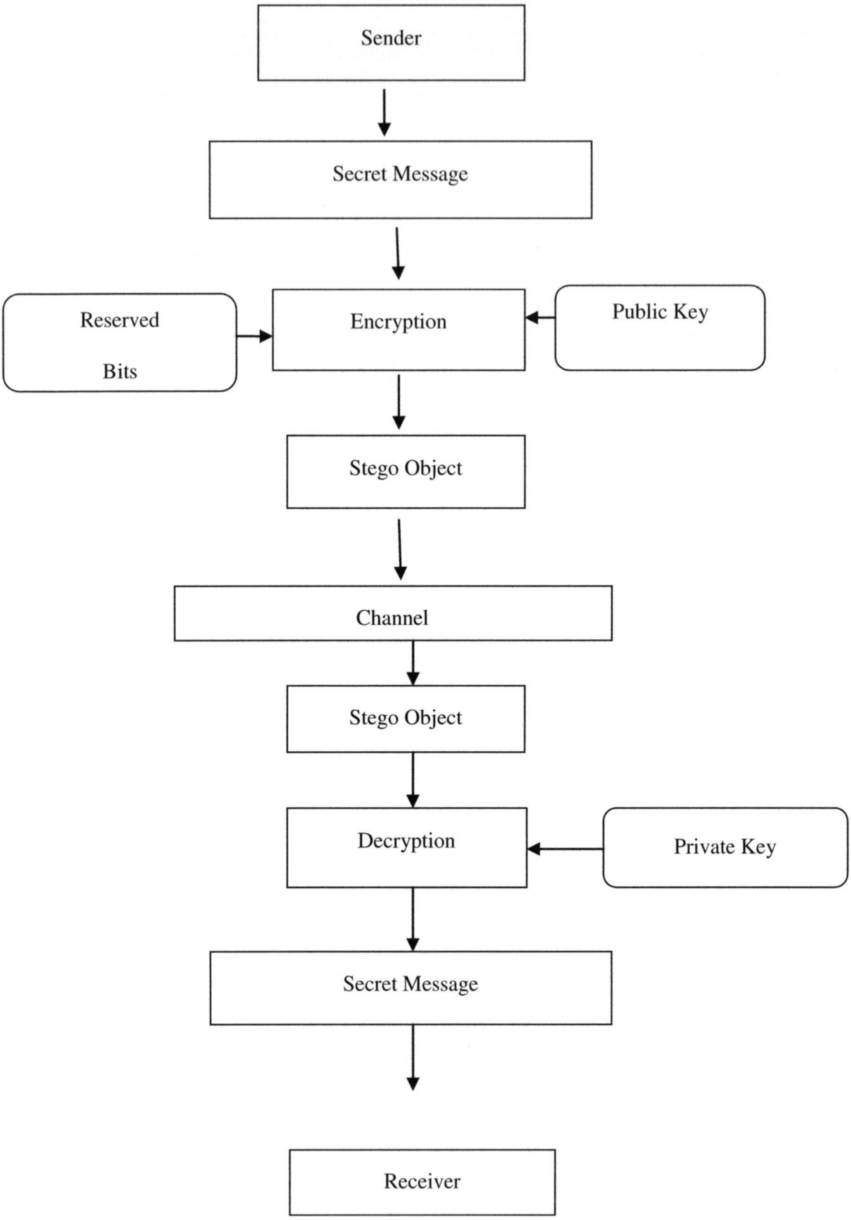

Fig. 8 ZigBee Architecture

And hence finally the sensitive information or data can be delivered to the other participant in the communication. The technique provides a way out to keep the sensitive information being transferred hidden from the third parties who illegally try to gain access to this sensitive information or data.

3.1 Proposition in Steps

Step 1: The sender initiates the covert text that is needed to be delivered to the recipient.
Step 2: Covert text is put in the reserved field bits accordingly in the frames.
Step 3: The data gets encrypted employing public together with associated private keys.
Step 4: The resultant obtained is a stego object now.
Step 5: The stego object passes through the channel to the other communicating participant.
Step 6: The Stego object is decrypted at the receiver end using the associated private key.
Step 7: The receiver receives the secret message or data accordingly.

4 Advantage

The Steganographic technique discussed above has an advantage over cryptography and hence provides a better way out in maintaining the security of communications, i.e. to be further secure. As in cryptographic techniques enciphered messages no matter how resistant they are, can provoke doubt or suspicion and can be adequate for an attacker that spies on the system to facilitate somewhat significant anything is sensed. For that reason of preserving safety in such set of connections we put forward an alternate approach to protect data, using asymmetric Steganography is a practice which permits a number of participants to dispatch covert texts through an unrestricted channel in a way so even an opponent will not be able to recognize as in covert texts are being forwarded. So like cryptography protects inside of a document alone, Steganographic technique could be generalized protecting equally, contents of message together even hiding this reality of covert texts being transferred. And once the fact is hidden that a secret message is being transferred in a communication network, there are very less chances of intrusions or any security attacks taking place along the communication.

5 Implementation

The technique finds an implementation in networks that are based on wireless connections sourced on the protocol ZigBee. The method provides better, secure and a reliable way to secure sensitive information or data and hence can be implemented in the following:

(1) It can be implemented in home safety monitoring and remote control systems that are based on ZigBee-GSM [8].
(2) It has an execution in lighting control systems that are operated remotely and sourced further on ZigBee expertise and SoC solutions [9].
(3) It can even find implementations in areas of designs and implementations in vehicle tracking sourced on ZigBee-RFID [10].
(4) Area of designs and intelligent home energy organizations system sourced on ZigBee expertise can employ the proposed technique to enhance security and reliability [11].
(5) The smart meters used presently that are developed with characteristics as in support to meters, response to demands and even supporting load controls, support in prices, message text supports, support in safety plus security, etc., can be developed on ZigBee-based applications and further this technique could be executed in these intelligent meters also.
(6) Any ZigBee application employing the technique could even find an implementation in several devices such as the sensors for smoking and heating controls, equipment used in medicine and science, controls for homes and industry units plus quite a few additional devices sourced on wireless communications.

6 Conclusion

This chapter puts forward a method for keeping the sensitive information being transferred over a network secure and reliable. The sole aim of this chapter is to put forward a method which enables passing a covert text in an outline of a steganographic object and even hiding reality of a covert message being transferred in communication networks. And, this can be achieved using the asymmetric steganographic technique.

References

1. Alliance, Z. http://www.zigbee.org. Accessed 12 July 2010.
2. Mahmood, N. R., Kufa University Education College, Azeez, A. A., Mohammad Kufa University Education College, & Rasool, Z. N., Kufa University Sciences College. *Public key steganography*.
3. Martins, D., Guyennet, H., & Computer Science Department LIFC, University of Franche-Comte Beasancon. *Steganography in MAC layers of 802.15.4 protocol for securing wireless sensor networks*.
4. Handel, T. G., & Sanford II, M. T. Hiding data in the OSI network model. In *Proceedings of the First International Workshop on Information Hiding, (London, UK)* (23–38). Springer.

5. Trabelsi, Z., El Sayed, H., Frikha, L., & Rabie, T. (2007). A novel covert channel based on the IP header record route option. *International Journal of Advanced Media and Communication, 1*(4), 328–350.
6. Murdoch, S. J., & Lewis, S. (2005). Embedding covert channels into TCP/IP. In *Information hiding: 7th international workshop, volume 3727 of LNCS* (pp. 247–261).
7. Backes, M., Cachin, C., & IBM Research, Zurich Research Laboratory, CH-8803 Rüschlikon, Switzerland. *Public-key steganography with active attacks.*
8. Ahmad, A. W., Hanyang University, Ansan, South Korea, Jan, N., Iqbal, S., & Lee, C. *Implementation of ZigBee-GSM based home security monitoring and remote control system.*
9. Sun, M., School of Electronics & Information Engineering, Tongji University, Shanghai, Liu, Q., & Jiang, M. *An implementation of remote lighting control system based on ZigBee technology and SoC solution.*
10. Anuradha, P., Network Engineering, Vel Tech Multi Tech Dr. Rangarajan Dr. Sakunthala Engineering College, Chennai, India, & Sendhilkumar, R. *Design and implementation of ZigBee-RFID based vehicle tracking.*
11. Han, D.-M., School of Computer Science & Engineering, Kongju National University, Kongju, South Korea, & Lim, J.-H. *Design and implementation of smart home energy management systems based on ZigBee.*

Implementation of Six Sigma Methodology in Syrian Pharmaceutical Companies

Yury Klochkov, Bacel Mikhael Alasas, Adarsh Anand and Ljubisa Papic

Abstract The increased competition in the global pharmaceutical market and the necessity to reach higher levels of quality of the pharmaceutical products force manufacturers to seek and adopt more effective and reliable quality management methods and techniques which allow them to introduce products with the highest possible quality level and reduced quality costs, while maintaining conformance to the pharmaceutical GMPs, technical and legislative requirements. One of the popular modern quality management methodologies is Six Sigma, which proved its high ability to increase business profits and competitiveness within more than 30 years of implementation in manufacturing and service sectors. Recently, Six Sigma methodology has been adopted by global pharmaceutical companies such as Baxter, Eli Lilly, Johnson & Johnson and Novartis and obtained considerable benefits from its abilities. This research aims at investigating the possibility to implement Six Sigma methodology in the Syrian pharmaceutical companies, and to find out what benefits a pharmaceutical company can get through the implementation of this methodology. We conducted a case study in a pharmaceutical company in Syria (Orient Pharma) in order to examine the effectiveness and advantages of Six Sigma methodology. For this purpose, a quality improvement project was conducted using DMAIC roadmap to enhance the quality for one of the main products of the company. The obtained results of DMAIC project showed an

---

Y. Klochkov (✉) · B. M. Alasas
Peter the Great St. Petersburg Polytechnic University, St. Petersburg, Russia
e-mail: y.kloch@gmail.com

B. M. Alasas
e-mail: bacel.mikhaelalasas@mail.ru

A. Anand
University of Delhi, New Delhi, Delhi, India
e-mail: adarsh.anand86@gmail.com

L. Papic
University of Kragujevac, Cacak, Serbia
e-mail: dqmcenter@mts.rs

P. K. Kapur et al. (eds.), *System Performance and Management Analytics*,
Asset Analytics, https://doi.org/10.1007/978-981-10-7323-6_19

enhanced process capability, an enhanced process Sigma level, decreased variability in the process outputs. The main difficulties that have been observed during the study are resistance to change, lack of training, lack of necessary resources, attitude toward quality in the company. As a conclusion, considerable benefits can be obtained through implementing Six Sigma methodology in the Syrian pharmaceutical companies.

Keywords Quality improvement · Six Sigma methodology · Pharmaceutical industry · DMAIC methodology

1 Introduction

Six Sigma methodology was developed at Motorola company in 1987 as a way to achieve business excellence. Many researchers attempted to provide a comprehensive description of Six Sigma concept. Six Sigma is a rigorous and high-efficiency application of the proved quality principles and techniques. It combines elements from the scientific works of quality management pioneers and different methodologies, and aims to reach a level of performance which does not contain any defects [1–4].

The term "Sigma" is symbolized by the Greek letter σ and used by statisticians to measure the variability in process outputs, so the company's performance is measured by the Sigma level of its processes. Historically, companies accepted performance level at three or four sigma standard in spite of the fact that the processes at that performance level produced 6200–67,000 defects per million opportunities (DPMO). However, the standard Six Sigma (equal to 3.4 DPMO) represents a response to the growing requirements of the customers and the growing complexity of manufacturing processes and new products [2, 5]. Six Sigma is defined as a strategy to improve the business performance of an organization as a whole. It is characterized by a high degree of organization and discipline, and a strong focus on the customers and the efforts toward improving organization's profitability. Six Sigma uses effective statistical methods and is based on quality principles used to improve processes and products through a framework known as DMAIC that consists of five consecutive phases (Define, Measure, Analyze, Improve, Control) [3]. Six Sigma also is a systematic data-driven methodology aimed at solving chronic problems, which business sectors encounter. It provides an excellent framework to manage improvement projects; and applies a lot of statistical and nonstatistical tools in a manner provides the best solutions for the investigated problems [4]. Six Sigma is a much-disciplined methodology that relies on statistics to eliminate the defects in products and processes, and depends on the full involvement of the company's personnel [6].

2 Six Sigma Project

This research aims at implementing Six Sigma methodology in a Syrian Pharmaceutical Company (Orient Pharma) to improve the perceived quality level of one of the main products manufactured by this company. The project goals include enhancement of process capability and decrease of its variability in order to achieve a higher Sigma level, while keeping the full fulfillment of GMPs and ISO 9001 requirements and other regulations [7–10]. To achieve this purpose, a DMAIC quality improvement project was designed in collaboration between the researchers and the interested departments in the company.

(1) **Define Phase**

Project Selection:
To determine the most produced product, the annual production records have been reviewed for all kinds in 2011 as shown in Fig. 1.

It is noticeable that the product "ORIENTOCIN—Tablets" is the main product of the company (45 batches per year). ORIENTOCIN—Tablets are a medicine used to treat gingivitis; its main API (active pharmaceutical ingredient) is 'Spiramycine'.

Project Charter:

The investigated stages of the process are illustrated in the following SIPOC diagram (Fig. 2; Table 1).

(2) **Measurement Phase**

To establish a general understanding of the investigated process (Preparing the formula), the project team developed a flowchart as shown in Fig. 3.

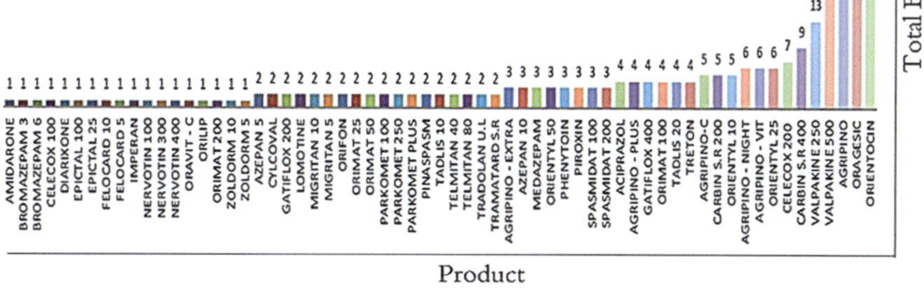

Fig. 1 Annual production for all kinds of products in 2011. *Source* Researcher

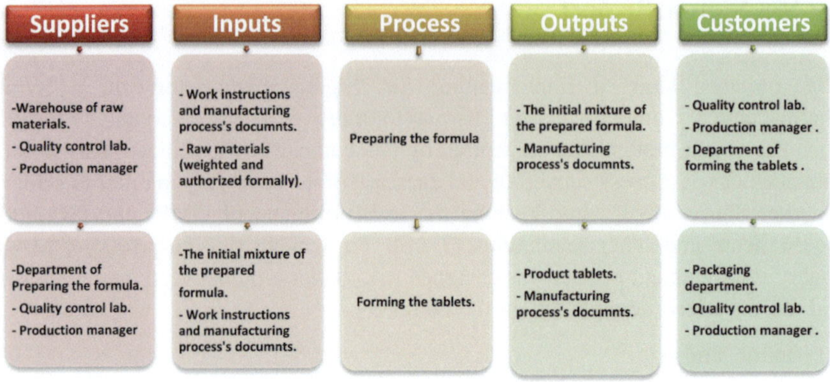

Fig. 2 High process level map (SIPOC) for the studied product. *Source* Researcher

Table 1 Project charter

Problem statement	The analytical data (supplied by the quality control lab in the considered company) demonstrated a significant variability in the concentration of API in the finished product ORIENTOCIN—tablets (the standard deviation value near 3500). This problem decreases the homogeneity of the produced units and affects the quality and effectiveness of the medicine negatively
Improvement goal	To decrease the variability in the concentration of API in the finished product ORIENTOCIN—Tablèts, to enhance process capability and to sigma level to the higher possible value
Project scope and limitation	The project covers the first two stages of the manufacturing process of the product stage 1 'formula preparation' and stage 2 'tablets formation'
Key performance indicators	Process capability index Ppk, process mean, standard deviation Std. D, sigma level
Project team	A cross-functional team includes members from production department, quality control laboratories, quality assurance
Critical-to-quality characteristic	API concentration in the finished product, each tablet should contain (774900 IU) $\pm$ 10%

Sampling and Measurement Plan:

- Investigated characteristic: API concentration in product tablets.
- Measurement procedure: The formal analytical method used in quality control lab in the company.
- Measurement techniques: Molecular absorption spectroscopy.
- Sampling: The samples were collected from the outputs of forming phase (tableting) as follows:
- Sample size: $n = 4$ tablets.

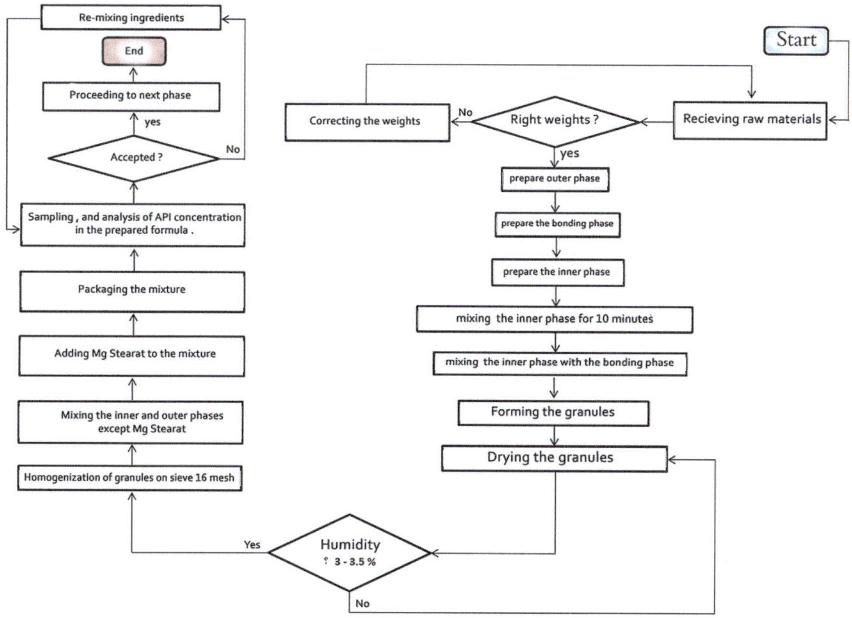

Fig. 3 Flowchart of the process at the stage 'formula preparation'

- Frequency: 1 sample each 5 min.
- Collected samples: 24.
- Sampling responsibility: Process operator.

Data Set:
According to the Sampling and Measurement plan, the data set had been collected (see Table 2).

Checking Process Stability:
By using software (Minitabv15), the project team created an Xbar-R control chart as shown in Fig. 4.

All points fall between UCL and LCL. No patterns were observed, so the process was under statistical control.

(3) **Analysis Phase**

To examine the normality of the data set, probability test was conducted as shown in (Fig. 5).

The test showed that the data follows the normal distribution law [11] and it is reliable to conduct a Process Capability Analysis as shown in Fig. 6.

The project team conducted a brainstorming to determine the possible causes of the studied problem and created cause-and-effect diagram as shown in Fig. 7.

The project team determined through brainstorming and technical expertise that the main causes affecting the investigated problem are as follows:

Table 2 Data set (before improvement)

Sample	Concentration of Spiramycine in each single tablet			
1	748983.36	786282.70	712326.35	814105.08
2	807598.64	738141.52	765466.22	751732.28
3	744493.58	784198.84	826338.23	820743.70
4	783988.56	782783.59	791388.32	729955.8
5	752702.08	755035.34	776623.88	726349.67
6	761741.15	767702.03	778985.67	778235.43
7	709386.50	746575.48	810163.70	745692.22
8	751653.11	777966.39	759402.64	712908.36
9	767159.78	823716.94	785748.68	748464.41
10	837044.65	736955.62	790873.70	782676.26
11	780268.57	729363.98	775379.26	782816.40
12	698562.66	776008.99	806399.59	783158.95
13	781179.10	809655.57	796241.47	731918.01
14	764158.61	758325.92	771724.85	767534.60
15	790165.9	765299.34	770208.46	766810.77
16	745395.27	756776.40	770734.25	765209.28
17	773001.15	743832.97	800763.52	722869.20
18	773984.201	785735.64	770083.07	774612.91
19	744419.41	776686.56	818007.63	803735.24
20	761741.15	767702.03	778985.67	762737.43
21	748131.50	769822.48	779167.70	776688.22
22	743912.78	777222.94	785748.63	748464.41
23	715232.71	787426.88	766404.75	773443.47
24	707498.15	752204.03	778985.67	778235.43

- Method of isolating inner phase components.
- Repetition of mixing the inner phase components of the product.
- Sequence of mixing inner phase components.
- Diameter of used sieves.

(4) **Improvement Phase**

New process activities were established considering the determined causes in the analysis phase; in addition, a new flowchart was prepared for the subprocess (preparing of inner phase) concerning the mechanism of preparing the inner phase of the studied product. The developed solution was applied on one batch. New measurements were conducted to collect the data set (Figs. 8, 9 and 10; Table 3).

The achieved results showed that process capability index Ppk has been increased from 0.86 to 1.60. Sigma level of the investigated process has been raised from 2.50 to 4.80. Process variability [12] decreased about 50% (Table 4).

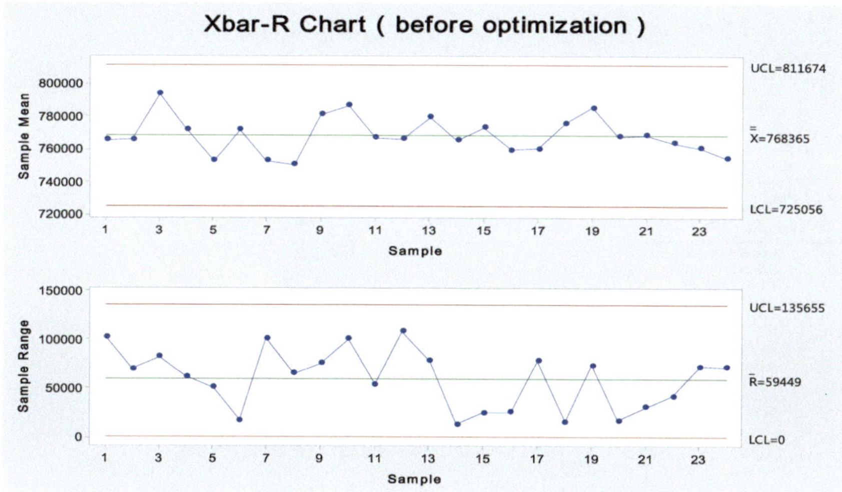

Fig. 4 Xbar-R control chart (before improvement) using software (Minitab v15)

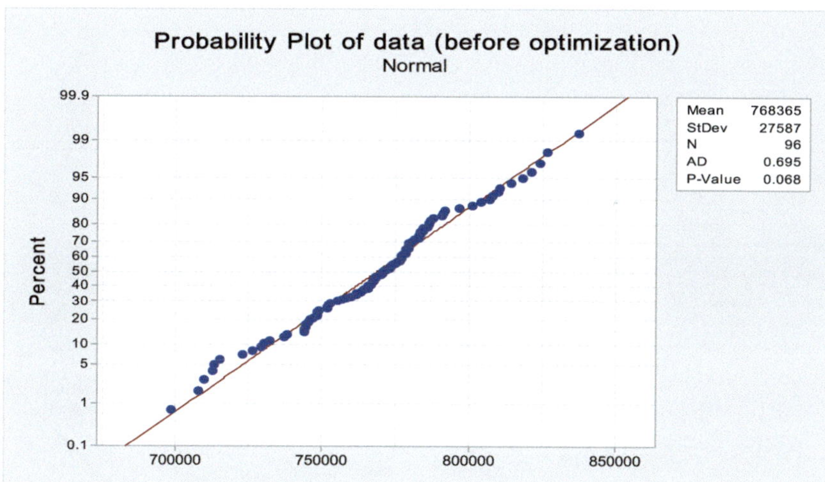

Fig. 5 Probability test (before improvement) using software (Minitabv15)

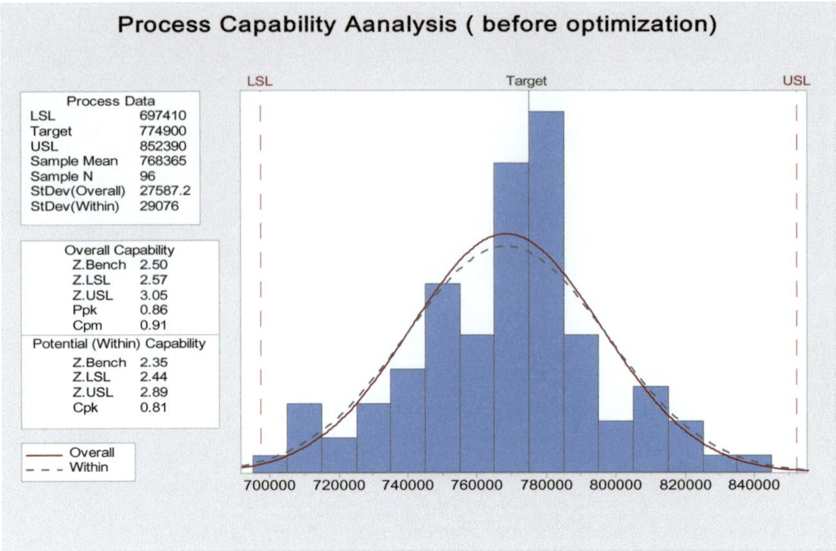

Fig. 6 Process capability analysis before improvement, using software (Minitabv15)

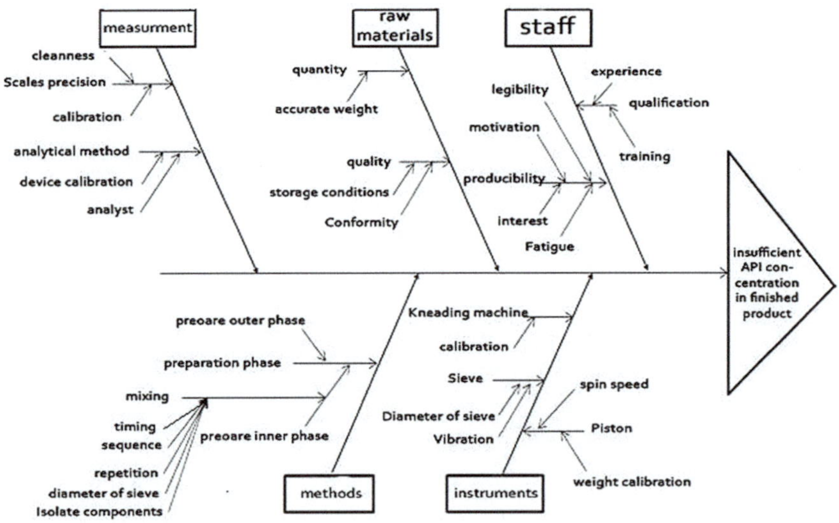

Fig. 7 Cause-and-effect diagram

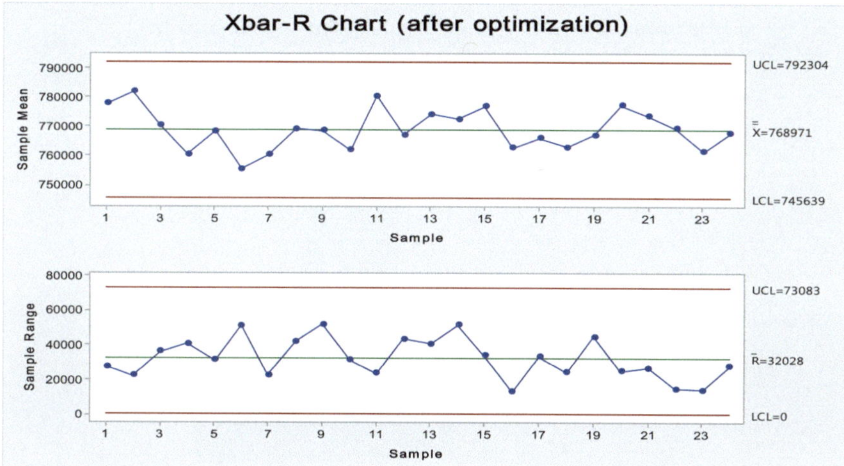

Fig. 8 Xbar-R control chart (after improvement) using software (Minitabv15)

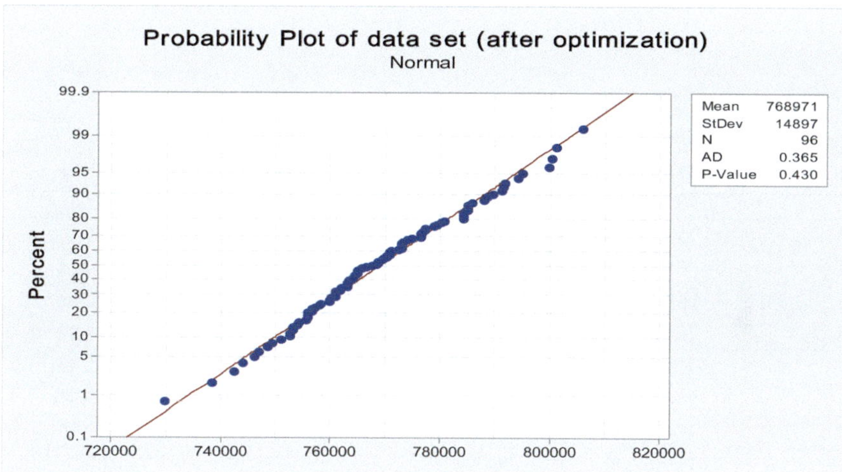

Fig. 9 Probability test (after improvement) using software (Minitabv15)

(5) **Control Phase**

In this phase, the project team accomplished the following activities:

- Validating the new process.
- Updating process documents.
- Training process operators on the new operation instructions.

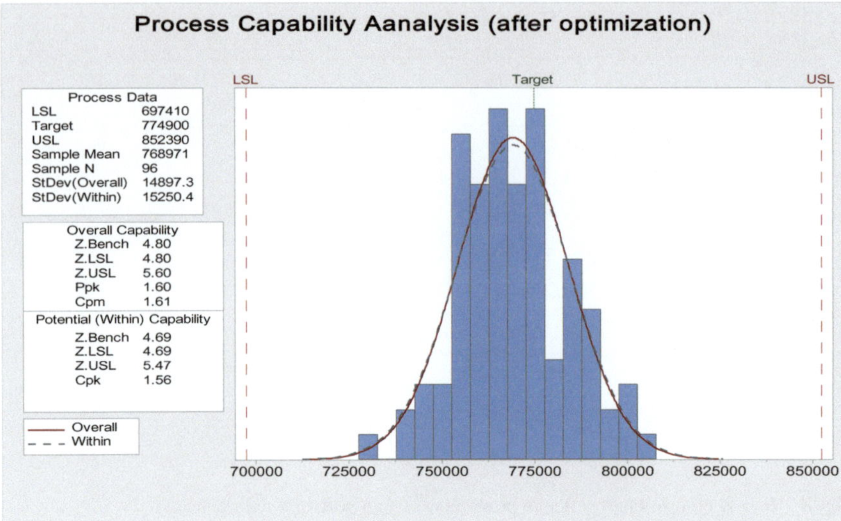

Fig. 10 Process capability analysis (after improvement) using software (Minitabv15)

Table 3 Data set (after improvement)

Sample	Concentration of Spiramycine in each single tablet			
1	791172.95	763853.35	770842.58	785748.61
2	785121.75	773079.26	774854.42	795047.41
3	755766.25	761008.78	773069.07	791553.31
4	755865.85	778774.51	768514.85	738479.71
5	753096.61	771215.23	784198.81	764972.63
6	764048.74	747003.61	729770.36	780767.82
7	768585.61	746228.71	766477.22	759842.83
8	784198.83	742354.21	774125.11	776348.34
9	754243.98	756644.191	805896.31	758072.35
10	756568.03	784198.81	753977.71	753202.81
11	779549.41	777224.71	794272.51	770900.84
12	791841.61	748553.41	763276.51	764972.62
13	760845.62	770250.62	801246.61	764972.63
14	785121.71	749328.31	800471.71	755402.84
15	769741.98	799696.82	772863.94	765821.34
16	763276.51	770250.61	760176.91	757223.63
17	760836.61	784198.82	750979.22	768700.81
18	764317.01	776682.27	758231.80	752648.96
19	773301.64	743904.01	763093.58	788437.45

(continued)

Table 3 (continued)

Sample	Concentration of Spiramycine in each single tablet			
20	785121.71	773079.26	788073.31	763151.86
21	777372.75	764593.83	789623.11	763151.84
22	767733.68	772575.31	761729.27	776348.34
23	760836.66	755711.86	769475.71	759842.83
24	780324.31	761998.99	776449.81	752648.91

Table 4 Comparison between KPI before and after improvement

	Standard Deviation	Ppk	Sigma level	Process mean (IU)
Before	27587.2	0.86	2.50	768,365
After	14897.3	1.60	4.80	768,971

- Controlling the process through the established Xbar-R chart to keep the process under statistical control.
- Updating the performance indices of the process to maintain the received enhancements.

3 Conclusion

Six Sigma is an advanced quality improvement methodology, which employs a wide variety of statistical and nonstatistical tools and techniques with highly qualified personnel to conduct quality improvement projects oriented to enhance the ability of the organization to achieve its strategic goals and to increase its competitiveness and market share. This research aims at investigating the possibility to implement Six Sigma methodology in the Syrian pharmaceutical companies, and to find out what benefits a pharmaceutical company can get through the implementation of this methodology. We conducted a case study in a pharmaceutical company in Syria (Orient Pharma) in order to examine the effectiveness and advantages of Six Sigma methodology. For this purpose, a quality improvement project was conducted using DMAIC roadmap to enhance the quality for one of the main products of the company. The obtained results of DMAIC project showed an enhanced process capability, an enhanced process Sigma level, decreased variability in the process outputs. The main difficulties that have been observed during the study were: personnel resistance to change, lack of training, lack of necessary resources, attitude toward quality in the company. As a conclusion, considerable benefits can be obtained through implementing Six Sigma methodology in the Syrian pharmaceutical companies.

References

1. Jernelid, M., & Roan, S. (2009). *Six sigma strategy applied to the pharmaceutical industry—How customers benefit* (MBA thesis, not published). School of Management, Blekinge Institute of Technology.
2. Pyzdek, T. & Keller, P. (2010). *The six sigma handbook: A complete guide for green belts, black belts, and managers at all levels*. Third Edition, McGraw-Hill.
3. Tang, L., et al. (2006). *Six sigma: Advanced tools for black belts and master black belts.* Wiley.
4. Sarkar, D. (2004). *Lessons in six sigma: 72 must-know truths for managers*. Response Books.
5. Muralidharan, K. (2015). *Six sigma for organizational excellence: A statistical approach.* Springer.
6. Hahn, G. J., Hill, W. J., Hoerl, R. W., & Zinkgraf, S. A. (1999, August). The impact of six sigma improvement—A glimpse into the future of statistics. *The American Statistician, 53*(3).
7. Glukhov, V., Turichin, G., Klimova-Korsmik, O., Zemlyakov, E., & Babkin, K. (2016). Document quality management of metal products prepared by high-speed direct laser deposition technology. *Key Engineering Materials, 684,* 461–467.
8. Djordjevic, A., & Cvetic, T. (2016). A business intelligence approach for choosing a optimal quality solution. *International Journal for Quality Research, 10*(2), 235–256.
9. Savovic, I., Bacovic, M., Pekovic, S., & Stanovcic, T. (2016). Impact of investment in quality and environmental protection on regional sustainability. *International Journal for Quality Research, 10*(3), 625–640.
10. Stanovcic, T., Bacovic, M., Pekovic, S., Jovanovic, J., & Savovic, I. (2016). The role of human resource practices on profits generated by the innovations: The role of top management support and regularity of employees meetings. *International Journal for Quality Research, 10*(4), 839–846.
11. Narasimhan, K. (2009). Six sigma: Basic tools and techniques. *Managing Service Quality, 19* (5), 631–632.
12. Papic, L., & Pantelic, M. (2014). Maintenance-oriented safety control charts. *International Journal of Systems Assurance Engineering and Management, 5*(2), 149–154.

Developing Plans for QFD-Based Quality Enhancement

Dmitriy Aydarov, Yury Klochkov, Natalia Ushanova, Elena Frolova and Maria Ostapenko

Abstract Quality Function Deployment (QFD) is a methodology for transforming customers' wishes into quality requirements for a product, service or process. QFD methodology was originally developed by Japanese researchers, who designed the approach for transforming customers' wishes (real or supposed) into detailed product characteristics using special matrices. QFD methodology provides better understanding of customers' expectations in the process of design and development of products, services, and processes and helps to consider real or supposed customers' requirements. House of Quality is used to show the relationship between customers' requirements and product characteristics. Product characteristics are realized using appropriate technological operations and equipment. If we know the methods for quality assessment of a separate operation (Cp indices, control charts, etc.), we can complete the House of Quality with the results of technological equipment analysis. Such data integration allows the complex solution of a problem of product competitiveness improvement. Using quality assessment methods for technological equipment, we acquire knowledge about defect probability at each separate production stage. Quality Function Deployment and integration of

D. Aydarov · N. Ushanova
Samara State Technical University, Samara, Russia
e-mail: aidarov.dm@mail.ru

N. Ushanova
e-mail: ushanovan@mail.ru

Y. Klochkov (✉)
Peter the Great St. Petersburg Polytechnic University, St. Petersburg, Russia
e-mail: y.kloch@gmail.com

E. Frolova
Buzuluksky Humanitarian-Technological Institute (branch)
of the federal state budgetary educational institution of higher
education «Orenburg State University», Orenburg region, Buzuluk, Russia
e-mail: fev_2004@list.ru

M. Ostapenko
Tyumen Industrial University, Tyumen, Russia
e-mail: ms_ostapenko@mail.ru

P. K. Kapur et al. (eds.), *System Performance and Management Analytics*,
Asset Analytics, https://doi.org/10.1007/978-981-10-7323-6_20

mentioned results (amount of defects, process stability) allow approaching assessment of each product characteristic with regard to its importance for a customer as well as with regard to the technical possibility to implement it.

Keywords Quality in engineering · QFD · Quality management
Characteristics of products

1 Introduction

Nearly, 80% of all defects detected during production and use of products are caused by quality problems at the stage of product development and design. About 60% of all failures within the warranty period are accounted for by inaccurate, hasty, or inadequate development. According to the data provided by the R&D department of General Motors Company (USA), the process of product development and production follows the "tenfold expenses" rule: if one of the quality cycle stages contains an error detected at the next stage, its correction requires ten times higher expenses compared to the case when this error is detected on time. If the error is detected at the stage after the next one, the expenses are 100 times higher, at the next stage after that the expenses are 1000 times higher, etc. Total quality management framework postulates changes in the approach to new product development, because it aims not only at supporting a high-quality level of products but also at meeting customers' requirements [1].

Activities for increasing the business culture level at an enterprise, which contribute to increasing the quality level at all operation stages, consider technologies for product design and production preparation. To decrease expenses, take into account customers' requirements and reduce development and production periods, special technologies for development and product analyses are applied:

- Quality Function Deployment (QFD) is a technology for the development of products and processes used to transform customers' requirements into technical requirements and production process parameters;
- Function cost analysis (FCA) is a method for analyzing expenses to achieve product functions or properties. FCA is applied to existing products and processes in order to reduce expenses, or to the product under development to reduce their production cost;
- Failure Mode and Effects Analysis (FMEA) is a method for analyzing defect probability and defect effect on customers; FMEA is carried out for products and processes under development to reduce potential defect risks;
- Functional physical analysis (FPA) is a method for analyzing the quality of suggested technical solutions, operation principles and elements of a product; FPA is carried out for products and processes under development.

Among all abovementioned technologies and methods, only QFD considers the interrelationship between customers' requirements and product properties, technical characteristics of its components and production parameters. QFD allows to transform customers' requirements into product functions and properties and production process parameters while distinguishing between consumer performance (primary quality parameters) and standard-specified product properties (secondary quality parameters), which can differ substantially.

QFD method was developed in the late 1960s in Japan and is used currently worldwide. QFD is a Delphi method (relying on a panel of experts) that uses tabular data presentation with tables of a specific form which is called House of Quality [2].

The central idea of the QFD method is that there is a difference between consumer performance (primary quality parameters, according to K. Ishikawa) and product parameters specified by standards and technical requirements (secondary quality parameters).

Secondary quality parameters are important for manufacturers, but not always significant for consumers. In an ideal case, a manufacturer is able to control product quality using primary quality parameters directly, but it is mostly impossible; therefore, a manufacturer should rely on secondary quality parameters [3].

QFD method is a sequence of actions which a manufacturer undertakes in order to transform real quality parameters of a product into technical requirements for a product, production process and equipment. QFD uses a series of matrices, i.e., tables of a specific form, which are called Houses of Quality. These tables help to relate customers' requirements with product properties, product properties with technical characteristics of its components, technical characteristics with production processes and production processes with production requirements. In practice, four Houses of Quality are usually used [4].

The first matrix presents customers' requirements in the rows and product/process properties in the columns. The main purpose of the first matrix is to establish correspondences between customers' requirements and product properties. The table cells contain indicators for the correspondence degree of these two parameters. Correlation degree is determined by experts. The scale contains the following values: 1—low correlation, 3—medium correlation, and 9—high correlation. If there is no correlation, the cell is empty.

The right part of the matrix allows to assess the degree of customers' satisfaction compared to that of the competitors. This part provides product benchmarking function. However, differently from conventional benchmarking, this stage includes a comparison of competing systems not in relation to technical characteristics, but to the degree of customers' satisfaction with product characteristics. Customer survey data are represented in numerical values on a five-point scale [5–7].

Analysis of competitors allows to draw conclusions about advantages and disadvantages of a product and to define the planned level of customers' satisfaction, which can be provided by a product to be developed. After a sequence of transformations in the right column, we obtain the value for the "required quality weight" which indicates the importance of a factor in various respects (importance for a customer, competitive advantages for a company, planned quality level, company priorities, etc.) Integrated index of improvement importance for each product property is calculated by serial addition of products of "required quality weight" and the corresponding values of property correlation with each customer' requirement. Conventional benchmarking does not provide such prioritization.

The second matrix is used to find out which system parameters are responsible for those properties. The right side of the matrix presents the comparison of competing products across their consumer properties. Similar to the first matrix, this comparison helps to determine if the improvements are necessary. As important, parameters-integrated indices of improvement importance (calculated in the first matrix) are used [8, 9].

The results provided by the second matrix are integrated coefficients reflecting the importance of product component improvements. Further, the matrix is constructed, which presents the interaction between component parameters and production processes; further matrix is constructed, which shows the relation between production processes and production requirements. Each matrix provides integrated importance indices. Thus, weighted coefficients for customers' requirements determined at the first stage undergo the full analysis process.

Construction of the second and all further matrices is similar to construction of engineering analysis tables; however, all QFD matrices are logically interrelated.

House of Quality method has the following advantages:

- It allows to establish the relation between customers' requirements, technical characteristics of a product, parameters of its components at all development stages (in other words, QFD has an algorithm absent in the engineering analysis).
- It provides the instrument used to transform customers' requirements into a set of controlled characteristics (required by conventional benchmarking) and requirements to the technical operations.

Thus, QFD method is a universal instrument for product development which integrates methods of marketing information processing, product benchmarking and engineering analysis. QFD method establishes a constant informational flow providing integration of all production systems and consideration of customers' requirements [10].

Various techniques for comparing competing products exist at present. Quality Function Deployment, or Houses of Quality method allows to transform customers' requirements into product properties and production process parameters. QFD is a Delphi method (relying on a panel of experts) that uses tabular data presentation with tables of a specific form which are called House of Quality. These tables present a sequence of actions to transform real quality parameters of a product into technical requirements to production processes and equipment. In the management aspects, QFD also helps:

- Identify and prioritize customer needs obtained from every possible source.
- Analyze the details of design and process improvement meeting the needs of engineers.

- Stimulate continuous improvement.
- Encourage communication and build teamwork within an organization.
- Reduce lead-time, optimize engineering resources and improve quality.
- Build partnerships with customers' participation [1–3].

2 Model Integration Results of the Assessment Technological Operations to Quality House

Using House of Quality method helps to find out the interrelation of customer's requirements and product characteristics. Product characteristics are implemented using certain technological operations and equipment. If we know which methods are used to assess the quality of each operation (calculation of Cp indices, control charts, etc.), we can add the results of analysis of equipment used to the House of Quality. Such integration of equipment quality data can result in complex solution of the problem of increasing competitive abilities of the products manufactured (Table 1).

The use of House of Quality method helps to find out the requirements to specific products (Fig. 1). Then, we select quality management instruments to assess the conformity of the defined requirements. For instance, in gearbox manufacturing methods such as Cp and Cpk as well as control charts are required. After technical drawings have been developed, we need to organize the technological process appropriate for customer's requirements realization. But before starting the production process, it is necessary to forecast the quality level in order to assess the risks related to potential defects. (3 Model predicts the level of quality) [5–7].

Quality function deployment and integration of the discussed results (defect rate, process stability) allow to assess the impact of a certain product characteristic both in the context of its importance for a customer and in the context of technological production possibility [8–11]. To achieve that, we can design the following diagram (Fig. 2).

The circle diameter depends on the impact of a characteristic: the higher the impact is, the larger the circle is. The higher the characteristic is represented, the

Table 1 Model integration results of the assessment technological operations to quality house

<table>
<tr><td>Customer's requirements</td><td>Requirement impact</td><td colspan="3">Product characteristic</td></tr>
<tr><td>Requirement 1</td><td>I_1</td><td colspan="3" rowspan="3">Relation between requirements and characteristics</td></tr>
<tr><td>...</td><td>...</td></tr>
<tr><td>Requirement N</td><td>I_N</td></tr>
<tr><td colspan="2">Technological operations</td><td colspan="3">Quality assessment methods used</td></tr>
<tr><td>N</td><td>Operation stage</td><td>Cp; Cpк</td><td>Control charts</td><td>...</td></tr>
<tr><td>i</td><td>...</td><td colspan="3" rowspan="2">Results</td></tr>
<tr><td>...</td><td>...</td></tr>
</table>

№			Requirements	relevance for the consumer	gear ratio	input shaft speed	output speed	transmitted power	transmitted torque	number of coils worm gear system	the number of gear teeth	the angular speed of the input shaft	the angular speed of the output shaft	wheel hardness	wheel worm gear system	endurance	angle screw turns	pitch diameter of the wheel	reference circle diameter	stiffer modulus a	rubbing speed	beam strength	oil viscosity	frictional heat dissipation	heat output	surface area
					Characteristics																					
					Worm gear system																		lubrication system			
					1	2	3	4	5	6	7	8	9	10	11	12	13	14	15	16	17	18	19	20	21	22
1	reliable and technically perfect worm reducer	Unconventional center-to-center spacing	diametral pitch	3,0							●						○	●	●							
2			housing removably	5,0																						●
3			lubrication control	4,3																					○	○
4			installation method	6,0																						○
5		requirements to engageable	lack of edge fin	7,2										○	○	▲				●	○		●			
6			continuity	9,0	○	●	●		○												●					
7			noiselessness	8,0										●			●				●					
8			exclusion repeated contact	6,0														○	○							
9			automatic braking	8,0	●	○												▲	▲				●			
10			reduced tooth crumble	10,0																			●			
11			lack of heating	7,5				○													●		○	●	●	○
12			material	6,3		○			▲							●					●	●				
13		composition	dimensions	6,5					▲		●					▲		●	●	○						
14			composition worm-gear drive	4,0																			●			
15			coating	3,0										▲	▲									●	●	
16			dimensions	8,0	○				●									●	●							
17		economy	cost of	7,6	●			●															▲			
18			weight	8,0					●	○	○			○	○			●	●							
19			efficiency	6,5						○							●									
20			total running time	5,0			▲					●	●		○											
	absolute importance				194	124	86	76	157	44	101	45	45	121	64	70	140	259	259	84	299	57	259	95	95	98
	the relative importance				7	4,5	3,1	2,7	5,7	1,6	3,6	1,6	1,6	4,4	2,3	2,5	5,1	9,3	9,3	3,0	11	2,1	9,3	3,4	3,4	3,5
	Operations																									
	first operation (Machine: 1K62)				Production «gear wheel», Quality control by Cp and Cpк																					
	...																									

Fig. 1 Sample quality houses

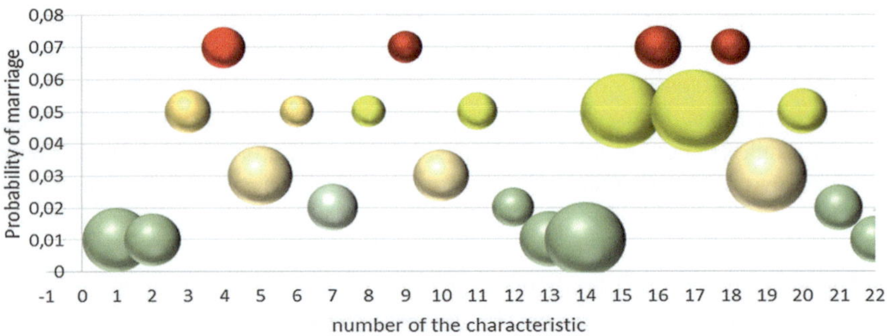

Fig. 2 Chart connection indicators "importance" of the characteristics of products, their number, and probability of default characteristics

higher the probability of its rejection in production. Thus, interpreting the results represented in Fig. 2, we can conclude that the first characteristic has the highest impact, and its production has a low probability of defect occurrence, whereas the fourth characteristic is the most problematic in this respect. Therefore, in the process of quality improvement program development more attention should be paid to the fourth characteristic. To simplify the decision-making procedure, we can suggest the following decision matrix (Fig. 3).

Cpk	The insignificance of the product characteristics	Few significant characteristic of the product	The significant characteristic of the product	The most significant characteristic of the product «importance»
1.67	Improving optional	Improving optional	Improvements are desirable	Improvements are desirable
1.33	Improvements are desirable	Improvements are desirable	Needed to improve	Needed to improve
1	Needed to improve	Needed to improve	Needed to improve	Needed to improve
0.87	Necessarily improve	Necessarily improve	Necessarily improve	Necessarily improve

Fig. 3 Matrix decision on the need to improve the characteristics of the product

3 Model Predict the Level of Quality

Cp, Cp and σ is usually calculated in reference to production [12–15], but if we calculate these indices in reference to operating equipment, we can forecast quality level at the preproduction stage. To know σ is most important.

If we have the results of constant assessment of quality indices Cp and Cpk, the value of mean displacement in reference to the center of the tolerance zone (Δ) and natural process variability (σ) for each piece of equipment, we can forecast the values of Cp and Cpk for a new technological operation at the planning stage of a new product; therefore, we can also assess probable defect rate.

For that assessment we need to calculate the predicted value of the mean using the formula:

supposing that the mean is displaced toward the upper tolerance limit

$$\overline{X}_n = \frac{UL_n + LL_n}{2} + \Delta,$$

where

UL_n — upper tolerance limit for a planned product;
LL_n — lower tolerance limit for a planned product;
Δ — mean displacement in reference to the center of the tolerance zone;

or
supposing that the mean is displaced towards the lower tolerance limit

$$\overline{X} = \frac{UL_n + LL_{n_H}}{2} - \Delta$$

Table 2 Example calculation Cp new and Cpk new

It is known from experience				We learn from the calculations		
Equipment	σ	Cp	Cpk	Production requirements	The upper limit (UL_n)	The lower limit (LL_n)
First operation					30.5	30.0
Machine: 1K62	0.05	2	1.8	The calculated values	Cp new	Cpk new
					(30.5–30.0)/(6*0.05) = **1.67**	1.67*(1.8/2) = **1.5**
Second operation					The upper limit	The lower limit
					29.8	29.5
Machine: Jet BD-7	σ	Cp	Cpk		Cp new	Cpk new
	0.025	1.8	1.6		(29.8–29.5)/(6*0.025) = **2**	2*(1.6/1.8) = **1.78**
…	…	…	…		…	…

To calculate $Cp_{\text{Н}}$ and $Cpk_{\text{Н}}$ for a new operation considering changes of tolerance zones and then to define ppm (number of defects per million of product units) or percentage of defects in reference to the indices (Table 2).

4 Conclusions

The central idea of the QFD method is that there is a difference between consumer performance (primary quality parameters) and product parameters specified by standards and technical requirements (secondary quality parameters).

Secondary quality parameters are important for a manufacturer, but not always substantial for a customer. In an ideal case, a manufacturer is able to control product quality using primary quality parameters directly, but it is mostly impossible; therefore, a manufacturer should rely on secondary quality parameters.

Our analysis of the methods used to establish the relation between technical characteristics of a product and customers' requirements shows that QFD is a universal instrument for product development which integrates methods of marketing information processing, product benchmarking, and engineering analysis. QFD method establishes a constant informational flow providing integration of all production systems and consideration of customers' requirements.

This effective application tool should be combined with TQM. The use of QFD method allows to simplify procedures involved in development of a new product. If House of Quality is supplemented with the results of statistical analysis of production equipment, it will become possible to forecast the quality level.

Acknowledgments The chapter is based on the research carried out with the financial support of the grant of the Russian Science Foundation (Project No 14-38-00009). Peter the Great St. Petersburg Polytechnic University.

References

1. Lee, S. F., Sai On Ko, A. (2000). Building balanced scorecard with SWOT analysis, and implementing "Sun Tzu's The Art of Business Management Strategies" on QFD methodology. *Managerial Auditing Journal, 15*(1), 68–76.
2. Hassana, A., Siadat, A., Dantan, J. Y., & Martin, P. (2010). Conceptual process planning—An improvement approach using QFD, FMEA, and ABC methods. *Robotics and Computer-Integrated Manufacturing, 26*(4), 392–401.
3. Luo, X., Tang, J., Kwong, C. K. (2013). A QFD-based optimization method for scalable product platform. In *Advances in product family and product platform design, part II* (pp. 343–365). New York: Springer.
4. Prasad, K. G. D., Subbaiah, K. V., & Rao, K. N. (2014). Supply chain design through QFD-based optimization. *Journal of Manufacturing Technology Management, 25*(5), 712–733.
5. Yazdani, M., Hashemkhani Zolfani, S., & Zavadskas, E. K. (2016). New integration of MCDM methods and QFD in the selection of green suppliers. *Journal of Business Economics and Management, 17*(6), 1097–1113.
6. Wu, C.-T., & Liu, N.-T. (2014). An extensive evaluation design approach to quality function deployment. In *International Applied Science and Precision Engineering Conference, Applied Mechanics and Materials*, pp. 1197–1201.
7. Karsak, E. E., & Dursun, M. (2014). An integrated supplier selection methodology incorporating QFD and DEA with imprecise data. *Expert Systems with Applications, 41*(16), 6995–7004.
8. Chun, J., & Cho, J. (2015). QFD model based on a suitability assessment for the reduction of design changes in unsatisfactory quality. *Journal of Asian Architecture and Building Engineering, 14*(1), 113–120.
9. Jin, J., Ji, P., & Liu, Y. (2014). Prioritising engineering characteristics based on customer online reviews for quality function deployment. *Journal of Engineering Design, 25*(7–9), 303–324.
10. Luo, X. G., Kwong, C. K., Tang, J. F., & Sun, F. Q. (2015). QFD-based product planning with consumer choice analysis. *IEEE Transactions on Systems, Man, and Cybernetics: Systems, 45*(3), 454–461.
11. Golubchik, E., Polyakova, M., & Gulin, A. (2014). Adaptive approach to quality management in combined methods of materials processing. *Applied Mechanics and Materials, 656,* 497–506.
12. Glukhov, V., Turichin, G., Klimova-Korsmik, O., Zemlyakov, E., & Babkin, K. (2016). Document quality management of metal products prepared by high-speed direct laser deposition technology. *Key Engineering Materials, 684,* 461–467.
13. Kasaei, A., Abedian, A., & Milani, A. S. (2014). An application of quality function deployment method in engineering materials selection. *Materials and Design, 55,* 912–920.
14. Polyakova, M. A., Rubin, G. S., Gun, G. S., & Danilova, Y. V. (2016). New approach to development methodology of requirements of standards for metal products. *CIS Iron and Steel Review, 12,* 45–49.
15. Rubin, G. S., Polyakova, M. A., Chukin, M. V., & Gun, G. S. (2013). Document protypology: A new stage in the standardization of metal products. *Steel in Translation, 43*(10), 666–669.

Modified Counting Sort

Ravin Kumar

Abstract There are various sorting methods in the literature, which are sequential in nature and have linear time complexity. But these methods are not preferred to use due to large memory requirements in specific cases. Counting sort is one, which lies in this domain. In this chapter, we have suggested an improvement on the counting sort. Due to this improvement, the memory requirement for counting sort is reduced up to a significant level. We have tested this modified counting sort on numerous data sets and the results obtained by these experiments are very much satisfactory. Results shows that this memory requirement is reduced at least 50% than traditional counting sort. So it opens up the opportunity of using this modified version in many sorting applications.

Keywords Sorting · Linear sorting · Counting sort · Modified counting sort Non-comparison sorting

1 Introduction

Sorting is the process of arranging numbers, alphabets or characters in a statistical order (in increasing or decreasing order) or in lexicographical order (alphabetical value like addressee key) [1–3]. These techniques are divided into categories based on their features. Research efforts are made to improve their performance by reducing their complexity. These algorithms are mainly divided into two categories, comparison-based sorting and non-comparison-based sorting. Comparison-based sorting algorithms involve comparing the values in the array to obtain the sorted sequence. Depending on the type of approach used to perform comparisons, few mainly used algorithms with their complexities are as stated:

R. Kumar (✉)
Department of Computer Science, Meerut Institute of Engineering and Technology, Meerut 250005, Uttar Pradesh, India
e-mail: mr.ravin_kumar@hotmail.com

P. K. Kapur et al. (eds.), *System Performance and Management Analytics*, Asset Analytics, https://doi.org/10.1007/978-981-10-7323-6_21

1.1 **Bubble sort**: The algorithm is slow compared to other algorithms. It has a complexity of $O(n^2)$ [4].
1.2 **Insertion sort**: Insertion sort is a sorting algorithm that works similar to sorting cards in hand. This algorithm also have the complexity of $O(n^2)$ [5].
1.3 **Merge sort**: This algorithm uses divide and conquer approach [6, 7]. It has a complexity of O(n log n).
1.4 **Quick sort**: Quick sort is more efficient than the preceding sorting algorithms. It has best-case complexity of O(n log n) and worst-case complexity of $O(n^2)$ [8].

Non-comparison-based sorting algorithms have less complexity because they do not require comparison operation among its elements to obtain the sorted sequence. Counting sort [9] is one classic algorithm used for non-comparison-based sorting. It has linear time complexity (i.e. O(n)), but is not preferred for sorting arrays having large numbers because it requires large amount of memory to obtain the sorted sequence. This paper presents a new approach of sorting which leads to a reduction of at least 50% of memory usage with a time complexity of O(n).

2 Proposed Algorithm

Traditional counting sort has a time complexity of O(n). The amount of memory required to perform sorting using counting sort depends primarily on the largest element present in the input array and less on the total number of elements in the sequence. In a sequence where '*T*' is the largest number, It will require an additional array of size '$T + 1$' with index ranging from '0' to '*T*'. For example, consider the case where we are sorting the values 1000 and 1 using traditional counting sort. Sorting these values require an additional array of size 1001 as the largest number in the input sequence is 1000.

Modified counting sort is designed to sort an array in the time complexity of O (n), and also reducing the memory requirement of additional array by at least 50 percent. In the proposed algorithm, the details of parameters and variables are described in Table 1.

Table 1 Description of the parameters and variables used in modified counting sort

Parameter	Description
RA	Input array that is to be sorted
min and max	Represents minimum and maximum value respectively
RB and RC	Represents two arrays
Ref1 and Ref2	Represents minimum value in RB and RC respectively
RS	Represents the sorted array

Modified_Counting_Sort(RA)
BEGIN

```
min = ∞
max = -∞
/* min and max are calculated from RA
  */
for i = 0 to Length[ RA ] -1
do
   if RA [ i ] > max
   then
      max = RA [ i ]
   end if
   if RA [ i ] < min
   then
      min = RA [ i ]
   end if
end for
div = (min + max) / 2
x = 0
y = 0
Ref1 = min
Ref2 = ∞
RBmax = -∞      // RA is divided in two arrays
for i = 0 to Length[ RA ] - 1
do
   if RA [ i ] < div
   then
      /
* Ref1 is subtracted from all the values in RB, and max value of RB is
        calculated and stored in RBmax */
      if RA [ i ] > RBmax
      then
         RBmax = RA [ i ]
      end if
      RB [ x ] = RA [ i ] - Ref1
      x ++
   else
      if RA [ i ] < Ref2
      then
         Ref2 = RA [ i ]   /* Ref2 of RC is calculated. */
      end if
      RC [ y ] = RA [ i ]
      y ++
   end if
```

```
end for
for i = 0 to Length[ RC ] -1
do
  RC [ i ] = RC [ i ] - Ref2
end for
```

/* Now applying the New_Counting _Sort on both RB and RC sub arrays. */

New_Counting_Sort(RB, RS, RBmax - Ref1, Ref1, 0)

New_Counting_Sort(RC, RS, max - Ref2, Ref2, Length[RB])

END

New_Counting_Sort(A, B, k, Ref, AddLen)

A is the input array, B is the sorted array and k is the maximum number, which is present in A array.

BEGIN

```
for i = 0 to k
do
  C[i] = 0
end for
for j = 0 to Length[A] -1
do
  C [ A [ j ] ] = C [ A [ j ] ] + 1
end for
for i = 1 to k
do
  C [ i ] = C [ i ] + C [ i - 1]
end for
for j = Length[A] - 1 down to 0
do
  B [ C [ A [ j ] ] + AddLen - 1] = A [ j ] + Ref
  C [ A[ j ] ] = C [ A [ j ] ] -1
end for
```

END

The time complexity of the modified counting sort is still O(n) and it reduces the amount of memory required to perform sorting operation.

3 Working Demonstration

Consider the following input data set on which we have to perform sorting operation using the proposed algorithm.

20	33	23	60	52	210	240	200	130	42

On applying the proposed algorithm over this data, we get min = 20, max = 240 and div = 130. Values less than div are as follows:

20	33	23	60	52	42

Similarly, values greater than or equal to div are as follows:

210	240	200	130

From the algorithm, we get Ref1 = 20 and Ref2 = 130.

Now after subtracting Ref1 from first array, store the obtained data in RB. Similarly, after subtracting Ref2 from second array, store the obtained data in RC.

After performing above operation, elements in RB become

0	13	3	40	32	22

Similarly, elements present in RC become

80	110	70	0

Now, our modified counting sort uses New_Counting_Sort on RB, which sorts RB and then adds the Ref1 value to the result of sorted RB:

0 + 20	3 + 20	13 + 20	22 + 20	32 + 20	40 + 20

Obtained result in RB:

20	23	33	42	52	60

And the size of C[] for doing this operation was 41 (i.e. from 0 to 40). Now, our modified counting sort uses New_Counting_Sort on RC, which sorts RC and then adds the Ref1 value to the result of sorted RC:

0 + 130	70 + 130	80 + 130	110 + 130

Obtained result in RC:

130	200	210	240

Now, final result of sorting is obtained in RS, and the sorted sequence becomes:

20	23	33	42	52	60	130	200	210	240

Total memory required by additional C[] is 111, (i.e. 0–110), while on applying traditional counting sort, it would require an additional array of size 241, indexed from 0 to 240 to sort this input sequence.

4 Result and Discussion

Modified counting sort is tested on various input sets, and a comparison is done with the traditional counting sort algorithm. Since both algorithms have time complexity of O(n), comparison is done on the basis of amount of memory required (Table 2).

To understand the effectiveness of the proposed algorithm the comparison using the above data set can be displayed graphically as shown in Fig. 1.

From the above data it is clear that in worst case, the memory used by the modified counting is half the memory used by the traditional counting sort algorithm. The time complexity of both the algorithms remains O(n). This can also be graphically represented as (Fig. 2).

Table 2 Sample data to demonstrate the comparison between traditional counting sort and modified counting sort

Input elements	Memory of C[] in Counting sort (i.e. length of C [])	Memory of C[] in Modified_Counting_Sort (i.e. length of C [])
1, 2, 3, 250, 251, 252	253	3
18, 21, 19, 23, 407, 401, 402, 400	408	8
1000, 1001, 1002, 1003, 1004, 1005	1006	3
900, 800, 2, 1, 700, 300, 4, 7, 9, 12, 27	901	300
9, 8, 11, 10, 1001, 1700, 1500, 2000	2001	1000
20, 100, 1, 0	101	21
1000, 999, 998, 997 …,3, 2, 1	1001	500

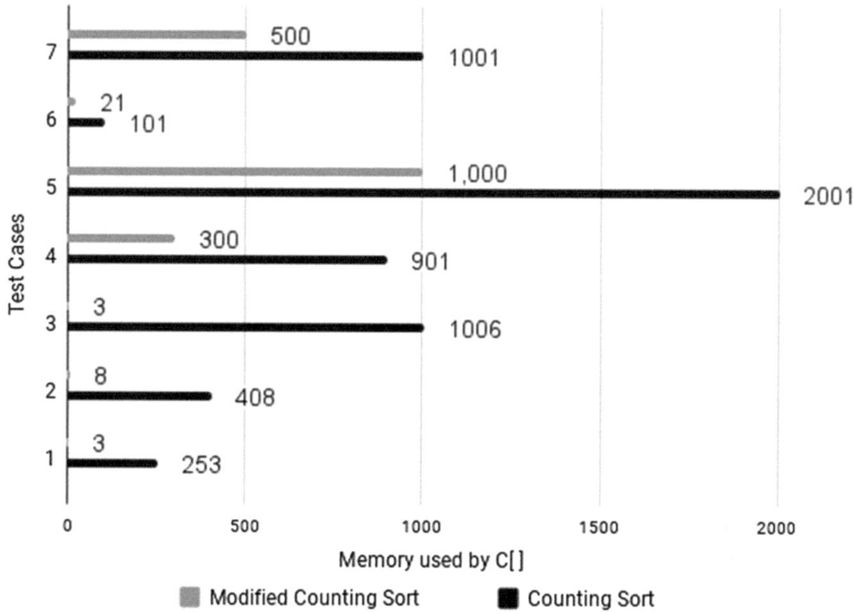

Fig. 1 Representation of memory used by each test data sequence in both traditional and modified counting sorts

Fig. 2 Representation of comparison of worst-case memory requirement by traditional and modified counting sorts

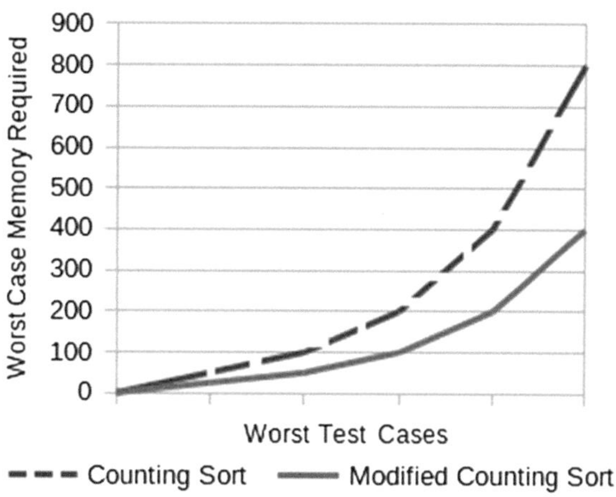

5 Conclusion

In this paper, an advanced method is introduced which reduces the memory requirements of the classic Counting sort algorithm by at least 50 percent. Modified Counting Sort can be used for improving performance of various existing sorting algorithms like radix sort [10], which requires an additional stable sorting algorithm to perform sorting. Using modified counting sort, It is possible to improve the current sorting applications like searching a value in a large array, Sorting in large databases and TCP/IP packet sorting [11].

References

1. Flores, I. (1960). Analysis of internal computer sorting. *ACM*, *7*(4), 389–409.
2. Franceschini, G., & Geffert, V. (2003). An in-place sorting with O(n log n) comparisons and O(n) moves. In *Proceedings of 44th Annual IEEE Symposium on Foundations of Computer Science,* pp. 242–250.
3. Knuth, D. (1998). T*he Art of Computer programming Sorting and Searching*, 2nd edn. Addison-Wesley.
4. Oyelami Olufemi Moses. (2009). Improving the performance of bubble sort using a modified diminishing increment sorting. *Scientific Research and Essay, 4*(8), 740–744.
5. Rupesh, S., Tarun, T., & Sweetes, S. (2009). Bidirectional expansion—insertion algorithm for sorting. In *Second International Conference on Emerging Trends in Engineering and Technology, ICETET-09*.
6. Radu, R., & Martin, R. *Automatic Parallelization of Divide and Conquer Algorithm" Laboratory of Computer Science*. Cambridge, MA, USA: Massachusetts Institute of Technology.
7. Dean, C. (2006). A simple expected running time analysis for randomized divide and conquer algorithms. *Computer Journal of Discrete Applied Mathematics, 154*(1), 15.
8. Friend, E. (1956). Sorting on electronic computer systems. *Computer Journal of ACM, 3*(3), 134168.
9. Rajasekhara Babu, M., Khalid, M., Sachin, S., Sunil, C., Babu, M. (2011). (IJCSIT) *International Journal of Computer Science and Information Technologies*, Vol. 2 (5) 2284–2287.
10. Andersson, A., & Nilsson, S. (1994). A new efficient radix sort. In *Proceedings of 35th Annual IEEE Symp. on Foundations of Computer Science*, pp. 714–721.
11. Meinel, C., & Sack, H. (2013). Internetworking. Berlin Heidelberg: X.media.publishing, Springer-Verlag. https://doi.org/10.1007/978-3-642-35392-5_2.

Analysis of Existing Clustering Algorithms for Wireless Sensor Networks

Richa Sharma, Vasudha Vashisht, Ajay Vikram Singh and Sushil Kumar

Abstract With the recent advancement in MEMS technology, researchers in academics as well as in industry are showing their immense interest in Wireless Sensor Networks (WSNs) since the past decade. WSNs are the networks composed of uniformly or randomly distributed autonomous low-cost nodes used for reliable monitoring of environmental parameters. These resource-constrained sensor nodes work in a synergetic manner to perform a sensing process. Wireless Sensor Networks have a significant role in different areas like habitat monitoring, health monitoring, intelligent and adaptive traffic management, military surveillance, target tracking, aircraft control, forest fire detection, air pollution monitoring, etc. These networks face some critical energy challenges while doing data aggregation, node deployment, localization, and clustering. This chapter presents the analysis of different clustering algorithms proposed so far to lengthen the network lifetime and to increase the network scalability.

Keywords Clustering · Scalability · Network lifetime · Evolutionary algorithms

R. Sharma (✉) · V. Vashisht · S. Kumar
Amity University, Noida, Uttar Pradesh, India
e-mail: richas193@gmail.com

V. Vashisht
e-mail: vvashisht@amity.edu

S. Kumar
e-mail: kumarsushiliitr@gmail.com

A. V. Singh
Middle East College, Muscat, United Arab Emirates
e-mail: asingh@mec.edu.om

P. K. Kapur et al. (eds.), *System Performance and Management Analytics*,
Asset Analytics, https://doi.org/10.1007/978-981-10-7323-6_22

1 Introduction

Wireless Sensor Networks are gaining a voluminous importance in every field with the increase in their demand in many real-life applications. WSNs are a network of densely deployed sensor nodes in a region of interest. WSNs are composed of a sensing unit to sense the environmental parameters, a processing unit with external memory, transceiver to collect and transmit the sensed data and a power supply. The three-sensor network elements are Nodes (also called Sensor nodes), Data Gatherer (Cluster head and Base station), and an External System. The sensor node is a basic unit of WSNs. These are resource-constrained devices with limited energy, small storage, less data processing capabilities, and limited bandwidth. Sensor nodes are randomly deployed with an objective to monitor, to sense, and explain a physical phenomenon. Sensor nodes act as a transducer as they convert the sensed physical phenomenon into an electrical signal. Sensor nodes are responsible for supervising the environmental conditions and periodically transmit that data to the base station. Since the sensor nodes are battery-powered devices and their batteries are irreplaceable (i.e., use and throw type of devices), so efficient energy utilization is a big challenge in WSNs. Long communication distances between the sensor nodes and the base station result in rapid draining of the node's limited energy available with them and reduce network's lifetime. Hence, lengthening the network lifetime, increasing the scalability of the network and balancing the load among sensor nodes are some challenges faced by WSNs [1].

2 Clustering Wireless Sensor Networks

Since energy conservation is the prime issue for the efficient performance of the WSNs, so a reduction in the communication distance between the sensor nodes and the sink is required. A prominent approach available for this purpose is clustering. Clustering is an effective means of organizing the sensor nodes into nonoverlapping sets called clusters with an intermediate node called Cluster Head (CH) acting as a head of other nodes. These cluster heads collect data from the nodes in the network and pass it to the base station.

Various advantages of clustering are summarized as follows:

- Balancing the load among the sensor nodes.
- Promote coverage maximization.
- Minimizing communication overhead of the nodes and the base station node.
- Reducing delay in data transmission.
- Promote energy conservation.
- Maximizing network lifetime.
- Fault-tolerant Networks.

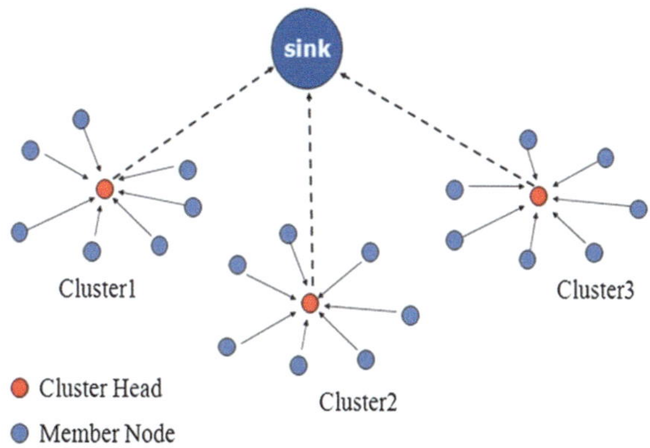

Fig. 1 Basic framework of clustered wireless sensor networks [2]

In WSNs, clustering is done mainly for two main reasons. First, since sensor nodes contain a large amount of redundant data sensed from an external environment so there is a need of data aggregation and data fusion to compress that large amount of ambiguous data into small meaningful information. With clustering, data aggregation and data fusion are done by cluster head at one centralized location. Second, if clustering is not done among the sensor nodes then all of the sensor nodes would have to communicate directly with base stations, which will result in excessive energy dissipation (Fig. 1).

3 Conventional Clustering Protocols and Their Performance Comparison

Prior to the concept of clustering in WSNs, routing was done through conventional protocols named direct transmission protocol, static clustering protocol, and MTE (Minimum Transmission Energy) protocol. All these protocols have pros and cons. In direct communication and transmission protocol, because of the direct data transmission between the base station and the sensor nodes, lots of energy was dissipated in transmission. In MTE protocol, multi-hop communication is employed for transmission and the base station which results in a large amount of energy dissipation while transmitting data through intermediate nodes acting as routers in between. In static clustering protocols, clusters are formed, within each cluster one node acts as a local base station. Sensor nodes relay their sensed data to the local base station that in turn passes that data to the global sink. This protocol suffers from a big problem that if the local base station of a cluster dies, then all the nodes in that cluster will also die as no intermediate will be left transfer their data to the global base station.

Heinzelman et al. [3] presented distributed cluster-based approach named LEACH. It focuses on the distribution of the energy load evenly among all the sensors nodes. This protocol has two phases named a setup phase and steady-state phase. In the first phase, sensor nodes organize themselves into clusters, by selecting one node as their CH. Each node elects itself to become a future CH. Between 0 and 1, a random number is selected by a node, if the threshold value *T* (*h*) is more than the randomly chosen number then in the current round that node will play the role of a cluster head, otherwise it joins the nearest cluster head in its neighborhood. Threshold *T*(*h*) is specified as

$$T(h) = \begin{cases} \frac{n}{1-n} * c \mod \frac{1}{n} \text{ if } h \in G \\ 0, \text{otherwise} \end{cases}$$

Here, n denotes the predetermined percentage of CH, *G* denotes the group of those nodes that had not been selected as CH in previous 1/*p* Rounds and c denotes the currently going round. Steady phase includes data aggregation and data fusion to be done by compressing the data before passing the sensed data to the sink. This results in minimizing the energy dissipation. LEACH has outperformed conventional static clustering algorithms, traditional protocols of direct data transmission, multi-hop routing and MTE by employing the rotation strategy among cluster heads.

LEACH has been considered as a benchmark to test various WSN clustering algorithms. LEACH-C, a centralized version of LEACH was introduced by Heinzelman et al. [4] later, in which selection of CHs is done at a centralized location by the base station (BS). Sensor nodes forward the information about themselves to the base station, i.e., their present location (where they are deployed) and their current energy. Base station finds out the average energy of the network and the nodes having high energy as compared to the calculated average energy are only considered eligible candidates for CH position. After this, the BS broadcasts this decision to every node in the network. The simulated annealing process is performed by the base station to search for the best solution out of all possible candidate solutions. This algorithm aims to decrease the total sum of the squared distance between the sensor nodes and their closest CH, resulting in minimum energy dissipation in transmitting data from sensor nodes to their cluster head. LEACH-C gives good output as compared to LEACH.

Khediri et al. [5] has proposed O-LEACH, a centralized and optimized approach, in which base station initiates the complete routing process. In each round the node with energy value, more than ten percent of the residual energy level of each sensor node is considered as the cluster head. Cluster formation and transmission phase are initiated after the CH selection process. If the energy is less than ten percent of the residual energy level then proceeds as such with the process of LEACH protocol. A simulation result shows that the stability of O-LEACH is longer than LEACH and LEACH-C.

Bandhopaya et al. devised EEHC (Energy-Efficient Hierarchical Clustering) [6] algorithm, which is a randomized clustering algorithm employing distributed approach. The algorithm generates a hierarchy of CHs which has shown that when the numbers of levels in the hierarchy are increased, the energy savings also increase. It is a multi-level hierarchical clustering algorithm that makes use of stochastic geometry to obtain parameters for minimum energy consumption.

Younis and Fahmy [7] proposed a protocol named HEED in which two parameters that is the currently available energy of the sensor node and the cost for intracluster communication is considered for CH selection. Based on probability function, cluster head selection is done as

$$n_{\text{prob}} = h_{\text{prob}} * e_{\text{residual}} / e_{\text{maximum}}$$

where e_{residual} denotes the current remaining energy of the sensor node, e_{maximum} is maximum initial energy same and fixed for all nodes and h_{prob} is the initial percentage of cluster heads among all n nodes. This results in a uniform distribution of CH nodes and better load balancing. The node will become CH according to n_{prob} or it can join a cluster according. PEGASIS [8] that was an extension of LEACH in which a chain of sensor nodes is made and nodes only transmit to their neighbors only and one randomly selected node will aggregate data from all other sensor nodes and transmit it to the BS (Table 1).

4 Metaheuristic Approaches for Clustering Wireless Sensor Networks

Most of the real-world problems are optimization problems under highly complex constraints, solving such problems is a challenging task. So, the current trend is to adapt nature-inspired metaheuristic approaches to efficiently tackle these problems. Nature-inspired algorithms are drawing the significant attention of the researchers in the designing of energy-aware algorithms to extend the lifetime of the WSNs. Evolutionary algorithms are algorithms inspired by nature. They are stochastic optimization algorithms working similar to the principle of natural evolution. The basic concept followed by these algorithms is to initialize a population of individuals and these individuals reproduce to generate new population only if they meet a certain selection criterion. These algorithms can search a large solution space to find out the best optimal solutions. These nature-inspired algorithms can efficiently solve various complex NP-hard problems due to their adaptable and flexible nature. Energy-Efficient Clustering in WSNs to keep distance to a minimum is one such NP-hard problem solved by these algorithms. To find an appropriate number of cluster heads in the network, metaheuristic approaches are adopted. If there are more CHs than the optimal number, more energy is dissipated in terms of the transmission of data from all these cluster heads to the base station, that is far away from the sensor nodes. If there are less CHs, more energy will be dissipated in relaying data from sensor nodes to the CHs, that are far away from the sensor nodes. In this chapter, a comprehensive survey of few of these algorithms is performed and the algorithms being focused in this chapter are Genetic Algorithm (GA), Particle Swarm Optimization (PSO), Firefly Algorithm, Honey Bee Mating Optimization Algorithm (HBMO) and Differential Evolution Algorithm (DE).

Table 1 Performance comparison of existing clustering protocols

Clustering algorithm	Clustering methodology	Model assumptions for sensor nodes and base station	Cluster head selection criteria	Simulation results	Conclusion
LEACH	Distributed	• Fixed BS (high-energy node), • Homogeneous and energy constrained nodes • Localized coordination among SNs • Incorporates data fusion • Number of clusters in network are fixed in priori	Assume predetermined number of CH and considers headcount of a node Random number selected by node should be lesser than $T(n)$ i.e. the threshold value	Reduction in energy dissipation up to 4–8 times when compared to MTE, and 7–8 times reduction in energy consumption than direct communication	Better than direct routing, static routing, and multi-hop routing
LEACH-C	Centralized	BS has information of geographical location and current energy of all sensor nodes beforehand	Depends on the residual energy of sensor nodes	40% more better throughput than LEACH	Efficient than LEACH protocol
EE-LEACH-C	Centralized	• Nodes are randomly distributed • Base station is at the central location • Clustering coordinators are being found	BS runs a sorting algorithm to sort nodes to be a future CH, according to descending value of their RE and node having maximum residual energy is selected as CH	Energy consumption reduces of whole network	Lengthen by 10% the lifespan of network in comparison to LEACH and by 5% in compared to LEACH-C

(continued)

Table 1 (continued)

Clustering algorithm	Clustering methodology	Model assumptions for sensor nodes and base station	Cluster head selection criteria	Simulation results	Conclusion
O-LEACH	Centralized	• Nodes are in static mode • Sensor nodes are uniformly dispersed • CHs forward data after aggregating it to the base station directly	Node having energy more than ten percent of the remaining energy value will be selected as CH in each iteration	Improve stability and lifetime of nodes, also more messages delivered as compared to LEACH and LEACH-C	Energy-efficient and achieve longer stability
LEACH-TLCH	Distributed	• Fixed BS in the center of region • SNs are homogeneous and uniformly distributed • SNs communicate multi-hop or single-hop • There is a fixed probability of 7% for a sensor node to become CH	Secondary cluster head is selected if the current energy of primary CH is less than the average energy of all the nodes	Running performance much better than LEACH	Improves energy efficiency and network lifetime
TL-LEACH	Distributed	CH instead of sending data directly to BS make use of another CH as an intermediate which lies between him and the BS	Two cluster heads are selected per cluster	Performance better than LEACH	Reduces the energy dissipation by decreasing the number of nodes communicating directly with BS

(continued)

Table 1 (continued)

Clustering algorithm	Clustering methodology	Model assumptions for sensor nodes and base station	Cluster head selection criteria	Simulation results	Conclusion
HEED	Distributed	• Sensor nodes are quasi-stationary • Nodes with no information about their location • All nodes have exactly same capabilities • After deployment, nodes are left unattended	Based on residual energy, the node is selected as a CH	Better than LEACH as energy consumption is less	HEED terminates in O(1) iterations thus increases overall lifespan of network

4.1 Clustering of WSNs Using Genetic Algorithm (GA)

GA is a well-known powerful and unbiased optimization search technique. It is one of the evolutionary algorithms that deal with solving constrained and unconstrained optimization problems. It is devised by Charles Darwin in 1858. The basis of the GA is the Darwin theory of natural selection that is "Survival of the Fittest". It works on the population of "individuals". For each "individual", a predefined fitness function is evaluated. The individuals with highest fitness values are considered closer to the optimal solution and thus are selected for crossover. Crossover operation generates new offsprings having some common features of their parent. For ensuring that the individuals generated should not have exactly the same features of their parents, the mutation operator is applied. The least fit individual is discarded at the end of the generation and newly generated offsprings are included in next generation. This complete process will iterate for a predefined number of generations or iterations.

4.1.1 Related Work

Khanna et al. [9], proposed a genetic algorithm based approach with reduced complexity to optimize multi-hop sensor networks. The author focused on two objectives: first is the creation of an optimal number of clusters with associated CHs and second is searching for a low-cost path using one or more hopes to the base station. For these two competing objectives, a multi-objective genetic algorithm with tradeoff analysis was employed to find a set of solutions towards the Pareto optimal front. The proposed GA-based work had two modules initialization and adaptation module. Sensor nodes are categorized into three categories CHs, sensor, intercluster router. Data fusion is done by the CH and it forwards data to other CH through intercluster route. Genetic Algorithm was employed to evaluate optimal route and the fitness function is calculated for each node. Fitness function for node selection is based on factors like Cluster Head Fitness, Node Communication Fitness, Router Load Fitness, Battery Status Fitness, and Total Node Fitness. Simulation results have shown a large reduction in the energy consumption and maximizing the objectives like node coverage and exposure.

Hussain et al. [10] used GA for intelligent hierarchical clustering technique to find the cluster heads, the number of clusters, the cluster members, and the transmission schedules. Fitness parameters considered are Cluster Distance-Standard Deviation (SD), Direct Distance (D) to sink, Transfer Energy (E), Number of transmissions (T), and Cluster Distance (C). The GA-based hierarchical protocol was found to perform better in comparison to LEACH and other similar cluster-based protocols. Heidari and Movaghar [11], presented a GA-based efficient approach for energy optimization of sensor networks. In this proposed work, base station employed a Genetic Algorithm to generate energy-efficient clusters. Randomly generated individuals constitute the initial population, GA is used to select CHs. One point crossover operation is applied on selected individuals and mutated operation is applied. Fitness function includes the total transmission distance (Total Distance) to sink, Cluster Distance (RCSD), Transfer Energy (E) as fitness parameters.

Bayrakli and Erdogan [12] proposed a method named GABEEC to optimize the lifetime of WSNs by means of rounds. The method consists of two phases: setup and steady-state phase just like LEACH protocol. In setup phase clusters are created and clusters will remain fixed with dynamically changing CHs for all rounds in future. The fitness function has three parameters: The round in which first sensor node is R_{FND}, the round in which last sensor node dies is L_{FND} and cluster distance is (C). The results are compared with LEACH and the proposed method performed better than LEACH. Kulia et al. [13] focused on a new evolutionary approach based on GA for load balanced clustering in WSNs. The fitness function is built with the objective to balance the load of each gateway(i.e., load balancing a Cluster Head) and for load distributing among all the gateways. The fitness function is calculated based on the standard deviation of the load of the cluster, in order to distribute the load evenly per cluster in the network. Barekatain et al. [14] proposed a collaborative approach based on k-means and improved GA to minimize the energy dissipation and for lengthening the lifetime of the network. In this work, the improved genetic algorithm is adapted to search for an optimal number of CHs and k-means algorithm is used to balance energy distribution. K-means methodology is included to overcome the limitation of GA of converging slowly for a large amount of data and getting stuck in local optimum solution in spite of finding the global optimum. Simulation result has shown that algorithm has prolonged the network lifetime by performing better than protocols like GABEEC, LEACH.

4.2 Clustering of WSNs Using Particle Swarm Optimization (PSO)

PSO is a well-known swarm-based technique inspired by the flocking behavior of the birds and is devised in 1995 by Kennedy and Eberhart. This technique is widely employed to solve various optimization problems. The set of candidate solutions to a given optimization problem constitutes a swarm. In a swarm, all particles contribute to a potential solution. Each P_i particle has velocity $G_{i,n}$ and $Z_{i,n}$, where n denotes the dimension of the search space. PSO optimizes a problem using several iterations, attempting to improve a particle solution. A fitness function evaluates the fitness of each particle. This algorithm aims to search for those positions of the particles, that would result in the finest evaluation of the fitness function. Initially, every particle in a swarm is accredited with a randomly chosen position and velocity for moving in the search space. Each particle calculate $\mathrm{pbest}_{i,n}$ that is its personal best position and also the global best position represented as gbest_n. Each particle updates its current velocity $V_{i,n}$ and position $X_{i,n}$ to achieve the best solution using the following equations:

$$G_{i,n}(j+1) = w.G_{i,n}(j) + k_1 r_1(j) * (pbest_{i,n} - Z_{i,n}) + k_2 r_2(j) * (gbest_n - Z_{i,n})$$
$$Z_{i,n}(j+1) = Z_{i,n}(j) + G_{i,n}(j+1)$$

where j denotes jth iteration, $G_{i,n}(j + 1)$ denotes particle's velocity in next iteration, $Z_{i,n}(j + 1)$ denotes particle's position in next iteration, w denotes inertia weight and k_1, k_2 are learning factors, usually $k_1, k_2 = 2$ and r_1, r_2 are random numbers between (0, 1).

4.2.1 Related Work

Latiff et al. [15] presented an energy-aware centralized clustering scheme (PSO-C) for WSN. Optimal nodes are chosen to be a CH using PSO-C. This protocol includes rounds and each round starting with setup phase to form clusters just as discussed in LEACH-C. Base selects all nodes to be eligible for CH position and then PSO is applied to find the best candidate which minimizes the cost function. Each particle in a swarm is evaluated through a cost function. The approach is to lessen the distance between CHs and cluster nodes. Singh and Lobiyal [16] has given a semi-distributed approach for energy-aware CH selection and analysis of retransmission of packets in the network. The objective is to find the optimized position of CHs and selecting optimal CHS by calculating their fitness function and is dependent on their current remaining energy, the average distance from the sensor nodes in the cluster and number of times the probable cluster heads have been CHs so far. The proposed scheme was discussed in comparison to LEACH-C and PSO-C to depict its effectiveness. Jana and Azharuddin [17] introduced a centralized, fault-tolerant PSO-based scheme. The main focus of this work was to deal with the hot spot problem arising due to multi-hop communication in WSNs. Their work consists of a routing phase and a clustering phase. In routing phase, the traffic load among CHs is distributed evenly and in clustering phase, the CHs whose energy is exhausting rapidly are allotted a few number of sensor nodes, this result in unequal clustering. The fitness function is devised based on an approximation of the lifetime of the gateways and energy dissipation for intercluster and intracluster activities. The first factor in the fitness function is considered for the efficient routing and second is for unequal clustering. This work mainly focuses on hotspots problem that the nodes which are closer to the BS die quickly because of the extra load of a large amount of data transmission to the BS. Solaiman and Sheta [18] presented a hybrid approach using K-means and particle swarm optimization algorithm named KPSO, for achieving energy management of WSNs. The proposed work was divided into three phases. Firstly, K-means algorithm is used for the partitioning of the whole network into k clusters. In the second phase, PSO algorithm is applied to find out the best CH within each cluster obtained and finally, cluster layout is evaluated.

4.3 *Clustering of WSNs Using Firefly Algorithm (FA)*

FA was first drafted at Cambridge University by Xin-She Yang [19]. It is inspired by the flashing behavior of fireflies. It is assumed that fireflies are unisex so each firefly will be attracted toward other fireflies regardless of its sex. The attractiveness is directly dependent on the brightness of the flash. With the increase in the distance between the fireflies, attractiveness decreases. The lesser bright one will always get attracted toward the brighter one.

4.3.1 Related Work

Sarma and Gopi [20] have proposed a centralized clustering algorithm for WSNs using Firefly algorithm. Best K CHs are determined that results in minimizing the cost function. A cost function is formulated as the sum of the two parameters, the first parameter is the maximum average Euclidean distance between the CHs and their cluster members and the second parameter is calculated by dividing the total initial energy of the sensor nodes in the network to the total cluster energy of the cluster heads in the current round. Simulation results have shown that using firefly algorithm to form clusters, increases the network lifetime in comparison to LEACH and SO-C. Nadeem et al. [21] presented an energy optimization method based on LEACH to increase WSN lifetime using firefly algorithm. They have defined fitness function as a fitness value which is energy consumed per round calculated fitness function using ABC algorithm. This algorithm performs better than LEACH and PSO-C (Centralized Particle Swarm Optimization) protocols.

4.4 *Clustering of WSNs Using Honey Bee Mating Optimization Algorithm (HBMO)*

HBMO is a recently developed nature-inspired swarm-based approach that resembles the mating behavior of the honey bees. The queen of the hive selects drone for mating during her flight and results into the generation of broods. Broods with more potential replace the weaker queens in the hive and the process repeats again to have potential generations.

4.4.1 Related Work

Sahoo et al. [22] presented LWTC-BMA, a trust-based and energy-aware clustering method using HBMO algorithm. Simulation analysis proved that the proposed algorithm outperformed the most popular LEACH with respect to following parameters that are, memory requirement, total network lifetime and the overhead arises due to communication. Clustering allows data aggregation and decreases the number of nodes taking part in transmission. An algorithm employs two phases, i.e., premiere phase and steady-state phase. In addition, the authors also calculated the trust value for every node to ensure that the selected CH not only have highest remaining energy but also is a trustworthy (reliable) node.

4.5 Clustering of WSNs Using Differential Evolution (DE)

DE is a robust and stochastic evolutionary technique used to calculate numerous complex optimization problems and is devised by Price and Storn. Similar to the genetic algorithm, it uses three operators: crossover, mutation, and selection. This method which utilizes NP d-dimensional vectors $X_{i,G}$ parameter where $i = 1, 2, 3, \ldots$ NP represents the population of generation G. The robustness and the effectiveness of the DE algorithm depend on three control parameter that are size of the population denoted by NP, amplification factor (F), and the crossover rate (Cr). Initially, population vectors are randomly chosen covering the entire parameter space. After initialization, three population vectors are randomly chosen to proceed with the algorithm. A mutated vector is produced by calculating the scaled difference of two randomly selected population vectors and adding it to the third randomly selected vector. This is known as Mutation. Then mutation vector's components are merged with the components of a preselected vector called target vector, to generate a new vector called trial vector. This process of blending the parameters is known as "Crossover". Finally, in the selection operation, if the fitness value of the trial vector comes out to be lesser than that of the target vector, then target vector will be replaced by the trial vector in the succeeding iteration.

4.5.1 Related Work

Kulia and Jana [19] have proposed an approach for clustering wireless sensor networks based on DE algorithm. Their work includes an additional phase named local improvement in comparison to the traditional DE. The idea is to achieve faster convergence and better performance. The main objective of this work is to balance the lifetime of the gateways. The principle followed to achieve this objective is that the gateways with less remaining energy should have minimum energy consumption as compared to those with higher residual energy. The fitness function computed for each individual considers two parameters, i.e., the standard deviation of lifetimes of the CHs and standard deviation of average cluster distance. The quality of newly produced trial vector is improved through local improvement phase. The algorithm performed better in comparison to basic DE and GA.

Potthuri et al. [23] have proposed an algorithm DESA, which aims at maximizing the network lifetime of the WSN by searching optimal cluster heads.

The proposed work makes use of DE for local search and SA for global optimal solutions. DESA consists of four phases that are initialization of the population, applying mutation operation followed by crossover operation, and at the end selection for next generation. The population is randomly initialized and further opposite point method is applied to generate another set of population called opposite population. From opposite population set '*n*' fittest individuals are selected for current generation. Mutation Strategy to be chosen depends on the value of random number chosen, if the random number is more than the threshold value it performs DE/rand/1 else it performs DE/current-to-best/1. For crossover operator, blending rate is used using the Gaussian Distribution. Selection phase uses Simulated Annealing Algorithm to select the fittest offspring to be included in next generation. The fitness function is to get best set of population vectors to be included in next generation. Hence, it is considered as a function of the distance of the node from their cluster head and the sensor node's energy. After finding the optimal cluster head and the group of the sensor nodes, the clusters are formed by cluster head distribution method. The results outperformed LEACH, conventional DE, HSA, and MHSA method.

Another algorithm named S-DE (Switching-Differential Algorithm) was proposed by Gaur and Kumar [24], to do clustering and select cluster heads using switching technique clubbed with differential evolution algorithm to improve the lifetime of network and to reduce its complexity. In this paper, switching technique is used to do CH election by switching the Cluster Head role from the node with highest remaining energy in one round to the other cluster node with highest remaining energy in next round. Fitness function to select the fittest individual is comprised of three parameters Remaining Energy (RE), Distance from the Cluster head to the Base Station (BSDist) and total intracluster communication distance (ICDist). According to the experimental studies done in recent few years, EA has performed very well for optimization problem as compared to other Evolutionary Algorithm (Table 2).

5 Performance Comparison and Analysis

The performance comparison of different protocols named LEACH, CHEF, LEACH-ERE, ECAFG is done with respect to the lifetime of the network. Table 3 shows the performance comparison with respect to a number of rounds until first node dead (FND). Table 4 presents the performance comparison on the basis of rounds until half node dead (HND). Table 5 presents the performance comparison on the basis of rounds until last node dead (LND). Different scenarios of different set of nodes 200, 250, 300, 350, and 400 [25]. Table 6 highlights the comparison of different protocols like DE, DECA, GA, LBC, EELBCA, and GLBCA for first gateway dead (FGD). FGD is also defined as a number of rounds from the beginning of the network until the first gateway runs out of its energy [19]. These performance comparisons of the different clustering protocols and algorithms have

Table 2 Performance analysis of evolutionary-based clustering algorithms

Evolutionary algorithm	Proposed method	Optimization criterion	Fitness parameters	Simulation tool	Performance evaluation
Genetic algorithm (GA)	GABEEC (Distributed)	By means of rounds, the network lifetime is lengthen.	Total transmission distance, cluster distance, transfer energy	MATLAB	Efficient than LEACH protocol
	LEACH-GA	Use optimal thresholding probability for cluster formation	Depending on the total energy consumption during data communication	MATLAB	Outperforms MTE, DT, and LEACH
	GAECH	To increase the first node die, last node die, half node die	Total energy consumption for single data collection round, CH dispersion, CHs energy consumption, standard deviation in energy consumption between clusters	MATLAB	Performed better than GCA, EAERP and LEACH
Particle swarm optimization (PSO)	PSO-C (Centralized)	To minimize energy consumption of network and to minimize the intra cluster distance	Maximum average Euclidean distance, ratio of sum of initial energies of the nodes to total present energy of CHs	MATLAB	Improved network lifetime in comparison to LEACH, GFCM, PSO-MV, CHEF
	KPSO	Hybrid clustering to achieve efficient energy management	Calculating Euclidean distance between nodes and associated cluster heads	MATLAB	49% betterment over the LEACH and 18% in comparison to the K-means clustering algorithm
	E-OEERP	Prevention of residual nodes in wireless sensor Networks and constructing optimal routing path using GSA	Distance between nodes and their associative CH number of nodes reachable from the cluster head	NS-2.32	Better performance in terms of lifetime

(continued)

Table 2 (continued)

Evolutionary algorithm	Proposed method	Optimization criterion	Fitness parameters	Simulation tool	Performance evaluation
Honeybee mating optimization (HBMO)	LWTC-BMA	To lengthen the lifespan of network by avoiding malicious nodes to become a cluster head	Remaining energy, direct trust of a node over expected CHs, distance between expected CHS and BS	MATLAB	Outperformed LEACH and TBCMA
Differential evolution (DE)	DECA	Preventing early death of highly loaded cluster heads	Deviation of the lifetime of CHs and Deviation of the average cluster distance	MATLAB	Outperforms traditional DE, GA, GLBCA and LBC
	S-DE	Switching technique applied to increase the life time	Remaining energy (RE), BSDist, ICDist	MATLAB	Performs better than GA
Fuzzy clustering	ECAFG	Uses static clustering of nodes by employing FCM and cluster head selection using genetic fuzzy system	Remaining energy of the sensor node, distance between the BS and the node, Distance of a node from their cluster center	MATLAB	Significantly lessen the energy consumption and prolonged the network lifetime
Global-simulated annealing genetic algorithm [26]	GSAGA	Formation of clusters for routing in WSNs that are energy efficient	Average Euclidean distance between sensor nodes and their respective cluster heads, ratio of the total initial energy of all the sensor nodes to the total current energy of CH node at present	NS2	Higher efficiency and better data delivery at the BS, increasing network lifetime

Table 3 Performance comparison on the basis of network lifetime (#Rounds until FND)

Protocols				
# of nodes	LEACH (#Rounds until FND)	CHEF	LEACH-ERE	ECAFG
200	1210	1320	1340	1950
250	1250	1420	1450	1930
300	1310	1440	1480	1920
350	1300	1450	1530	1910
400	1260	1320	1410	1905

Table 4 Performance comparison on the basis of network lifetime (#Rounds until HND)

Protocols				
# of nodes	LEACH	CHEF	LEACH-ERE	ECAFG
200	1500	1730	1750	1990
250	1500	1750	1760	1990
300	1500	1750	1790	2010
350	1470	1750	1780	2010
400	1450	1730	1760	2020

Table 5 Performance comparison on the basis of network lifetime (#Rounds until LND)

Protocols				
# of nodes	LEACH	CHEF	LEACH-ERE	ECAFG
200	1600	1780	1860	2070
250	1630	1790	1810	2050
300	1620	1810	1800	2050
350	1600	1840	1860	2090
400	1580	1800	1850	2060

Table 6 Performance comparison on the basis of network lifetime (#Rounds until FGD)

Protocols						
# of nodes	DECA	GA	DE	EELBCA	LBC	GLBCA
100	2700	2430	2380	2250	2170	2020
200	1570	1500	1480	1300	1290	1270
300	1290	1170	1190	1100	1000	970
400	1070	900	870	800	730	720

shown that the performance of the algorithms based on metaheuristic approaches is better than the traditional hierarchical protocols and above that the working of hybrid clustering algorithms is better than single approach algorithms in terms of network lifetime.

6 Conclusion

From the performance analysis of the existing clustering algorithms, we have concluded that the performance of the nature-inspired evolutionary algorithms-based clustering algorithms is superior to traditional clustering algorithms in terms of energy efficiency and network lifespan. Among these evolutionary algorithms, Differential Evolution outperforms other EA due to its effectiveness, efficiency, and robustness. Our next research work will concentrate on improving the performance of traditional DE by using the different mutation strategies or using the concept of self-adaptive nature of control parameters to design new energy-efficient clustering algorithms.

References

1. Yick, J., Mukherjee, B., & Ghosal, D. (2008). Wireless sensor network survey. *Computer Networks, 52*(12), 2292.
2. Kumar, S. N. (2014). A new approach for traffic management in wireless multimedia sensor network. *International Transaction of Electrical and Computer Engineers System, 2*(5), 128–134.
3. Heinzelman, W. B., Chandrakasan, A. P., & Balakrishnan, H. (2000). Energy efficient communication protocol for wireless microsensor networks. In *Proceedings of the 33rd annual Hawaii international conference on system sciences, 2000* (Vol. 2, pp. 3005–3014).
4. Heinzelman, W. B., Chandrakasan, A. P., & Balakrishnan, H. (2002). An application-specific protocol architecture for wireless microsensor networks. *IEEE Transactions on Wireless Communications, 1*(4), 660.
5. Khediri, S. E., Nasri, N., Wei, A., & Kachouri, A. (2014). A new approach for clustering in wireless sensors networks based on LEACH international workshop on wireless networks and energy saving techniques (WNTEST). *Procedia Computer Science, 32*(2014), 1180–1185.
6. Bandyopadhyay, S., & Coyle, E. J. (2003, April). An energy efficient hierarchical clustering algorithm for wireless sensor networks. In *INFOCOM 2003. Twenty-Second Annual Joint Conference of the IEEE Computer and Communications. IEEE Societies* (Vol. 3, pp. 1713–1723). IEEE.
7. Younis, O., & Fahmy, S. (2004). HEED: A hybrid, energy-efficient, distributed clustering approach for ad hoc sensor networks. *IEEE Transactions on Mobile Computing, 3*(4), 366–379.
8. Lindsey, S., & Raghavendra, C. S. (2002). PEGASIS: Power-efficient gathering in sensor information systems. In *Aerospace conference proceedings, IEEE* (Vol. 3, pp. 3–3). IEEE.
9. Khanna, R., Liu, H., & Chen, H. H. (2006). Self-organisation of sensor networks using genetic algorithms. *International Journal of Sensor Networks, 1*(3–4), 241–252.
10. Hussain, S., Matin, A. W., & Islam, O. (2007, April). Genetic algorithm for energy efficient clusters in wireless sensor networks. In *ITNG '07. Fourth International Conference on information Technology, 2007* (pp. 147–154). IEEE.
11. Heidari, E., & Movaghar, A. (2011, March). An efficient method based on genetic algorithms to solve sensor network optimization problem. *International Journal on Applications of Graph Theory in Wireless Ad Hoc Networks and Sensor Networks (GRAPH-HOC),* 3(1).
12. Bayraklı, S., & Erdogan, S. Z. (2012). Genetic algorithm based energy efficient clusters (gabeec) in wireless sensor networks. *Procedia Computer Science, 10*, 247–254, Conference on Ambient Systems, Networks and Technologies (ANT).

13. Kuila, P., Gupta, S. K., & Jana, P. K. (2013). A novel evolutionary approach for load balanced clustering problem for wireless sensor networks. *Swarm and Evolutionary Computation, 12,* 48–56.
14. Barekatain, B., Dehghani, S., & Pourzaferani, M. (2015). An energy-aware routing protocol for wireless sensor networks based on new combination of genetic algorithm & k-means. *Procedia Computer Science, 72,* 552–560.
15. Latiff, N. A., Tsimenidis, C. C., & Sharif, B. S. (2007, September). Energy-aware clustering for wireless sensor networks using particle swarm optimization. In *PIMRC 2007. IEEE 18th International Symposium on Personal, Indoor and Mobile Radio Communications, 2007* (pp. 1–5). IEEE.
16. Singh, B., & Lobiyal, D. K. (2012). A novel energy-aware cluster head selection based on particle swarm optimization for wireless sensor networks. *Human-Centric Computing and Information Sciences, 2*(1), 13.
17. Azharuddin, M., & Jana, P. K. (2016). Particle swarm optimization for maximizing lifetime of wireless sensor networks. *Computers & Electrical Engineering, 51,* 26–42.
18. Solaiman, B. (2016). Energy optimization in wireless sensor networks using a hybrid k-means PSO clustering algorithm. *Turkish Journal of Electrical Engineering & Computer Sciences, 24*(4), 2679–2695.
19. Kuila, P., & Jana, P. K. (2014). A novel differential evolution based clustering algorithm for wireless sensor networks. *Applied Soft Computing, 25,* 414–425.
20. Sarma, N. V. S. N., & Gopi, M. (2014). Implementation of energy efficient clustering using firefly algorithm in wireless sensor networks. *International Proceedings of Computer Science and Information Technology, 59,* 1.
21. Nadeem, A., Shankar, T., Sharma, R. K., & Roy, S. K. (2016). An application of firefly algorithm for clustering in wireless sensor networks. In *Proceedings of the International Conference on Recent Cognizance in Wireless Communication & Image Processing* (pp. 869–878). Springer India.
22. Sahoo, R. R., Singh, M., Sahoo, B. M., Majumder, K., Ray, S., & Sarkar, S. K. (2013). A light weight trust based secure and energy efficient clustering in wireless sensor network: Honey bee mating intelligence approach. *Procedia Technology, 10,* 515–523.
23. Potthuri, S., Shankar, T., & Rajesh, A. (2016). Lifetime improvement in wireless sensor networks using hybrid differential evolution and simulated annealing (DESA). *Ain Shams Engineering Journal*, 6 March 2016.
24. Gaur, A., & Kumar, T. (2016). *Switching-differential evolution (S-DE) for cluster head election in wireless sensor network*, IJARIIE-ISSN(O)-2395-4396 (Vol. 2 Issue 5).
25. Shokrollahi, A., & Mazloom-Nezhad Maybodi, B. (2017). An energy-efficient clustering algorithm using fuzzy C-means and genetic fuzzy system for wireless sensor network. *Journal of Circuits, Systems and Computers, 26*(01), 1750004.
26. Zhang, J., Lin, Y., Zhou, C., & Ouyang, J. (2008, December). Optimal model for energy-efficient clustering in wireless sensor networks using global simulated annealing genetic algorithm. In *IITAW '08. International Symposium on intelligent information technology application workshops, 2008* (pp. 656–660). IEEE.

Process Mining for Maintenance Decision Support

Adithya Thaduri, Stephen Mayowa Famurewa, Ajit Kumar Verma and Uday Kumar

Abstract In carrying out maintenance actions, there are several processes running simultaneously among different assets, stakeholders, and resources. Due to the complexity of maintenance process in general, there will be several bottlenecks for carrying out actions that lead to reduction in maintenance efficiency, increase in unnecessary costs and a hindrance to operations. One of the tools that is emerging to solve the above issues is the use Process Mining tools and models. Process mining is attaining significance for solving specific problems related to process such as classification, clustering, discovery of process, prediction of bottlenecks, developing of process workflow, etc. The main objective of this paper is to utilize the concept of process mining to map and comprehend a set of maintenance reports mainly repair or replacement from some lines on the Swedish railway network. To attain the above objective, the reports were processed to extract out time related maintenance parameters such as administrative, logistic and repair times. Bottlenecks are identified in the maintenance process and this information will be useful for maintenance service providers, infrastructure managers, asset owners and other stakeholders for improvement and maintenance effectiveness.

Keywords Process mining · Maintenance · Inductive visual miner
Decision support structure

A. Thaduri (✉) · S. M. Famurewa · U. Kumar
Luleå University of Technology, Luleå, Sweden
e-mail: adithya.thaduri@ltu.se

S. M. Famurewa
e-mail: stephen.famurewa@ltu.se

U. Kumar
e-mail: uday.kumar@ltu.se

A. K. Verma
Western Norway University of Applied Sciences, Haugesund, Norway
e-mail: AjitKumar.Verma@hvl.no

P. K. Kapur et al. (eds.), *System Performance and Management Analytics*,
Asset Analytics, https://doi.org/10.1007/978-981-10-7323-6_23

1 Introduction

Process mining is a newly developing research area that combines the process modeling, data mining, analysis and computational intelligence to determine the process effectiveness. The core aspects of process mining are to "*discover, monitor and improve real processes by extracting knowledge from event logs*" accessible in today's systems. Process mining comprises of process discovery (i.e., mining process models from an event log), conformance checking (i.e., monitoring deviations by comparing model and log), social network/organizational mining, automated simulation modeling, case prediction, and history-based recommendations [1, 2].

Process mining can be implemented as shown in Fig. 1. By using the software system for data acquisition/condition monitoring, one needs to extract the process from the real world such as business process, industrial process, customer's interactions, etc. This process is recorded in the event logs, transactions, messages, etc. From the event logs, various types of methods are available to develop the process model. These are as follows:

Discovery: This technique generates a process model from the generated event log from the process systems. For many organizations and industries, the existing technologies can discover real processes from event logs based on simulations at the various stages of a process.

C*onformance*: In this step, the developed process model from Discovery is evaluated with a new set of event logs generated from the same process for conformance model's accuracy. The different types of models of conformance checking are procedural, organizational, and declarative models that can be operated in business rules/policies, laws, etc.

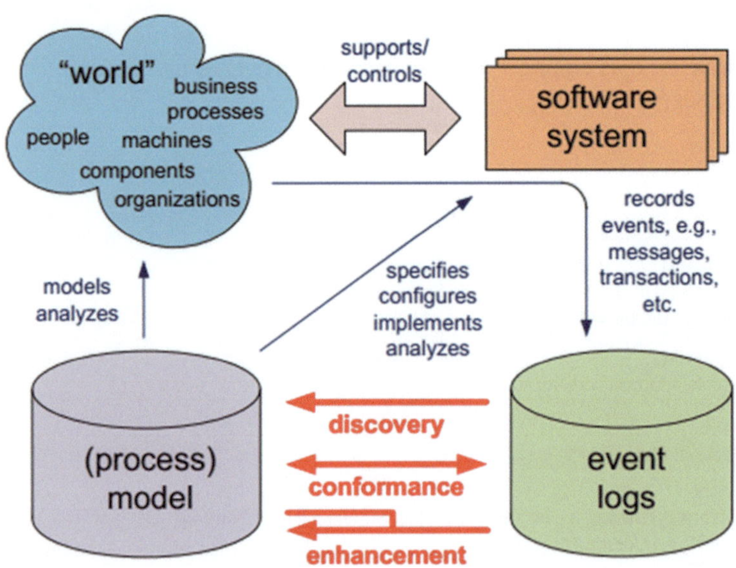

Fig. 1 Three main groups of process mining: **a** process discovery, **b** process conformance checking, and **c** process enhancement [3]

Enhancement: This step is to amend or transform the previously developed process model using evidence obtained from the real process logged in event log. The main difference between conformance checking and enhancement is that the former measures the alignment between model and reality, whereas later aims at modifying or advancing the priori developed model.

Figure 2 defines the three groups of process mining methods. Some of the examples of discovered models are Business Process Model and Notation (BPMN), Event-driven Process Chain (EPC), Petri net or Unified Modeling Language (UML) activity diagram. Conformance checking techniques are meant for diagnostic purposes. Enhanced techniques are needed there might be a change in the behavior of the process [3] (Fig. 3).

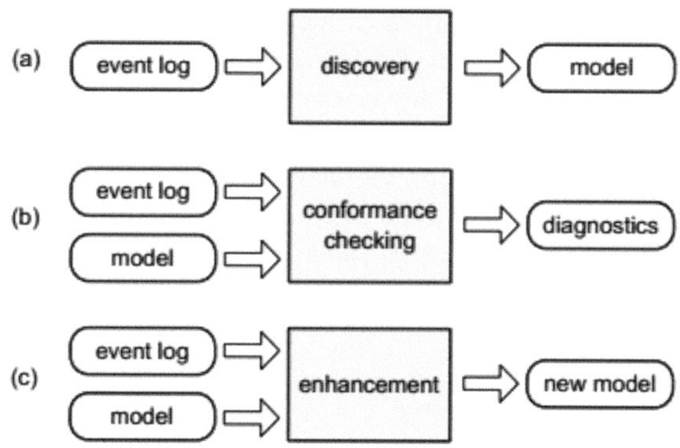

Fig. 2 The three kinds of process mining **a** process discovery, **b** process conformance checking, and **c** process enhancement [3]

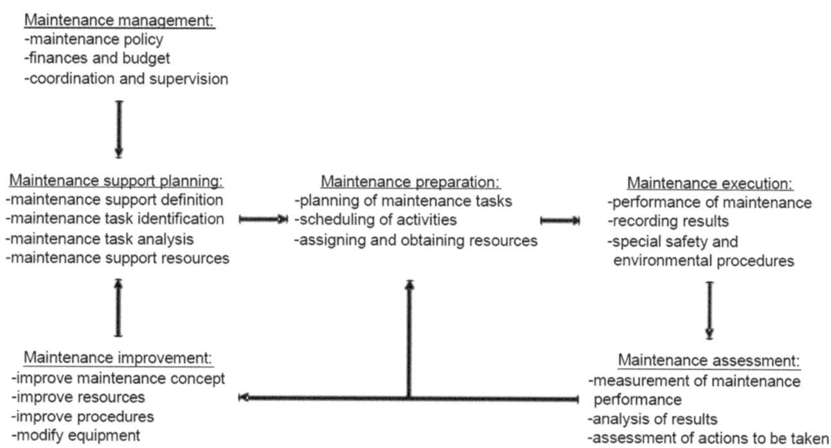

Fig. 3 Maintenance process for the railway industry

2 Data Mining Versus Process Mining

Data mining and process mining have a lot in common. Both techniques are part of business intelligence, viz., the analysis of large volumes of data to achieve greater insights. Both approach things in a similar way. Both data mining and process mining apply specific algorithms to data to uncover hidden patterns and relationships. The goal of data mining and process mining is to provide insight and to let users come to better decisions.

We use data mining to analyze data and to detect or predict patterns. For example, which target groups buy which products, where does my marketing campaign have the greatest effect, etc. Data mining has no direct link with business processes, as opposed to process mining. The latter focuses on discovering, controlling and improving actual business processes. By analyzing data derived from the IT systems that support our processes, process mining gives us a true, end-to-end view of how business processes operate.

Data mining analyzes static information. In other words, data that is available at the time of analysis. Process mining on the other hand looks at how the data was created. Process mining techniques also allow users to generate processes dynamically based on the most recent data. Process mining can even provide a real-time view of business processes through a live feed [4].

3 Process Mining Problems

Since this area is relatively new, there are few suboptimal current methods. To get the best out of this concept, it is essential to have accessible, sufficient, and relevant data that can provide a high payoff for the right decisions. To make even best use, it is also required to have data with changing environment. The main issues related to business process mining are [5]:

Noise: Noise can be generated by the incorrect or incomplete data and stored without data cleaning.

Hidden tasks: Tasks that not logged by the user but still affect the process.

Duplicate tasks: Two or three process resembles the same process.

Mining loops: A process may have loops that iterate continuously without exiting the step.

Different perspectives: Process events recorded from different personnel lead to different process models.

Visualizing results: It needs to be understood by the top management.

Heterogeneous results: Information extracted from multiple sources on different platforms.

Concurrent processes: Process that is occurring simultaneously.

Local/global search: Local process model is simple but needs a clear understanding of neighboring process where global process model is complex but shows a holistic picture of the process to find the best optimal solution.

Process re-discovery: The process algorithm rediscovers process models with the complete event log.

4 Process Mining Metrics

There are several metrics or features required as an input to the process mining. These are differentiated based on perspectives like process or control, resource, and operational perspective.

Process/control-flow perspective: Flow time, waiting time, processing time and synchronization time.

Resource perspective: Frequencies, time, utilization, and variability.

Operations perspective: Queue length, arrival rate, departure rate, dwell/sojourn time.

5 Other performance characteristics

5.1 Applications of Process Mining

The application of the process mining can be segregated per the way of types [3]:

5.2 Applications of Process Discovery

The procedures that handle the cases are compulsory by the information system to extract the knowledge of the process. These procedures are informal, and hence, it is not properly recorded. Even though they are recorded, they are of poor quality and not well represent the real process. Thus, it is quite significant to know the underlying process to discover the bottlenecks

for discoursing problems in the process among various investors,

for improving the process with new techniques and refining them,

for enhancing the model,

for rearranging a system.

5.3 Applications of Conformance Checking

These techniques compare the detected behavior with the developed model by relating the events that are mapped to change over dismissals in the Petri net. This can be achieved by quantifying and diagnosing deviations. This technique can be used for the following:

1. for verifying the confidence of developed processes,
2. for recognizing the cases that are deviated in the process by understanding through other cases,
3. for identifying process subparts with most deviations,
4. for auditing the financial process,
5. for qualifying the discovered process model,
6. for supervising evolutionary process discovery algorithms like genetic algorithms, machine learning, and other computational intelligence algorithms.

Conformance checking mainly can be used for evaluating a process discovery algorithm. The auditors, further, the above information need to be validated by assessing whether to execute business processes execute within certain restrictions established by different stakeholders.

6 Applications of Enhancement

Due to changes in the contextual parameters, it is necessary to implement computational intelligence tools for effectiveness [6]. Hence, it is necessary to enhance or improve the developed models for the event logs based on the application. This can be achieved by diagnosing the aberrations from the event log and fed to the model. It is possible to analyze different timings between activities and calculate statistics such as mean, averages, standard deviation, variances, and confidence intervals for recognizing the bottlenecks. Here, the typical clustering techniques can be utilized for the following:

for constructing social networks workflow and analyzing resource performance,
for analyzing the decision points by traditional classification techniques,
for constructing a decision tree elucidating the observed behavior
for online predictions and recommendations.

The applications of process mining are mostly like the data mining. They are categorized as follows:

Process discovery,
Social network/organizational mining,
Classification, clustering, estimation, visualization,
Automated building of simulation models,
Case prediction, and history-based suggestions,
Customer relationship management,
Job planning and scheduling,
Challenges in process mining.

There is an increase in demand for the growth of data acquisition and collection of event logs. To cater the needs of the business goals, there is a need for processing, analyzing and extracting information to meet requirements related to acquiescence, efficacy, and customer satisfaction. The application of process mining methods is in nascent stages and there are inherent challenges that need to be addressed. The main challenges to be addressed for process mining are [1–3] as follows:

Data acquisition and cleaning: Data is extracted from disparate data sources and event data needs to be generated but it may contain incomplete, missing and uncleaned that requires supervision.

Complex event logs with various characteristics: Event logs may consist of different characteristics. There might be complexity in the type of event logs; large data with several event logs, unnecessary data, data from complex systems, etc., which is difficult to manage.

Demonstrative benchmarks: For conformance and enhancement of process models, it is customary to get sample data sets for calibration and there might exist quality issues which make the improvement disarrayed.

Concept drift: The process, in meantime, can change due to various internal and external process. Understanding and analyzing those processes need major concentration for business decisions.

Representational bias: The modeling might tend to representational bias and care should be taken to ensure the high quality.

Quality Control: Due to the complexity of the process, quality factors such as fitness, simplicity, precision, and generalization are needed to be considered for processing. The main challenge is to find models that are best optimized for above quality factors.

Cross-organization: In most of the infrastructures and process industries, often, data comes from different sources not also within the organization but also from multiple outside sources. Due to the managerial collaborations, some organizations work together to share the knowledge of event logs but for some organizations, it is quite difficult to manage due to data security issues.

Operational support: Process mining can be useful for both offline and online support. The main operational support activities are detection, prediction, and recommendation. During online support, the processing of above activities is a challenge to consider the dynamic environment.

Process mining with data mining: There are several challenges at various stages to combine the process mining techniques with other existing analysis approaches in the data mining such as optimization techniques, data mining, simulation, visual analytics, etc., to extract useful insights from event data.

Usability: The developed models and visualization needs to be more intuitive and user friendly for the decision support.

Awareness: If the process mining is not understood properly, this might provide incorrect conclusions and wrong decisions. Hence, the results to be presented with clear indications.

7 Maintenance Process

Process mining was studied on few applications with respect to maintenance [7], especially for knowledge maintenance. Dongen [8] reported that process models is laborious for installation of the systems, and maintaining them was beneficial for identifying to extract useful information. Buijs et al. [9] also emphasized on aggregation of the maintenance data for effective planning and scheduling. Van Der Aalst et al. [10] applied the process mining to analyze and improve the flexibility of the process. Karray et al. [11] studied the transient changes in the process and services on a maintenance platform.

The maintenance of infrastructure is an inclusive process of repair, replacement, and renewal process to meet the demands of the infrastructure owner/manager such as higher asset availability, safer, and better quality and reduced cost. To attain these demands, maintenance strategies are essential at different hierarchical levels of organization. Table 1 encapsulates the accounts of maintenance process from different perspectives. The improvement in maintenance process can be achieved by adapting the issues with advanced techniques, models, methods, methodologies. The key issues highlighted are the allocation of resources to work orders, planning, and scheduling of different maintenance actions, control, and organization process in maintenance. Hence, the major factors that will increase the improvement in performance of railway infrastructure are maintenance execution process, maintenance need analysis, and maintenance planning and scheduling [12].

The maintenance execution can be further elaborated in the Fig. 4. First, the maintenance process can be controlled and assessed by the observation and inspection of the activities. These activities are reported for any failures or maintenance actions of the ongoing operation. The activities can be categorized as follows:

Administrative Activities,
Logistics Activities,
Active Repair/Replacement Activities.

Table 1 Maintenance process from different perspectives

Generic maintenance process	Maintenance process in railway industry	Maintenance process in Trafikverket
Maintenance budgeting Setting maintenance objectives Formulating strategy Establishing responsibilities	Budget determination	Budget allocation Identifying objectives from regulation and white paper Establishing strategy from existing handbook Contract procurement
Planning	Long-term quality prediction and diagnosis Project prioritization and selection Project identification and definition	Condition assessment Maintenance need analysis
Scheduling	Possession allocation and timetabling of track possession	Track possession schedule
Execution	Implementation	Execution
Assessment	Work evaluation	Assessment and verification
Improvement	Feedback loop	Follow up of contract

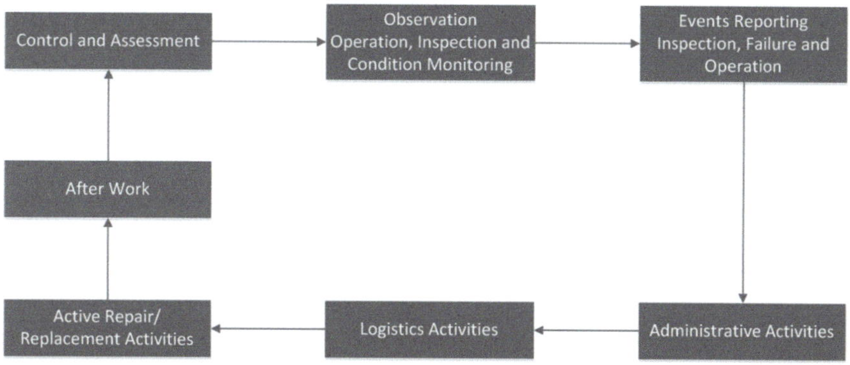

Fig. 4 Maintenance execution process

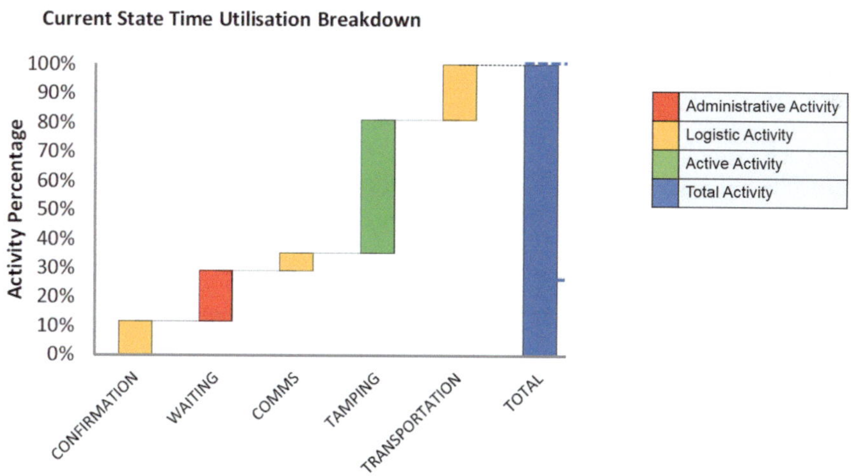

Fig. 5 Example of maintenance action in maintenance execution

An example of the above activities is shown in Fig. 5. These are delays that are recorded by the maintenance personnel. If we consider the maintenance process in Fig. 3, the concept of process mining can be well fitted into the maintenance assessment by recording the maintenance execution activities.

8 Inductive Visual Miner

There are several miners developed for process mining to retrieve valuable information from event logs from the process and Inductive visual Miner (IvM) is one of them [13, 14]. The architecture of IvM represents the binding of event analysis and

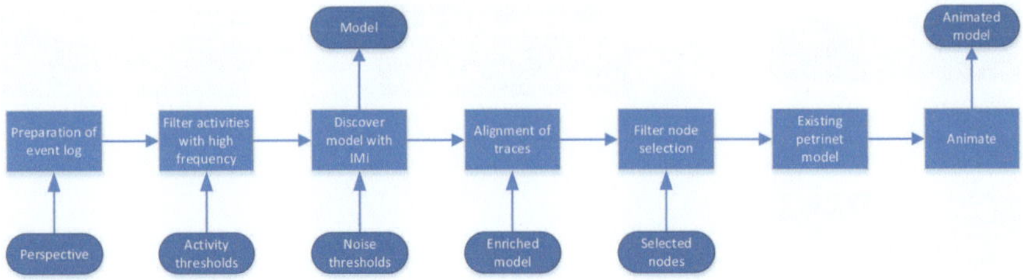

Fig. 6 Inductive visual miner tasks [15]

visualization, shown in Fig. 6. The important feature of IvM is that user can change the different parameters online for finding variations without the need for restarting the chain. Though the animation might take some time to rebind, the Miner makes sure that visual effects can be shown to the user until finishing the next task.

While preparing the log task, the events logs are classified by one of the classifiers called perspective classifier. After classifying, the events with high frequency of activities, task is filtered. To discover the task, Inductive Miner-infrequent (IMi) algorithm is applied. This algorithm produces a process tree by noise filtering approach with thresholds for most frequent paths. Alignments of the tasks are carried to align the best matching runs to observe the deviations within the process to develop an enriched model. The filter node selection filters take selected nodes to align the traces accordingly. The last task is mainly for visualization for animation by computation of aligned traces to show quick preview. This model can be used as an input to the existing developed model to carry out animations. A user can change the different parameters for fine-tuning to make analysis to find bottlenecks to improve the process [15].

9 Case Study

The Iron Ore Line (Malmbanan) in northern Sweden starts in Luleå and ends in Narvik in Norway (see Fig. 7). The traffic on the line consists of both passenger and freight trains. The freight traffic in this line consists of heavy haul trains with axle load of 30 MGT and annual traffic volume more than 25MGT. The trains normally operate in very severe climate conditions with mountains, high snow and long winter season with extreme temperatures ranging from −45 °C to +25 °C that can lead to accelerated wear and damage rate [16].

The Swedish iron-ore mining company LKAB is the main freight operator on this line, transporting iron-ore with an axle load of 30 tonnes and at speed of 60 km/h. In this study, the concept of maintenance process mining was applied on Iron Ore line shown in Fig. 7. The route has a total length of 400 Km. 3 main track sections namely; 111, 113, and 118. The number of work orders or maintenance reports on the three sections for the year 2012 was 1656. The analysis has been done with

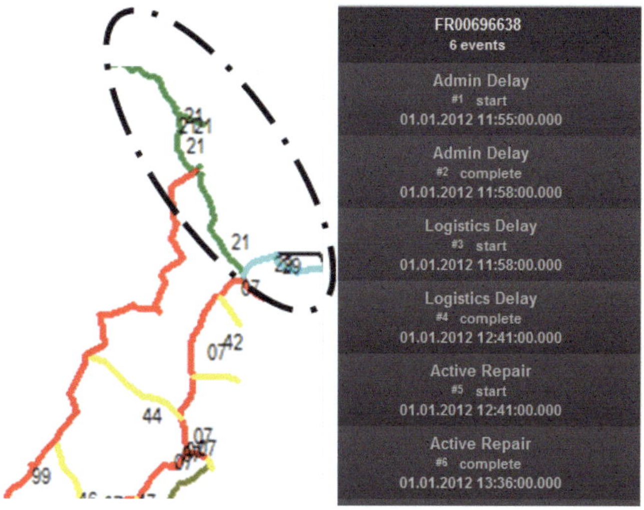

Fig. 7 Northern part of Sweden

different complexities and varies three factors are considered for investigating the bottleneck; traffic disruption, traffic section, and asset system.

10 Results and Discussion

Using ProM tool [17], the Inductive visual Miner was applied on the Maintenance Execution part. The developed Petri net model is shown in Fig. 8.

By looking into Fig. 8, it was concluded from the process mining that the active repair replacement took a lot of time and also queue lengths. This step can be considered as a bottleneck in this process. If we are interested, the work orders are

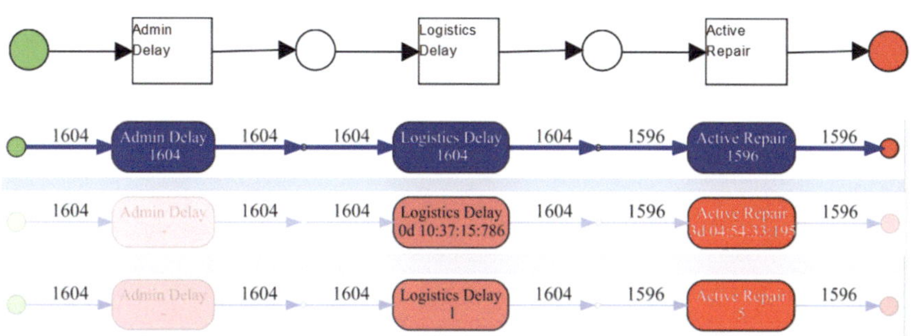

Fig. 8 Petrinet for whole track section

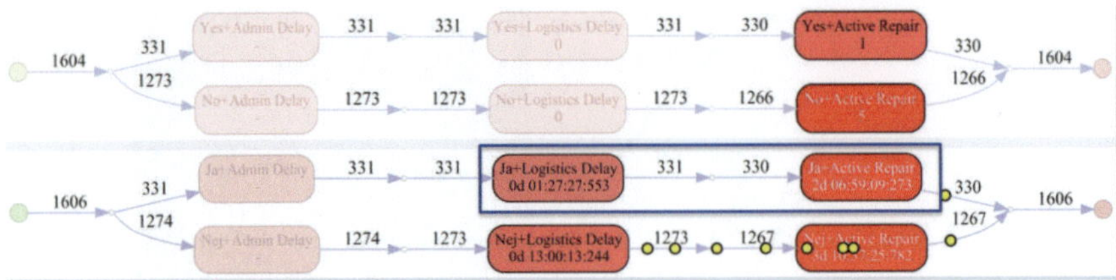

Fig. 9 IvM with traffic disruption

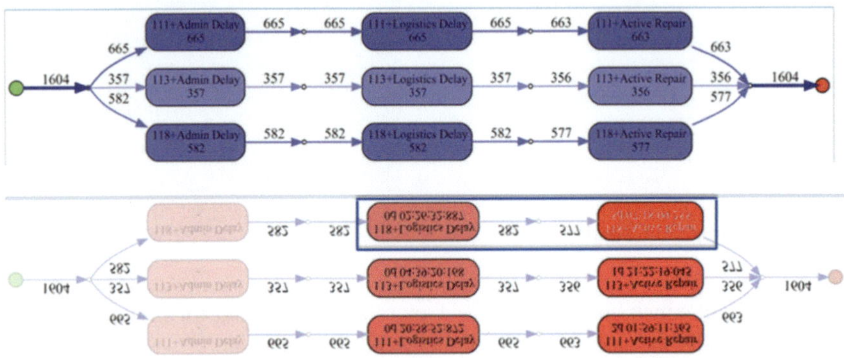

Fig. 10 IvM for different track sections

due to the traffic disruption as illustrated in Fig. 9. This part we are further investigated in the below figures.

Figure 10 implied that Section 118 has large number of maintenance actions during this period. It is also interesting to note that the length of the track section is also large. To filter further, the information from Fig. 10 can be classified based on the traffic disruption; yes or no shown in Fig. 11. It was concluded that the 118 section without traffic disruption was longer than the traffic disruption. It was because the priority for the maintenance engineers is mostly focused on the traffic

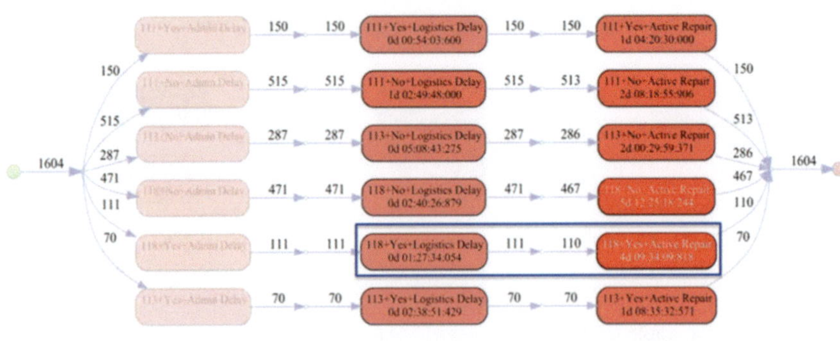

Fig. 11 IvM for 118 track section with traffic disruption

section type. Hence, we are looking to the combination of 118+ traffic disruption to find out the main asset type that was creating the bottleneck.

Figure 12 illustrates the comprehensive list of asset type that was affected due to the traffic disruption. Out of all the work orders, the track maintenance took lot of time to recover to the working condition. The critical asset types were further filtered in Fig. 13. The criticality of the asset types is chosen based on the number of work orders, total time taken to do active repair and by expert judgment. These asset types were switches and crossings, signaling system, positioning system, interlocking system, track, animals, balise and the alternative power.

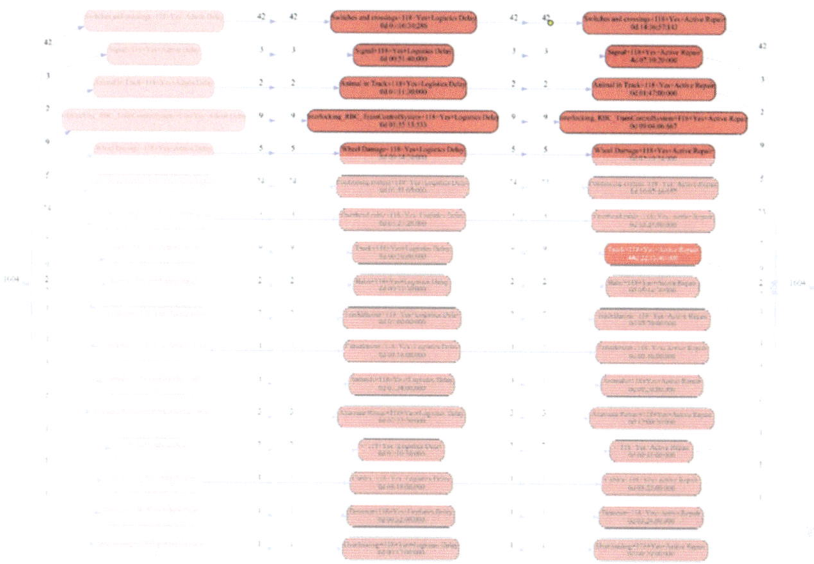

Fig. 12 IvM for 118 track section with traffic disruption for several systems

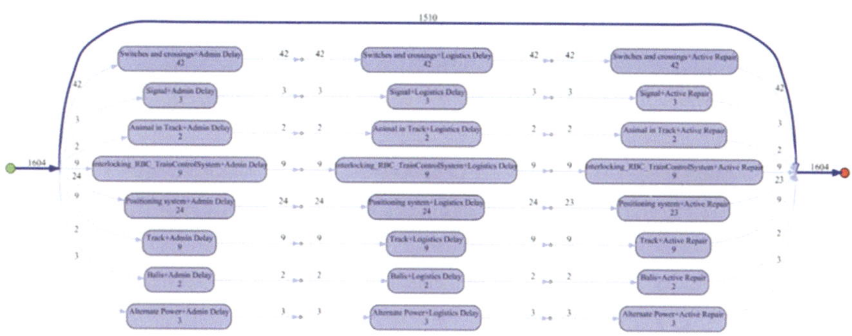

Fig. 13 IvM for 118 track section with traffic disruption for critical systems

11 Conclusion

The process mining model presented is useful for extracting knowledge from maintenance events logs. The model identifies critical maintenance activities for each track section and system in the case study. The PM model can be used for future prediction of events after conformance checking and enhancement. The PM model can be integrated with OR techniques to optimize maintenance resources by cost required on the iron ore line. Maintenance PM can be merged with lean optimization techniques to eliminate or reduce non-value-adding events in railway infrastructure maintenance.

References

1. Van der Aalst, W. (2011). Getting the data.
2. van der Aalst, W. M. P. (2011). Process mining: Overview and opportunities. *Process Mining, 5,* 301–317. https://doi.org/10.1007/978-3-642-19345-3.
3. Van Der Aalst, W., Adriansyah, A., De Medeiros, A. K. A., et al. (2012). Process mining manifesto. *Lecture notes in business information processing* (pp. 169–194). Berlin, Heidelberg: Springer.
4. Houthoodf, D. (2015). Data mining vs. process mining: what's the difference? In: Linkedin. https://www.linkedin.com/pulse/data-mining-vs-process-whats-difference-dennis-houthoofd. Accessed 13 Apr 2017.
5. Tiwari, A., Turner, C., Majeed, B., B.b M (2008). A review of business process mining: State-of-the-art and future trends. *Business Process Management Journal*, *14*, 5–22. https://doi.org/10.1108/14637150810849373.
6. Thaduri, A., Kumar, U., & Verma, A. K. (2014). Computational intelligence framework for context-aware decision making. *International Journal of System Assurance Engineering and Management, 5*, 1–12. https://doi.org/10.1007/s13198-014-0320-8 U6 - http://link.springer.com/article/10.1007/s13198-014-0320-8 M4.
7. Santos, I. H. F., Machado, M. M., Russo, E. E., et al. (2015). Big data analytics for predictive maintenance modeling: Challenges and opportunities. In: OTC Brasil. Offshore Technology Conference.
8. Van Dongen, B. F. (2005). A meta model for process mining data. In Proceedings of the CAiSE Workshops (pp 309–320).
9. Buijs, J.C.A.M., Dongen, B.F., & Aalst, W. M. P. (2012). Towards cross-organizational process mining in collections of process models and their executions. Bus Process Manag 2–13.
10. Van Der Aalst, W., Günther, C., Recker, J., & Reichert, M. (2006) Using process mining to analyze and improve process flexibility—Position paper. In: CEUR Workshop Proceedings (pp. 168–177).
11. Karray, M. H., Chebel-Morello, B., & Zerhouni, N. (2014). PETRA: Process Evolution using a TRAce-based system on a maintenance platform. *Knowledge-Based Syst, 68,* 21–39. https://doi.org/10.1016/j.knosys.2014.03.010.
12. Famurewa, S. M. (2015). Maintenance analysis and modelling for enhanced railway infrastructure capacity. Lulea University of Technology.
13. Buijs, J. C. A. M. (2014). Flexible evolutionary algorithms for mining structured process models.
14. Gottschalk, F., van Der Aalst, W. M. P., & Jansen-Vullers, M. H. (2008). Mining reference process models and their configurations. *Lect Notes Comput Sci (including Subser Lect Notes*

Artif Intell Lect Notes Bioinformatics), 5333, 263–272. https://doi.org/10.1007/978-3-540-88875-8_47.
15. Leemans, S., Fahland, D., & van der Aalst, W. (2014). Process and deviation exploration with inductive visual miner.
16. Larsson, D. (2004) A study of the track degradation process related to changes in railway traffic. Lulea University of Technology.
17. Dongen, B. F., de Medeiros, A., Verbeek, H., et al. (2005). The ProM framework: A new era in process mining tool support. *Applications and Theory of Petri Nets, 2005,* 444–454.

Software Release Time Problem Revisited

Nitin Sachdeva, P. K. Kapur and A. K. Shrivastava

Abstract With technological advancements in the Information Technology (IT) world, Software Reliability Growth Models (SRGMs) have been extensively made use of by both researchers and practitioners. To withstand the challenges posed by this exponential growth in the IT sector, researchers have propagated the need to obtain optimal software release time by optimizing overall testing cost. In this chapter, the authors suggest a novel approach to optimize release time considering the cost of fault detection and correction as distinct cost and treat them separately in the cost modeling framework. We develop testing effort-dependent SRGMs in a unified framework and thus provide for the proposed cost model validation based on real-life data.

Keywords Software reliability · Release time · Detection/correction process Unified framework

1 Introduction

Every software development organization today strives to produce high-quality software with an objective of minimum development cost and delivered in the market much sooner than its competitors. To achieve this dual objective of optimal cost and timely delivery, developers employ software reliability engineering

N. Sachdeva
Institute of Management Technology, Ghaziabad, Uttar Pradesh, India
e-mail: nitin.sach@gmail.com

P. K. Kapur
Amity Centre for Interdisciplinary Research, Amity University, 201313 Noida, Uttar Pradesh, India
e-mail: pkkapur1@gmail.com

A. K. Shrivastava (✉)
Fortune Institute of International Business, New Delhi, Delhi, India
e-mail: kavinash1987@gmail.com

P. K. Kapur et al. (eds.), *System Performance and Management Analytics*, Asset Analytics, https://doi.org/10.1007/978-981-10-7323-6_24

(SRE) and perform software testing. SRE delivers the desired product with required characteristics keeping in mind the pressing issues of reliability and timely delivery. Thereby, SRE ensures a balance between all the customer needs from obtaining a reliable software which is made available to them at a predetermined delivery time with effective life cycle cost.

Within the branch of SRE, numerous software reliability growth models (SRGMs), have been proposed with the greater thrust being kept on identifying the relation between the number of failures occurring and the execution time [1–3]. Much of the research carried out in this area considers SRGMs with diverse testing environments by distinguishing failure and correction processes, training and learning phenomenon of the testers, possibility of error generation and/or perfect/ imperfect debugging, whether the fault detection rate (FDR) is fixed, increasing/ decreasing with time, etc. A unique characteristic common to almost all these SRGMs has been that either they tend to follow exponential [4] distribution or S-shaped [5] distribution and therefore, accordingly provide a fit on several different datasets [1, 3, 6]. Lately, attempts have been made and flexible SRGMs considering both types of failure data sets have been proposed [6, 7], but no single SRGM can be treated to be the best due to numerous internal and external factors. This all tends to make the process of model selection really tough and in order to reduce this difficulty, a unified modeling framework has been a popular solution [1, 3, 8–15]. These unified models help developers with insightful investigations without making multiple assumptions about their specific testing environments. Yet another unification framework is based on the understanding of fault detection process (FDP) and fault correction process (FCP) with these FCPs being described by detection process with time delay [16]. This concept of FCP being considered as a separate process was put forward by Schneidewind [17] with an assumption of fixed time lag which was further relaxed by Xie and Zhao [18] with a time-dependent lag. Later, Xie et al. [16] suggested a more general framework in this direction with FDP being described by NHPP-based SRGMs and FCP as a delayed detection process with random delay function. Lastly, Kapur et al. [1] proposed a unification scheme based on the concept of hazard rate function. Alongside the concept of time lag exists yet another insightful concept of testing effort often applied as testing effort function (TEF) in modeling software reliability scenarios [8–12, 14, 15, 19–22]. These testing effort functions have been quite accurately described by using Logistic and Weibull TEFs [1, 3]. Kapur et al. [10] made use of learning phenomenon with testing effort in developing SRGMs leading to the birth of flexible SRGMs with testing effort-based learning process. Further, a unified framework for modeling testing effort-based SRGM was also proposed by Kapur et al. [11]. Later, Inoue et al. [23] suggested a testing effort-based lognormal SRGM. Testing effort-based SRGM in an imperfect debugging environment has also been talked about by Zhao et al. [22]. Peng et al. [13] for the very first time suggested to differentiate the detection/correction process by proposing two-staged SRGM under testing effort-based imperfect debugging environment and lately, Zhang et al. [15] with testing effort under the imperfect debugging environment demonstrated a unified framework for modeling such SRGMs. To add to the literature, S-shaped testing effort functions were proposed by Li et al. [12] for modeling SRGMs with imperfect debugging. Kapur et al. [9] proposed a generalized framework for a software upgradation model with testing effort and two types of imperfect debugging.

In this chapter, the proposed work accounts for a novel approach to modeling debugging cost of detection and correction separately with respect to testing effort expenditure based on the unified framework for SRGMs. Interestingly, we demonstrate that already existing Non-Homogeneous Poisson Process (NHPP)-based SRGMs are a good fit for our proposed modeling framework. So in Sect. 2, a testing-effort-dependent SRGM under unified framework is discussed, wherein we consider fault detection and correction processes separately. In Sect. 3, the classical release time problem is discussed by highlighting the literature and existing gap therein. In Sect. 4, we present the numerical illustration for the proposed cost model while concluding the proposed work in Sect. 5 by providing important research contributions, limitations and future course of action of this work

2 Unified Framework for Developing Testing Effort-Based Fault Detection and Correction Process

2.1 Notations

$m_d(E_t)$	Average number of faults detected by time T with Testing Effort E_t
$m_c(E_t)$	Average number of faults corrected by time T with Testing Effort E_t
a	Total number of faults
b_1/b_2	Fault detection/correction rate
$F(E_t)$	Cumulative probability distribution function for fault removal process
$f(E_t)$	Probability density function for fault removal process

2.2 Assumptions

The proposed models are based upon the following fundamental assumptions of software reliability modeling:

1. The fault removal phenomenon is assumed to ollow NHPP.
2. Number of faults remaining in the software may lead to Software failures at a later stage of execution.
3. Failure rate based on testing effort intensity is proportional to the remaining software fault content, where

$$F(E(t)) = \int_{0}^{E(t)} f(x)\mathrm{d}x$$

4. There is a time lag (delay) between the fault detection and correction process.
5. We assume perfect software debugging process with no additional bug added during the fault removal process.

2.3 Model Development

Based on Assumptions 2 and 3, we have the following differential equation for fault detection process

$$\frac{\mathrm{d}m_d(E_t)}{\mathrm{d}t} / \frac{\mathrm{d}E_t}{\mathrm{d}t} = b(E_t)(a - m(E_t)) \quad (1)$$

where $b(E_t) = f(E_t)/(1 - F(E_t)$ denotes the s the fault removal rate with the testing effort function E_t.

With, $m(t = 0) = 0$ and $E(t = 0) = 0$, on solving the above equation, we get

$$m_d(E_t) = aF(E_t) \quad (2)$$

Considering a lag between the time a failure is observed and the time to correcting the fault. This time lag can be construed as lag caused due to varied software testers skills, high/low severity of the faults, alteration to the defect density, etc. In such a scenario, use of one-stage process to model both detection/correction process is not feasible and hence such a process can be given by

$$\frac{\mathrm{d}m_c/\mathrm{d}t}{\mathrm{d}E_t/\mathrm{d}t} = \frac{(f * g)(E_t)}{1 - (F \otimes G)(E_t)}[a - m_c(E_t)] \quad (3)$$

Assuming, $m(0) = 0$ we get

$$m_c(E_t) = a(F \otimes G)(E_t) \quad (4)$$

Here, $F(E_t)$ is the testing effort-dependent probability distribution function for fault detection time meeting the necessary criteria of being probability distribution function:

1. At $t = 0, E_t = 0$ and $F(E_t) = 0$. We make use of Weibull type testing effort function due to its better fit in modeling SRGMs with testing effort than Exponential and Rayleigh-type testing effort function.

2. Then, $t > 0, E_t > 0$ and $F(E_t) > 0$.
3. With $F(E_t)$ described by Weibull-type testing effort function, increase in W_t indicates an increase in $F(E_t)$. Similarly, the continuity of $F(E_t)$ can also be explained.
4. With $t \to \infty, E_t \to \overline{E}$, thereby, $F(E_t)$ is $F(\overline{E})$. Here $\overline{W}$ is a very large positive number and, therefore, $F(\overline{E}) \approx 1$

The two-stage fault correct process under perfect debugging environment is given by Eq. (4) with $m(E_t)$ *as the mean value function.* We consider two different rates of exponential distribution functions $F(E_t)$ and $G(E_t)$ as shown in Table 1.

3 Optimal Release Time—Testing Effort-Based Cost Modeling

The three main quality attributes considered by software developers include scheduled delivery, software reliability, and the overall development cost. With this threefold objective in mind, developers tend to attain them at their best possible values to sustain their bottom lines and business for as long as possible. Reliability has always been considered as the prime quality measure due to the importance bestowed upon it by the end users. They consider reliability as the safest quality measure for the software. Apart from this, other reasons being lack of standard software implementations in the various domains worldwide, the perilous dependency of various systems on computing systems, global competition and trades and unprecedented growth in IT. Even with such advancements, there is still no standardized way to test whether a software is 100% fault free in order to attain maximum reliability. Further, users' requirements often conflict with the developer's deliverables. Users demand shorter delivery time, less expensive and highly reliable or quality software whereas the development aims at minimizing their development cost so as to maximize profits and remain competitive. Such a situation often calls for a trade-off for developers to manage these conflicting objectives. Therefore, the solution lies in optimizing the software release time and total testing duration in order to minimize overall testing cost. Such a problem in software development environment is typically called, Software Release Time Decision Problem (SRTD) and has been widely dealt with in the literature of software reliability engineering [1, 3].

Optimizing software release and testing stop time provides for not only with the possibility of maximizing ROI for developers by reduced cost, defeating

Table 1 Mean value function of SRGM considered

Model	**F**(E_t)	**G**(E_t)	**m**(E_t)
SRGM	$E_t \sim \exp(b_1)$	$E_t \sim \exp(b_2)$	$a\left[1 - \left\{\frac{1}{b_1 - b_2}\left(b_1 e^{-b_2 E_t} - b_2 e^{-b_1 E_t}\right)\right\}\right]$

competition by capitalizing on market opportunity and increasing organizational goodwill but also helps achieve user's requirements of early delivery and reliable software at a lower cost. The idea is to provide an economically priced and reliable software at a faster rate to the users. Interestingly, on one side delay in software release cause penalty\revenue loss on developers, while on the other hand, a premature release may lead to an unreliable product in the user's hands, which eventually means goodwill loss to the developer. Hence, it is imperative to optimize both the release and testing stop time to counter both the market opportunity, customer requirements and minimize overall testing cost and also to avoid the possibility of dual losses arising out of the early or late release. This problem in the software reliability engineering literature has been widely formulated as an optimization problem under well-defined constraints.

These two conflicting objectives of achieving higher performance to be derived from prolonged software testing versus cost reduction with early release, numerous SRGMs have been proposed in the literature considering the debugging cost during testing and operational phase [1, 3, 24–28]. A testing effort-based SRGM to optimize release time and effort in a minimizing cost environment has been proposed by Yamada et al. [29]. Huang et al. [30] on the other hand proposed a framework by simultaneously considering testing effort and efficiency to optimize two important time points. Peng et al. [13] developed testing effort based cost model considering fault detection and correction process. Recently, Tickoo et al. [28] proposed a cost model to optimize software release and patching time using testing effort function. Also, Kapur et al. [24, 25] proposed an optimal scheduling policy to determine optimal release and testing stop time under reliability and budgetary constraints to minimize overall testing cost. Almost invariably, all this existing work considers detection/correction model as one. Lately, Kapur et al. [26] developed a framework wherein the detection and correction processes are detached and treated separately and in our proposed work here, we extend this concept by incorporating testing effort function to determine testing stop time of a software optimal testing effort considering detection and correction process together in the cost model. Through our research, we propose five major costs as follows:

1. **Testing Cost Per Unit**—Testing cost per unit refers to the testing planning, test case generation, test case execution and analysis of testing cost. It also includes CPU hours consumed in the process of testing by the developer.
2. **Bug Detection Cost (Testing Phase)**—Cost of bug detection in the testing phase directly relates to cost associated with bug detection process. This cost is assumed to be linear with the total number of bugs lying in the system. In our proposed modeling framework, we consider that this cost varies when we move from testing to operational phase of the software based upon varying team resources and efforts in each of the phase.
3. **Bug Detection cost (Operational Phase)**—Cost of bug detection during the operational phase arises in case of software failure occurring during the operational usage of the software and the bug getting reported. Such a cost includes bug detection cost, loss of revenue due to system downtime, liability cost, and

customer's dissatisfaction cost. As already proposed in the literature [1, 3], we assume that the cost of software failure in the field is significantly higher in magnitude than the debugging cost during the testing period for the same fault detection process.

4. **Bug Correction Cost (Testing Phase)**—Cost associated with bug correction during testing phase directly relates to cost associated with bug removal process. In our proposed modeling framework we consider that this cost is dependent on the bug detection cost too as with an increase in the bug detection cost higher is the bug correction cost.
5. **Bug Correction Cost (Operational Phase)**—Bug correction cost during operational phase is due to the bugs, which got detected while software is in the field and got corrected during the same time.

After the functional forms of each of these proposed costs are obtained, optimal testing stops time of the software can be obtained by minimizing the total of all the five costs. In the following sections, we formulate the proposed cost model. All other costs of testing and debugging process which is involved in the testing process are considered to be negligible in this study.

The proposed cost models are based on the failure observation/fault removal phenomenon as modeled using Non-homogenous Poisson Process (NHPP), [1] discussed in Section II. The entire fault detection/correction process follows typical NHPP criterions like the independence of bug detection process, the occurrence of failures due to remaining number of faults in the system, detection/correction to be an instantaneous process, perfect debugging, a finite number of faults to be removed in the finite lifecycle of the software.

$$\begin{aligned} C(E_t)_{\text{total}} = {} & C(E_t)_{\text{per unit testing cost}} + C(E_t)_{\text{bug detection cost in testing phase}} \\ & + C(E_t)_{\text{bug detection cost in operational phase}} + C(E_t)_{\text{bug correction cost in testing phase}} \\ & + C(E_t)_{\text{bug correction cost in operational phase}} \end{aligned} \tag{5}$$

$$\begin{aligned} C(E_t)_{\text{total}} = {} & c_1.E_t + c_2.m_d(E_t) + c_3.[m_d(E_\infty) - m_d(E_t)] + c_3.[m_d(E_t) \\ & - m_c(E_t)] + c_4 m_c(E_t) \end{aligned} \tag{6}$$

Here c_1, c_2, c_3 and c_4, denote per unit testing cost, bug detection cost during the testing phase, bug detection/correction cost during operational phase and bug correction cost during the testing phase, respectively. In the literature, debugging process has always been considered to include both detection and correction process simultaneously, whereas seldom, in reality, it is observed that the detection and correction take place simultaneously rather they happen successively. Working on these lines, we consider that the bug detection cost may be different than the bug correction cost both during the two software phases of testing and operational. Further, we have considered the following four constraints Viz., Reliability®, Ratio of effort and max. effort to be utilized and ratio of correction and correction(correction efficiency p_c) and budget which are given as follows.

$$R = \frac{m_d(E_t)}{\theta}, \quad \text{where}\, \theta = m_d(E_\infty), \quad \frac{E_t}{\overline{\overline{E}}}, \quad p_c = \frac{m_c(E_t)}{m_d(E_t)}.$$

Therefore, the final objective can be written as

$$\min C(E_t)_{\text{total}} = c_1.E_t + c_2.m_d(E_t) + c_3.[m_d(E_\infty) - m_d(E_t)] + c_3.[m_d(E_t) - m_c(E_t)] + c_4 m_c(E_t)$$

subject to

$$p_c = \frac{m_c(E_t)}{m_d(E_t)} \geq p_0, R = \frac{m_d(E_t)}{\theta} \geq R_0, \frac{E_t}{\overline{\overline{E}}} \leq E_0, C(E_t)_{\text{total}} \leq C_B \tag{7}$$

where p_0 *and* R_0 are the aspiration level of correction efficiency and reliability, whereas E_0 *and* C_B are the constraints on effort utilized and budget.

4 Numerical Illustration

For illustration purpose, we make use of the data set provided by Obha [6]. With 328 total faults removed in 19 weeks by a total testing effort of 47 CPU hours, the data set was considered for our model validation purpose. The parameter estimation values of the mean value function based on the Yamada SRGM, *m(t)* is given in Table 3, These estimates are obtained using SPSS. We make use of Weibull testing effort function in this SRGM and the estimated values of testing effort function parameters are given in Table 2, due to its high prediction ability. The Weibull function is given as$E(t) = \overline{E}\left(1 - e^{-vt^k}\right)$.

The optimal software release time based on the proposed cost model is obtained here. For numerical illustration purpose, we have taken the cost coefficient as $c_1 = 5, c_2 = 1, c_3 = 6\,\text{and}\, c_4 = 12$. Also, the aspiration level of correction efficiency and reliability are $p_0 = 0.99\,\text{and}\, R_0 = 0.98$ and the value of $E_0 = 0.3\,\text{and}\, C_B = 6000$. Considering all the cost coefficient together with the aspiration level of correction efficiency, reliability, and constraints on effort utilized

Table 2 Estimation of Weibull testing effort function parameters

Parameters	$\overline{E}$	V	K	R^2
	799.3	0.002	1.115	0.998

Table 3 Parameter estimates of Yamada SRGM

Parameters	a	b_1	b_2	R^2
	411.4	0.068	0.132	0.988

and budget we optimized the cost model given in Eq. (7) using MAPLE software to obtain the optimal results. We obtain the optimal software release time under the given constraints as 27 weeks and the effort utilized before the release time is 60.72 CPU hours.

5 Conclusion

Numerous problems pertaining to finding optimal software release time have been proposed in the literature. But in all the existing works either detection or correction of software fault is considered to model the cost function. Considering either detection or correction to obtain the optimal release and cost does not give the idea exact idea of release time of the software as well the budget required. In this work, we proposed a unified framework considering testing effort-based cost model wherein, the cost of detection and correction are taken to be separate and their influence is studied software release time. We provide numerical analysis on the proposed model by considering exponential detection and correction process with Weibull testing effort function under three constraints, viz. reliability, ratio of corrected and total number of detected faults and consumed testing effort. Currently, we have optimized cost function under the above four constraints and obtained the optimal results under different aspiration level. In future, we will develop an optimization model considering all the three attributes together using multi-attribute utility theory (MAUT). The above model can be extended by considering the two-dimensional aspect of software reliability, i.e., time and effort or time and coverage.

References

1. Kapur, P. K., Pham, H., Gupta, A., & Jha, P. C. (2011). *Software reliability assessment with OR application*. Berlin: Springer.
2. Musa, J. D., Iannino, A., & Okumoto, K. (1987). *Software reliability: Measurement*. Applications, McGraw Hill: Prediction.
3. Pham, H. (2006). *System software reliability, reliability engineering series*. Berlin: Springer.
4. Goel, A. L., & Okumoto, K. (1979). Time-dependent errordetection rate model for software reliability and other performance measures. *IEEE Transactions on Reliability, R-28*, 206–211.
5. Yamada, S., Ohba, M., & Osaki, S. (1983). S-shaped reliability growth modelling for software error detection. *IEEE Transactions on Reliability, 32*(5), 475–478.
6. Ohba, M. (1984). Software Reliability Analysis Models. *IBM Journal of Research and Development, 28*(4), 428–443.
7. Kapur, P. K., Garg, R. B., & Kumar, S. (1999). Contributions to hardware and software reliability. *Series of Quality, Reliability & Engineering, 3*, 145–187.

8. Kapur, P. K., Pham H., Anand, S., & Yadav, K. (2011). A unified approach for developing software reliability growth models in the presence of imperfect debugging and error generation reliability. *IEEE Transactions on Reliability, 60*(I), 331–340.
9. Kapur, P. K., Singh, O., Shrivastava, A. K., & Singh J. N. P. (2015). A software up-gradation model with testing effort and two types of imperfect debugging. In *IEEE Xplore Proceedings of International Conference on Futuristic Trends in Computational Analysis and Knowledge Management*, held at Amity University Greater Noida Campus, UP on 25–27 Feb., pp. 613–618.
10. Kapur, P. K., Goswami, D. N., & Bardhan, A. (2007). A general software reliability growth model with testing effort dependent learning process. *International Journal of Modelling and Simulation, 27*(4), 340–346.
11. Kapur, P. K., Ompal, S., Aggarwal, A. G., & Kumar, R. (2009). Unified framework for developing testing effort dependent software reliability growth models. *WSEAS Transactions on Systems, 4*(8), 521–531.
12. Liu, Y., Li, D., Wang, L., & Hu, Q. (2016). A general modeling and analysis framework for software fault detection and correction process. *Software Testing, Verification and Reliability.*
13. Peng, R., Li, Y. F., Zhang, W. J., & Hu, Q. P. (2014). Testing effort dependent software reliability model for imperfect debugging process considering both detection and correction. *Reliability Engineering and System Safety, 126,* 37–43.
14. Wang, L., Hu, Q., & Liu, J. (2016). Software reliability growth modeling and analysis with dual fault detection and correction processes. *IIE Transactions, 48*(4), 359–370.
15. Zhang C., Gang, C., Liu, H., Meng, F., & Wu, S. (2014) A unified and flexible framework of imperfect debugging dependent SRGMs with testing-effort. *Journal of Multimedia, 9*(2), 310–317.
16. Xie, M., Hu, Q. P., Wu, Y. P., & Ng, S. H. (2007). A study of the modeling and analysis of software fault-detection and fault-correction processes. *Quality and Reliability Engineering International, 23*(4), 459–470.
17. Schneidewind, N. F. (1975). Analysis of error processes in computer software. *Sigplan Notices, 10*(6), 337–346.
18. Xie, M., & Zhao, M. (1992). The Schneidewind Software reliability model revisited. In *Proceedings of the 3rd Int'l Symposium on Software Reliability Engineering, Research Triangle Park*, NC, USA 1992, pp. 184–192.
19. Huang, C. Y., Kuo, S. Y., & Lyu, M. R. (2007). An assessment of testing effort dependent software reliability growth models. *IEEE Transactions on Reliability, 57,* 198–211.
20. Kumar, M., Ahmad, N., & Quadri, S. M. K. (2005). Software reliability growth models and data analysis with a pareto test effort. *RAU Journal of Research, 15*(1–2), 124–128.
21. Li, Q., Li, H., & Lu, M. (2015). Incorporating S-shaped testing-effort functions into NHPP software reliability model with imperfect debugging. *Journal of Systems Engineering and Electronics, 26*(1), 190–207.
22. Zhao, Q., Zheng, J., & Li, J. (1998). Software reliability modeling with testing-effort function and imperfect debugging. *TELKOMNIKA, 10*(8), 1992–1998 (2012).
23. Inoue, S., & Yamada, S. (2013). Lognormal process software reliability modeling with testing-effort. *Journal of Software Engineering and Applications,* 6, 8–14.
24. Kapur, P. K., Shrivastava, A. K., & Singh, O. (2017). When to release and stop testing of a software. *Journal of the Indian Society for Probability and Statistics, 16*(1), 19–37.
25. Kapur, P. K., & Shrivastava, A. K. (2015). *Release and testing stop time of a software: A new insight.* 978-1-4673-7231-2/15/$31.00 © IEEE.
26. Kapur, P.K., & Shrivastava A. K. (2016). A new dimension to software release time problems. *SRESA Journal of Life Cycle Reliability and Safety Engineering.*
27. Shrivastava, A. K., & Kapur, P. K. (2017). *Development of software reliability growth models with time lag and change-point and a new perspective for release time problem, mathematics applied in information systems.* USA: Bentham Science.

28. Tickoo, A., Kapur, P. K., & Shrivastava, A. K. (2016). Testing effort based modeling to determine optimal release and patching time of software. *International Journal of System Assurance Engineering and Management, 7,* 427–434.
29. Yamada, S., Ohtera, H., & Narihisa, H. (1986). Software reliability growth models with testing effort. *IEEE Transactions on Reliability, R-35*(1), 19–23.
30. Huang, C.-Y., Lyu, & M. R. (2005). Optimal release time for software systems considering cost, testing-effort, and test efficiency. *IEEE Transactions on Reliability, 54,* 583–91.

Diffusion Modeling Framework for Adoption of Competitive Brands

Adarsh Anand, Gunjan Bansal, Arushi Singh Rawat and P. K. Kapur

Abstract In order to survive in today's competitive market, every brand/company is altering and refining its offerings at a fast pace. Market thus sees a variety of products available almost at the same time. In the midst of all the major aspects, firms need to look at how customers respond to products, which are similar looking, equally priced, and even have similar features. To cater to this understanding, the present proposal deals with the concept of brand preference. The objective of our modeling framework is to observe the shifting behavior of customers and to predict the sales level in the presence of various brands available together. Today's market provides the customers with multiple options to choose from, thereby, taking this ideology into account, the current study is able to identify all the possible variations that might impact the overall sales of a particular product because of inter- and intra-shifting of customers amongst various brands available at the time of purchase. Validation of the model has been done on real-life car sales data for the automobile industry.

Keywords Brand preference · Competition · Diffusion

A. Anand · G. Bansal (✉) · A. S. Rawat
Department of Operational Research, University of Delhi,
New Delhi 11007, Delhi, India
e-mail: gunjan.1512@gmail.com

A. Anand
e-mail: adarsh.anand86@gmail.com

A. S. Rawat
e-mail: arushisinghrawat92@gmail.com

P. K. Kapur
Centre for Interdisciplinary Research, Amity University,
Noida 201313, Uttar Pradesh, India
e-mail: pkkapur1@gmail.com

P. K. Kapur et al. (eds.), *System Performance and Management Analytics*,
Asset Analytics, https://doi.org/10.1007/978-981-10-7323-6_25

1 Introduction

"*Which airline do you fly? Unless your answer is 'cheapest flight possible,' you have a* ***brand preference***." [21]. As it has been perfectly described here that it is all about the choice a customer makes of a brand irrespective of the price, quality, etc. It is more like a perception that an individual has of a particular brand which makes them choose a brand over the others in the market. Often it is the image of a brand that becomes way more important than the other features of the products that the brand offers. The existence of brand preference as a concept is completely dependent on the brand-related queries such as measurement of a brand, functions of a brand, interaction of a brand with individuals in the market, etc. It is to be noted that the main aim of branding is to make a brand valuable for individuals. Value basically refers to the preferential position that a brand can expect among the distributors and customers in the market. Brand value translates into demand for a brand which in turn, increases the willingness of a customer to pay a higher price for well-known brands and hence increases the overall sales of the brand [9, 25].

In today's market, one of the distinctive features of modern marketing is the extensive use of brands by manufacturers and distributors and the general preference for branded items by consumers. In our marketing process, brands are an important communicator of economic information; they aid in product identification and tend to protect buyers and sellers from uncertainty regarding product quality. Hence, it can be said that diffusion of brands in the market is dichotomous; as in it is an important aspect for both buyers and sellers. The seller can depend on the rate of diffusion for planning their production and supply, whereas the buyers can choose a suitable brand (with a higher rate of diffusion) out of several others available in the market. The rate of diffusion can be explained as the spread of a brand over its customers, it happens to play an essential role in preference of a brand in the market. **Adoption** (the reciprocal process as viewed from a consumer perspective rather than the distributor) is similar to diffusion except that it deals with the psychological processes an individual goes through, rather than an aggregate market process.

The importance of information concerning consumer attitudes toward acceptance of, preferences for, and loyalty toward various brands within a product class has long been recognized [5, 10]. There are several brands in the market which offer similar features but still have varying prices. Hence the concept of brand preference comes in. *Brand preference* can be formally defined as the choice that a customer makes of a specific company's product or service, when he/she is provided with other, equally priced options. This concept ponders upon several factors behind making such a choice of product or the brand [22, 24]. Thus, it helps companies in evaluating the loyalty of the customer towards a particular brand which in turn, builds up the overall brand image in the market. There has been a study wherein, the shift from the attributes of a brand to the overall experience derived from a brand has been talked about [23]. Hence, it has been observed that the experiential data obtained an important aspect for deriving brand preference. A total of four major constructs related to the model have been established as mentioned below:

1. **Brand Experience**: It can be explained as sensations, feelings, and behavioral responses towards brand-related stimuli that are a part of brand's design, identity, environment, etc. Brand preference is learning and a time-taking process for any individual provided brand experience is the initial source of learning. Since brand preference would change when brand experience changes. Therefore, it can be said that they are directly related to each other.
2. **Brand Personality**: It is a set of human characteristics that are credited to the brand name. It involves everything that a customer can relate to. Being closely related to brand associations, brand personality can be termed as a symbolic representation of brand associations [14].
3. **Brand Associations**: It can be defined as anything that is deep rooted and forms the base of a customer's mentality about a brand. Thus, it becomes extremely essential for a brand to develop a positive association in the minds of the customers. According to human associative theory (HAM) [3], brand associations are formed by direct and indirect experiences. Brand associations basically give reasons for choosing a brand. Thus, the brand association is directly proportional to brand preference.
4. **Human-Brand Personality Congruence**: This concept basically revolves around how a customer perceives a particular brand and what exactly he/she feels would go with his/her personality. A brand always has a certain image in the market and if it matches with the personality of an individual (i.e., brand congruity) then the customer will not only go for that brand but can also prove to be a loyal customer in future. Therefore, the impact of brand image on brand preference is extremely high when there is a balance between human personality and the brand image.

The concept of brand preference has always been a concerned and noteworthy field for researchers, organizations, managers, etc. Characteristics' of market dynamics have been studied by using Markov process that captures brand preference information mathematically [7, 16]. Brand preference has also been evaluated using attitude model wherein, a consumer's attitude toward a brand is hypothesized to be a function of the relative importance of each of the product attributes and the beliefs about the brand on each attribute [6, 19]. D'Souza and Rao have discussed how frequent repetitions of advertisement have a positive influence in generating brand preference among customers [11]. Manufacturers continuously put extra efforts to build a healthy and long-term relationship with their existing and new customers. For this purpose, they always analyze purchase intensions of customers, improve their customers' services and also work to maintain their brand image [8].

In other words, brand preference can be stated as a perfect blend of economic psychology and marketing science. Therefore, in this chapter, our objective to study the influence of brand preference that directly affects consumer's behavior such as purchasing, saving, brand choice, etc., and induce consumers to deviate from one brand to another brand. Also after using the derived information from the analysis, sales have been forecasted for different brands. Rest of the chapter has been structured as follows.Modeling framework based on brand preference when more than one brand exists in the marketplace has been discussed in Sect. 2. In Sect. 3, a numerical illustration based on automobile industry dataset has been demonstrated. Results have been interpreted and discussed in Sect. 4. Further, the significance of this study for managers has been specified in Sect. 5, i.e., Managerial Implication. The study has been concluded in Sect. 6.

2 Concept of Brand Preference Modeling Under Competitive Scenario

2.1 *Literature Review*

Modeling framework is an integral part of studying any concept in detail. Here, in this study, we have formulated a mathematical framework of brand preference, which is derived using well-known diffusion model known as Bass Model (1969). This model captures the diffusion process that is based on two types of influences, i.e., external and internal influences. Externally influenced are adopters who are influenced through advertisements and promotions of the new product and those adopters who are influenced via word of mouth by individuals are under the category of internal influence. The rate by which both the groups impact the leftover individuals are p and q, respectively; where p is termed as the coefficient of innovation and q is read as the coefficient of imitators. This well-established model was based on hazard rate function which says the probability of purchasing a new product by a new purchaser at time t given that no purchase has been made. The model has been widely used and well accepted in many fields like *multi-generational of a particular product*: When a particular product with its newer and updated versions have been launched in some time interval is known as multi-generational of a particular product. Norton and Bass in 1987 had introduced this concept of successive generations of a product and then further extensions have been made [1, 12, 13, 17, 20]

Product Line of a particular brand: Concept of Product line emerges when a particular brand offers more than one type of the products in the marketplace [4, 27], *Product Services*: [2, 15] and many more.

2.2 *Mathematical Modeling of Brand Preference Under Competitive Scenario*

Brand preference as it is described as a concept, which becomes even more interesting to study when we have an extremely competitive market. In such a market, we have several brands existing together and giving each other a tough competition of survival. This, in fact, helps the customers to choose from a wide range of products which have similar features but different prices. Thus, it becomes difficult and also interesting to analyze the deviating behavior of customers in the presence of many brands competing with each other. Successful diffusion of a new product directly depends upon the image of a brand subsist in the marketplace. It is the concept of adoption of new products in the market which led to the idea of multiple brands and, in turn, brand preference. To understand brand preference and deviating behavior of customers mathematically, we have considered only two product categories which have similar but distinguished features and each category consist of two competitive brands such as Maruti Suzuki and Hyundai Motors.

The modeling framework is based upon some assumptions which have been listed below:

(1) The number of potential customers, m is a constant.
(2) There are two broad categories of potential buyers, i.e., Purchasers and Deviators.
(3) The monopolistic aspect of Bass model has been explained in this chapter as we are considering that the competition between the two brands coexists.
(4) Only one unit is being purchased by any new buyer. No repeat purchase is possible as we have taken cars as our products which happen to be durable products.
(5) Product categories are comparable to each other.

In order to explain a mathematical model, we have considered all the possible types of customers that may exist when two product categories consist of two competitive brands in the marketplace.

(Note: For easy representation, we have used acronyms for Maruti Suzuki as M, Hyundai Motors as H, first Product category as P1 and with its respective brands: P1H for Hyundai and P1M for Maruti. Similarly, for second Product category P2 and with its respective brands: P2H for Hyundai and P2M for Maruti). Categorization of all types of customers can be explained as below and, categorization of all the types of customers has been demonstrated using Fig. 1:

(a) Potential purchasers of P1H—Type 1
(b) Potential purchasers of P1M—Type 2
(c) Potential purchasers of P2M—Type 3
(d) Potential purchasers of P2H—Type 4

The market under consideration can have several products existing simultaneously and hence one cannot be sure of the choice of customers as they can divert from a particular product to another. Further, we have defined all those purchasers who would have adopted Hyundai car (P1H) but may divert to other products (P1M, P2H, P2M). And, also there are chances of deviation of customers from other brands to P1H. Hence, these customers can be categorized as

(e) Potential customers of P1H may prefer car of other brands of the same product category, i.e., P1M over P1H—Type 5,
(f) Potential customers of P1H who would like to go the same brand but of different category, i.e., P2H over P1H—Type 6,
(g) Potential customers of P1H who would prefer other brand along with different product category, i.e., customers may adopt P2H instead of P1H—Type 7,

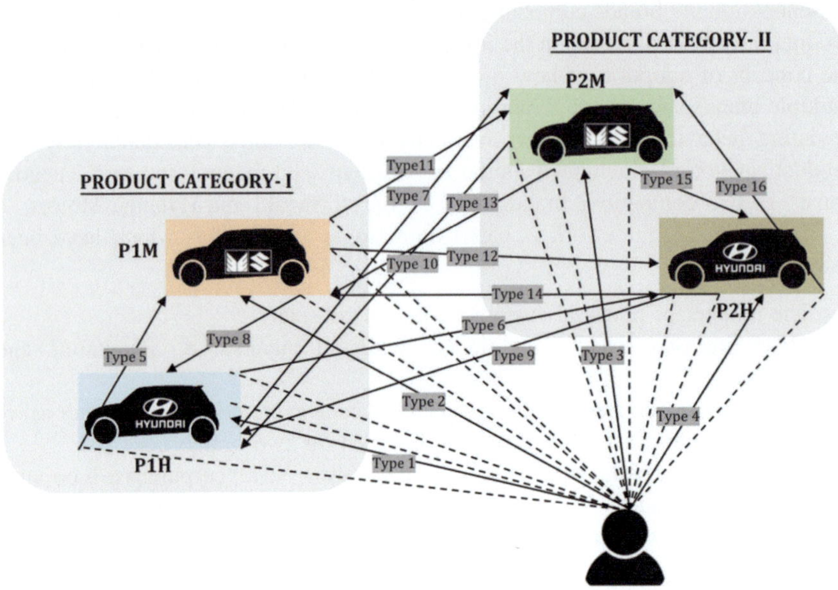

Fig. 1 Understanding of potential adopters

(h) Potential customers of the P1M product, who may prefer other brand of the same product category and deviate to P1H—Type 8,
(i) Potential customers of the P2H category who would prefer same brand but deviate to another product category, i.e., P1H of the same brand—Type 9,
(j) Potential customers of P2M may prefer another brand of other product category and deviate to P1H—Type 10.

Similarly, we have those customers who would deviate from one product category to another product category and/or shift to another brand. These categories are given below:

(k) Potential customers of P1M prefer the same brand but go for another product category, i.e., P2M—Type 11
(l) Potential customers of P1M who would like to deviate to another brand of another product category, i.e., P2H—Type 12
(m) Potential customers of product category P2 who would choose another product category (i.e., P1) but like to prefer the same brand M.—Type 13
(n) Potential customers of P2H prefer another brand of another product category and deviate to P1M—Type 14
(o) Potential customers of P2M may prefer to deviate to another brand of the same category, i.e., P2H—Type 15
(p) Potential customers of P2H who may prefer another brand, i.e., P2M—Type 16

In order to understand the deviating behavior of the customers, we have defined types of adopters from Type 1 to Type 4 who are those potential adopters of respective product categories that consist of two brands. Whereas, Type 5 to Type 16 are those adopters who would act as deviators and may deviate to another product with a motive of purchasing a best available product on the basis of their brand preference.

Notations

λ_j	Rate that classifies overall potential customers of product category into numbers of brands available
m_i	Potential adopters of the *ith* product category
$F_{ij}(t)$	Cumulative likelihood of purchasing product of *jth* brand belonging to the *ith* product category. p_{ij} Coefficient of innovation of *ith* product category of *jth* brand
q_{ij}	Coefficient of imitation of *ith* product category of *jth* brand
α_k	Brand preference factor that deviates from P1H product to other products
δ_k	Brand preference factor that deviates from P1M product to other products
ρ_k	Brand preference factor that deviates from P2H product to other products
γ_k	Brand preference factor that deviate from P2M product to other products
i	Represents the number of product categories
j	Represents the number of brands available in each *ith* product category
k	Represents the number of competitive products available corresponding to a particular product, i.e., $(n-1)$; when total numbers of products are n

We have explained the expression of the aggregated expected sales of the first product category of Hyundai car and Maruti Suzuki car (i.e., P1H and P1M):

$$\begin{aligned}\text{P1H} = {} & (\lambda_1.m_1.F_{11}(t)) - (\alpha_1.\lambda_1.m_1.F_{11}(t).F_{12}(t)) - (\alpha_2.\lambda_1.m_1.F_{11}(t).F_{21}(t)) \\ & - (\alpha_3.\lambda_1.m_1.F_{11}(t).F_{22}(t)) + (\delta_1.(1-\lambda_1).m_1.F_{12}(t).F_{11}(t)) + (\gamma_1.\lambda_2.m_2.F_{21}(t).F_{11}(t)) \\ & + (\rho_1.(1-\lambda_2).m_2.F_{22}(t).F_{11}(t))\end{aligned} \tag{1}$$

$$\begin{aligned}\text{P1M} = {} & ((1-\lambda_1).m_1.F_{12}(t)) - (\delta_1.(1-\lambda_1).m_1.F_{12}(t).F_{11}(t)) - (\delta_2.(1-\lambda_1).m_1.F_{12}(t).F_{21}(t)) \\ & - (\delta_3.(1-\lambda_1).m_1.F_{12}(t).F_{22}(t)) + (\alpha_1.\lambda_1.m_1.F_{11}(t).F_{12}(t)) + (\gamma_2.\lambda_2.m_2.F_{21}(t).F_{12}(t)) \\ & + (\rho_2.(1-\lambda_2).m_2.F_{22}(t).F_{12}(t))\end{aligned} \tag{2}$$

In Eqs. 1 and 2, $F_{i,j}(t) = \left[\frac{1-e^{-(p_{ij}+q_{ij})t}}{1+\left(\frac{q_{ij}}{p_{ij}}\right)e^{-(p_{ij}+q_{ij})t}}\right]$ represents the distribution fraction for the adoption process. m_1 is the potential adopters of the single product category (i.e., P1) which consist of similar features of products but since it consists of different brands (Hyundai Motors and Maruti Suzuki). And, λ_1 is the rate that classifies the overall potential customers of P1 category into respective numbers of brands. Hence,

$(\lambda_1.m_1.F_{11}(t))$ and $((1-\lambda_1).m_1.F_{12}(t))$ have become the potential customers of brand Hyundai and Maruti Suzuki respectively. The expected count of customers may get changed (either increase or decrease), when potential customers influence from the other products available in the marketplace. Since we have considered two different product categories which are comparable to each other and have two brands each. Therefore, for potential customers of P1H have three available options to choose from such that either to prefer the same brand but of a different category (P2H) or to prefer a different brand of any category, i.e., (P1M and P2M). For this purpose, we have considered α_1, α_2 and $\alpha_3(=1-\alpha_1-\alpha_2)$ brand preference factors which are termed as the rate of diversion from P1H to P1M, P2H and P2M respectively. Since, these are the adopters who are deviating from P1H to P1M, P2H, and P2M; therefore, we have subtracted them from the total sales of P1H. Similarly, there is a possibility of certain additions of potential adopters in some proportions to the overall sales of P1H, which may come from the other available products such as P1M, P2H, and P2M with the rate of δ_1, γ_1 and ρ_1 as the brand preference factors respectively. Therefore, we can say that, potential customers of P1H, i.e., $\lambda_{1.}m_{1.}F_{11}(t)$ who would prefer P1M with the rate of $\alpha_{1.}$ would add up to the P1M and get subtracted from P1H with the rate of $F_{12}(t)$. Hence, collectively $\alpha_{1.}\lambda_{1.}m_{1.}F_{11}(t)_{.}F_{12}(t)$ is the factor that acted as deviated customers of Type-5. Similarly, we can say for $\alpha_{2.}\lambda_{1.}m_{1.}F_{11}(t)_{.}F_{21}(t)$ and $\alpha_{3.}\lambda_{1.}m_{1.}F_{11}(t)_{.}F_{22}(t)$ are the deviators who would be considered as Type-6 and Type-7 respectively. Since these are the terms which refer to the customers deviating from P1H to other products; we have subtracted them from the expression of the overall sales of P1H. Likewise, $(\delta_{1.}(1-\lambda_1).m_{1.}F_{12}(t)_{.}F_{11}(t))$, $(\gamma_{1.}\lambda_{2.}m_{1.}F_{21}(t)_{.}F_{11}(t))$ and $(\rho_{1.}(1-\lambda_2).m_{1.}F_{11}(t)_{.}F_{22}(t))$ are representing as Type-8, Type-9, and Type-10 of deviators respectively. And, $(\delta_{1.}(1-\lambda_1).m_{1.}F_{12}(t)_{.}F_{11}(t))$, $(\gamma_{1.}\lambda_{2.}m_{1.}F_{21}(t)_{.}F_{11}(t))$ and $(\rho_{1.}(1-\lambda_2).m_{1.}F_{11}(t)_{.}F_{22}(t))$ are the terms which refer to the customers deviating to P1H from P1M, P2H, and P2M, respectively, therefore, these factors have been added up in the expression of the overall sales of P1H. Similarly, we have written a similar expression for other product category follows:

$$\begin{aligned}\text{P2H} = {} & (\lambda_2.m_2.F_{21}(t)) - (\gamma_1.\lambda_2.m_2.F_{21}(t).F_{22}(t)) - (\gamma_2.\lambda_2.m_2.F_{21}(t).F_{11}(t)) \\ & - (\gamma_3.\lambda_2.m_2.F_{21}(t).F_{12}(t)) + (\alpha_2.\lambda_1.m_1.F_{11}(t).F_{21}(t)) + (\delta_2.(1-\lambda_1).m_1.F_{12}(t).F_{21}(t)) \\ & + (\rho_3.(1-\lambda_2).m_2.F_{22}(t).F_{21}(t))\end{aligned} \tag{3}$$

$$\begin{aligned}\text{P2M} = {} & ((1-\lambda_2).m_2.F_{22}(t)) - (\rho_1.(1-\lambda_2).m_2.F_{22}(t).F_{11}(t)) \\ & - (\rho_2.(1-\lambda_2).m_2.F_{22}(t).F_{12}(t)) - (\rho_3.(1-\lambda_2).m_2.F_{22}(t).F_{21}(t)) \\ & + (\alpha_3.\lambda_1.m_1.F_{11}(t).F_{22}(t)) + (\delta_3.(1-\lambda_1).m_1.F_{12}(t).F_{22}(t)) \\ & + (\gamma_3.\lambda_2.m_2.F_{21}(t).F_{12}(t))\end{aligned} \tag{4}$$

With the help of the above equations, we can understand the other types of deviators as well. Type-11 would fall in the class of $\delta_{3.}(1-\lambda_1).m_{1.}F_{12}(t)_{.}F_{22}(t)$, which would indicate those adopters who have deviated from P1M $\rightarrow$ P2M.

Type-12 are the adopters who may deviate from P1M → P2H and mathematically represented as $\delta_{2}.(1-\lambda_{1}).m_{1}.F_{12}(t).F_{21}(t)$. Adopters of Type-13 are those who deviated from P2M → P1M and represented as $\rho_{2}.(1-\lambda_{2}).m_{2}.F_{22}(t).F_{12}(t)$. Type-14 are the adopters who are deviated from P2H → P1M and defined as $\gamma_{2}.\lambda_{2}.m_{2}.F_{21}(t).F_{12}(t)$. Adopters of Type-15 are those who deviated from P2M → P2H and represented as $\rho_{3}.(1-\lambda_{2}).m_{2}.F_{22}(t).F_{21}(t)$ and Type-16 are those adopters who are deviated from P2H → P2M and mathematically represented as $\gamma_{1}.\lambda_{2}.m_{2}.F_{21}(t).F_{22}(t)$. The proposed modeling framework can be generalized up to "x" product categories consisting of "y" brands in each category. In the next section, we have validated these equations and analyzed the deviating behavior of customers with respect to the sales of the cars considered.

3 Data Analysis and Numerical Illustration

Here, in this study, we have considered two product categories and within each product category, we have taken two car brands (Hyundai and Maruti). The product categories and its products are comparable with each other in terms of features and prices. This way, the only parameter that differentiates the cars from each other is their brand and any kind of choice made by the customer will imply his/her inclination toward that particular brand because of brand preferences. For the purpose of validation of our proposed modeling, we have used car sales data from online sites [18, 26]. Here, we have considered overall sales of two product categories with comparable attributes. These product categories are slightly different (depending on features, price, etc.) but are still comparable to each other given below in Table 1.

Here, our objective is to establish how two product categories with similar features and prices are differentiated by the customers just on the basis of the brands they belong to and to predict the sales of all different brands that are present in each category as per customers' brand preferences. Hence, brand preference playing a significant role in choosing a suitable car. Here, first product category includes two cars, one from each brand, namely P1H (For Hyundai) and P1M (for Maruti). Similarly, for the second category, we have P2H (for Hyundai) and P2M (for Maruti). Actual data of these two product categories can be seen in Table 2.

Table 1 Product categories with their respective features

Attributes	Product category1 (PC1)	Product category2 (PC2)
Car type	Hatchback	Estate/Hatchback
No. of seats	5 seater	5-seater
Engine displacement	1 L	1.4 L
Engine length	3600 mm	3600–4000 mm

Table 2 Sales of product categories (P1 and P2)

Time	Product category P1 (in '000)	Product category P2 (in '000)
Jan. 11	66.589	51.826
Feb. 11	128.174	95.052
Mar. 11	194.999	139.498
Apr. 11	244.876	178.208
May 11	293.658	218.301
June 11	338.955	252.661
July 11	382.392	276.215
Aug. 11	426.847	306.813
Sep. 11	479.755	343.338
Oct. 11	514.22	372.588
Nov. 11	565.681	412.069
Dec. 11	615.412	446.445
Jan. 12	679.618	489.013
Feb. 12	745.126	532.08
Mar. 12	815.811	576.385
Apr. 12	859.381	617.891
May 12	900.814	656.815
June 12	945.253	693.292
July 12	982.295	722.939
Aug. 12	1012.876	743.01
Sep. 12	1061.682	776.736
Oct. 12	1115.613	817.241
Nov. 12	1164.654	858.633
Dec. 12	1208.677	893.421

Further, for estimating parameters of the proposed model, we have used the simultaneous nonlinear least square methodology and solved it in SAS software package. Table 3 is representing the estimated values of diffusion parameters such as external factors (p_{ij}), internal factors (q_{ij}) and brand preference factors (α_k, δ_k and σ_k(where $k = 1, 2, 3$)) of all 4 products, i.e., P1H, P1M, P2H, and P2M, respectively, that have been obtained using above proposed Eqs. (1 to 4). And, comparison criteria's of these equations are shown in Table 4.

Table 3 Parameter estimation results

	$m_{1,2}$	$\lambda_{1,2}$	p_{ij}	q_{ij}	α_k	δ_k	γ_k	σ_k
P1H	1250120	0.78	0.000335	0.299	–	0.25	0.21	0.32
P1M			0.194219	0.536741	0.158665	–	0.201	0.716019
P2H	1265913	0.123	0.183155	0.521151	0.45	0.2	–	0
P2M			0.0101	0.301	0.391335	0.55	0.589	–

Table 4 Comparison criteria

Equations	SSE	MSE	Root MSE	R-Square	Adj R-Sq
P1	2.1969E9	1.1562E8	10752.9	0.9992	0.9991
P2	1.1481E9	6.04E7	7773.4	0.9992	0.9991

Using Fig. 2, clearly the graph represents an excellent fit between actual and predicted values of the two product categories. And, Fig. 3 represents the predicted sales behavior of four products of different two brands, i.e., P1H, P1M, P2H, and P2M, respectively, as per deviating behavior of customers due to brand preference factors.

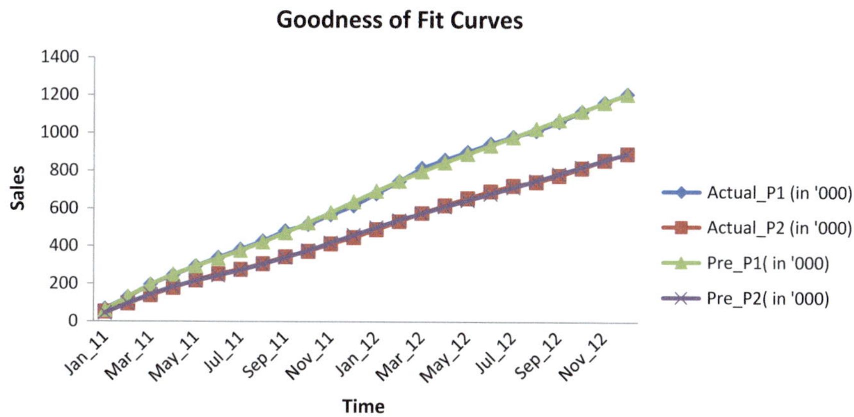

Fig. 2 Goodness of fit curves

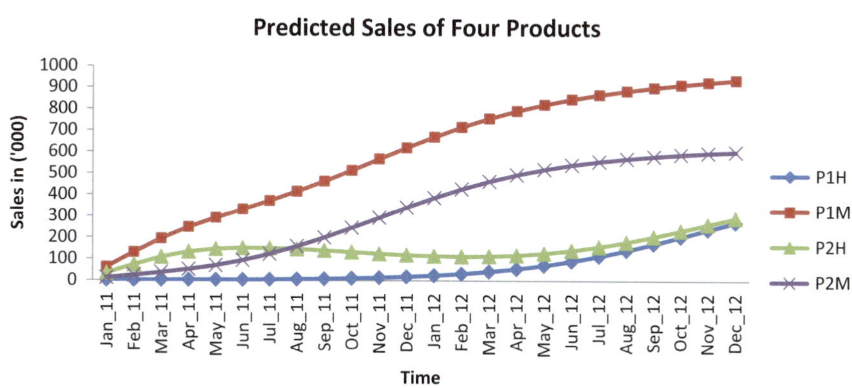

Fig. 3 Predicted sales of four products

4 Interpretation and Findings

Table 3 and Fig. 3 demonstrated that potential customers of P1 and P2 are observed as $m_1 = 1250.120$ and $m_2 = 1265.913$. And, these potential markets can be classified using the parameter $\lambda_1 = 0.78$ for P1H and $(1 - \lambda_1) = 0.22$ P1M and similarly, $\lambda_2 = 0.123$ for P2H and $(1 - \lambda_1) = 0.877$ for P2M. External and internal factors for product P1M is highest among all other products which represent that this product is well accepted by the customers. Further, brand preference factors have been considered individually such that α_k that represent the deviators moving from P1H to other products (P1M, P2H, and P2M), here it can be observed that maximum amount of Hyundai customers are deviating to different category of the Hyundai cars only (i.e., 0.45 to P2H) and 0.39 are preferring Maruti brand of another category (P2M). Whereas, in case of δ_k and σ_k factors, maximum proportions of deviation has occurred within the product category and preference has given to its already preferred brand, i.e., (from P1M to P2M with 0.55 and from P2M to P1M with 0.716) and also there is no deviation from P2M to P2H. Moreover, in case of γ_k, which is representing a deviation from P2H, has the highest value for P2M (i.e., 0.589). Hence, it can be analyzed that most of the customers prefer Maruti over Hyundai as per the values obtained. And, also the predicted values of these 4 products demonstrated that sales of product P1M and P2M are higher than the Hyundai products.

Table 4 and Fig. 2, as per the goodness of fit criterion, values of SSE, MSE, Root MSE are low for both product categories and also the value R-Square and Adjusted R-square corresponding to each of the product category PC1 and PC2 is 0.9992 and 0.9991, that is, both values are approaching 1 which signifies that the proposed model is well fitted with the actual sales data.

5 Managerial Implication

Leading brands are perceived to undergo tough competition in order to sustain in today's ever-changing market scenario. Usually, when a couple of firms lead the market, they are no longer eager to increase their sales by capturing the market of the rival firms. Rather, customer retention becomes an unsaid priority. Although, the competition between the brands stays alive forever, but the preference for a brand among customers always increase the sales of a brand and results in losses of the other brands under consideration. Customers, being the significant entity, always require a product that not only satisfies their basic requirements but also feels connected and loyal toward their preferred brand.

Here, in this study, we endeavor to deduce that it is extremely important for a firm to identify all possible deviations of the customers that may impact the overall sales of a particular product when it belongs to either of the two forums; between brands as well as within the brand. Consideration of all possible scenarios of deviations; whenever customers have multiple choices to choose from, it can be understood that when a customer is specific to a particular brand then he/she may deviate to similar products but only of that preferred brand. Therefore, it is significant for managers to study the impact of brand preference factors which are persuading customers to deviate from one product to other products. Furthermore, it can help the managers to thoroughly look into the sales produced from each of the deviations and choose the most profitable scenario for the firm.

6 Conclusion

Brand preference, as the name implies, is all about choices that one makes when he/she is provided with products which are not just equally priced but also have similar features. Here, we have considered a similar scenario where we have cars of similar features, prices, etc., and brand being the only major difference (viz., Hyundai Motors, Maruti Suzuki), we have been able to comprehend that by applying the concept of brand preference as a factor in the methodological framework; deviating behavior of customers can be observed and also analyzed that which particular product is cannibalizing the sales of other products. With the help of consolidated data of car sales (viz., Hyundai Motors and Maruti Suzuki), this study has shown that if a customer has a bond with the specific brand then he will choose the different product but of the same brand otherwise, he may convert to the other brand. Hence, we have concluded that brand preference plays a very significant role in the shift of the customers between and within the brands.

Acknowledgments The research work presented in this chapter is supported by grants to the first author from DST, via DST PURSE phase II, India.

References

1. Aggrawal, D., Anand, A., Singh, O., & Kapur, P. K. (2015). Modelling successive generations for products-in-use and number of products sold in the market. *International Journal of Operational Research, 24*(2), 228–244.
2. Anand, A., Aggrawal, R., Singh, O., & Aggrawal, D. (2016). Understanding diffusion process in the context of product dis-adoption. *240*(2), 7–18.
3. Anderson, J. R., & Bower, G. H. (2014). *Human associative memory*. Psychology press.
4. Bansal, G., Anand, A., Aggrawal, D., & Agarwal, M. (2016). Competitive diffusion modeling framework for adoption of product lines. *Communications in Dependability and Quality Management An International Journal, 19*(1), 43–54.
5. Bass, F. M., & Wilkie, W. L. (1973). A comparative analysis of attitudinal predictions of brand preference. *Journal of Marketing Research*, 262–269.
6. Bass, F. M., & Talarzyk, W. W. (1972). An attitude model for the study of brand preference. *Journal of Marketing Research*, 93–96.

7. Blin, J. M., & Dodson, J. A. (1980). The relationship between attributes, brand preference, and choice: A stochastic view. *Management Science, 26*(6), 606–619.
8. Chen, C. F., & Chang, Y. Y. (2008). Airline brand equity, brand preference, and purchase intentions—The moderating effects of switching costs. *Journal of Air Transport Management, 14*(1), 40–42.
9. Christodoulides, G., & De Chernatony, L. (2010). Consumer-based brand equity conceptualization and measurement: A literature review. *International Journal of Research in Marketing, 52*(1), 43–66.
10. Cobb-Walgren, C. J., Ruble, C. A., & Donthu, N. (1995). Brand equity, brand preference, and purchase intent. *Journal of advertising, 24*(3), 25–40.
11. D'Souza, G., & Rao, R. C. (1995). Can repeating an advertisement more frequently than the competition affect brand preference in a mature market? *The Journal of Marketing*, 32–42.
12. Jiang, Z. (2010). How to give away software with successive versions. *Decision Support Systems, 49,* 430–441.
13. Jiang, Z., & Jain, D. C. (2012). A generalized Norton–Bass model for multigeneration diffusion.
14. Keller, K. L. (1998). Branding perspectives on social marketing. *Advances in Consumer Research, 25,* 299–302. http://www.acrwebsite.org/search/view-conference-proceedings.aspx?Id=7887, Accessed March 18, 2017.
15. Libai, B., Muller, E., & Peres, R. (2009). The diffusion of services. *Journal of Marketing Research, 46*(2), 163–175.
16. Maffei, R. B. (1960). Brand preferences and simple Markov processes. *Operations Research, 8*(2), 210–218.
17. Mahajan, V., & Muller, E. (1996). Timing, diffusion and substitution of successive generations of technological innovations: The IBM mainframe case. *Technological Forecasting and Social Change, 51,* 109–132.
18. Motorbeam, http://www.motorbeam.com/. Accessed July 03, 2015.
19. Nakanishi, M., & Bettman, J. R. (1974). Attitude models revisited: An individual level analysis. *Journal of Consumer Research, 1*(3), 16–21.
20. Norton, J. A., & Bass, F. M. (1987). A diffusion theory model of adoption and substitution for successive generation of high-technology products. *Management Science, 33*(9), 1069–1086.
21. Davis, O. (2016). http://study.com/academy/lesson/brand-preference-definition-lesson-quiz.html.
22. Padberg, D. I., Walker, F. E., & Kepner, K. W. (1967). Measuring consumer brand preference. *Journal of Farm Economics,* 723–733.
23. Schmitt, B. (2010). Experience marketing: Concepts, frameworks and consumer insights. *Foundations and Trend in Marketing, 5*(2), 55–112. https://www8.gsb.columbia.edu/sites/globalbrands/files/Experience%20Marketing%20-%20Schmitt%20-%20Foundations%20and%20Trends%202011.pdf. Accessed March 18, 2017.
24. Sriram, S., Chintagunta, P. K., & Neelamegham, R. (2006). Effects of brand preference, product attributes, and marketing mix variables in technology product markets. *Marketing Science, 25*(5), 440–456.
25. Stanton, J. L., & Lowenhar, J. A. (1974). A congruence model of brand preference: A theoretical and empirical study. *Journal of Marketing Research,* 427–433.
26. Team-bhp. http://www.team-bhp.com/. Accessed July 03, 2015.
27. Wilson, L. O., & Norton J. A. (1989). Optimal entry timing for a product line extension. *Marketing Science, 8*(1), 1–17.

Two-Dimensional Vulnerability Patching Model

Yogita Kansal and P. K. Kapur

Abstract In this paper, we develop a vulnerability patching model based on the nonhomogeneous Poisson process (NHPP) with different dimensions. Here, first, we assumed that the patching of discovered vulnerabilities can also cause patching of some additional vulnerabilities without causing any patch failure. This patching model is known as one-dimensional vulnerability patching model (1D-VPM) as it is only dependent on the time at which the vulnerabilities are patched. Further, we extend the one-dimensional vulnerability patching model by introducing the number of software users as a new dimension for software patching resources. In this two-dimensional model, we assume that the effort spent by users in installing the patches plays a major role in remediating the software vulnerabilities. It does not matter how quickly the vendor releases the patch until the users installed them correctly. Hence, we develop two-dimensional vulnerability patching model with patching time and software users as a two-dimensional vulnerability patching model (2D-VPM). Cobb–Douglas production function is used to create the two-dimensional patching model. An empirical study is performed on the vulnerability patching data (for Windows 8.1) to validate and compare the proposed models.

Keywords Vulnerability · Patching · One-dimensional · Two-dimensional vulnerability patching model · Cobb–Douglas production function

Y. Kansal (✉)
Amity Institute of Information Technology, Amity University, Noida, India
e-mail: ykansal35@gmail.com

P. K. Kapur
Amity Centre for Interdisciplinary Research, Amity University, 201313 Noida, Uttar Pradesh, India
e-mail: pkkapur1@gmail.com

P. K. Kapur et al. (eds.), *System Performance and Management Analytics*, Asset Analytics, https://doi.org/10.1007/978-981-10-7323-6_26

1 Introduction

Evolution of new and diverse business network infrastructure increases the potential security threats (also known as software vulnerabilities) that make patch management as the topmost priority for the vendors. Due to some external constraints like development cost, time to deliver, and unexpected changes in specifications, the developers are not able to penetrate and debug the vulnerabilities efficiently in testing phase. For such situations, software patching is considered to be a feasible solution during operations [1]. Software patch is a piece of code that helps the programmer in fixing the discovered vulnerability against exploitation. It keeps all the software users up to date with the enhanced features. For developing patches, a sufficient amount of resources are required to stay ahead of the hackers [2]. Most importantly, managers are supposed to allocate an enough time to developers for creating subsequent patches. Patches with short development time may cause misconfiguration and catastrophic errors. Thus, operational time/patch release time becomes the most crucial factor that was solely mathematically analyzed by researchers till now.

In general, the main thrust of research for vulnerability modeling literature focused on determining the optimal patch release time. Beattie et al. [3] proposed a mathematical cost model that extrapolates the best time to release the patches and covered the cost of vulnerability discovery due to attack and the cost of destruction due to flawed patches over time. Cavusoglu et al. [4] developed a game theoretic model for patch management and considered that the patch release and update policy is either time-driven or event-driven. Later, Okamura et al. [5] extended the work done by Cavusoglu et al. [2] and developed a patching model through nonhomogeneous vulnerability discovery process that evaluates the optimal patch release time. However, these authors have developed cost models for optimizing the patch release time. Recently, Kansal et al. [6] proposed a time-based vulnerability patching model that is driven by two factors: unsuccessful patching rate and successful patching rate.

From discussion, vulnerability patching models (VPMs) developed in past few years are solely depends on one factor, i.e., patching time thus it can be defined as one-dimensional vulnerability patching model (1D_VPM). However, these researchers have not contributed towards the vulnerabilities which are patched due to the presence of strong regression between the vulnerabilities. Moreover, no researcher has discussed the impact of software users on software patches. Since the success of the patches depends on its release time and the effort spent by users in installing the patches, it becomes utmost important to combine these two dimensions together in vulnerability patching model [7].

The main objective of this research paper is to propose vulnerability patching models under different assumptions and dimensions. Despite the availability of patches, the severity of the discovered vulnerability is compounded as the users are remained at risk due to the delay in patch deployment. Thus, we develop a time-based vulnerability patching model also known as one-dimensional vulnerability patching model (1D-VPM) that solely depends on patching time (the time at which vulnerabilities are patched). This model operates on an assumption that the patching of discovered vulnerabilities can also cause patching of some additional vulnerabilities without causing any patch failure. Since, knowing when your organization is in the sights of cyber-attackers is a difficult challenge, regular patching or time-based patching is most important.

Vulnerabilities can appear in almost any type of software, but the most attractive to targeted attackers is software that is widely used [8]. Software such as Internet Explorer and Adobe Flash are the applications in which majority of vulnerabilities are discovered because of the vast number of consumers. It has also been observed that when people have not moved quickly enough to apply patch (though the patch is available), at that point, hundreds of thousands of interconnected systems get infected. Therefore, we extend the proposed one-dimensional vulnerability patching model (1D_VDM) with the introduction of a new dimension, i.e., software users. Here, we have assumed that vulnerability discovery performed on a mass scale is substantially forcing the vendors to release a patch. It is just a matter of fact that if a small number of known or less critical vulnerabilities are reported then the probability of releasing a successful patch increases as the developers are then able to create patches with lesser resources. On the contrary, if large number of vulnerabilities or zero-day exploits are reported then the patching rate may decrease as large amount of resources are needed. Thus, here we scrutinized how the collaborative approach of software users and time can improve the success rate of patches.

For accomplishing the mentioned goals, we have used the Cobb–Douglas production function [9]. The modeling explains the behavior of software patches under the mentioned aspects and helps in decision-making problems. The proposed model measures the impact of new factors on software patches. Furthermore, we demonstrate the implications of our proposed one-dimensional approach by comparing it with existing models from literature, the Okamura model and Kansal et al. model. In addition, the results of two-dimensional vulnerability patching model are also compared with 1D-VPM. The research methodology begins with the proposition of one-dimensional and two-dimensional vulnerability patching models that focuses on finding the intensity at which the vulnerabilities are patched based on time and software users with time, respectively. Next, we obtain parameter estimates based on vulnerability data set for the two proposed models. Subsequently, various performance measures have been calculated for each of the two models.

Table 1 Notations used

Notation	Description
$\hat{\rho}(t)$	Expected number of patches released with respect to patching time t
$\hat{\rho}(r)$	Expected number of patches released with respect to patching resource r
A	Proportion of vulnerabilities patched independently
r	Patching resources
t	Patching time or patch release time
u	Number of software users
B	Actual number of patched vulnerabilities
C	Proportion of vulnerabilities patched dependently
Δ, δ	Intermediate variables

For numerical illustration purpose, we have manually extracted the vulnerability data set of Windows 8.1 from Common Vulnerability Exposure [10] database. We begin with the description of notations used throughout the paper discussed in Table 1.

The rest of the paper is organized as follows: Sect. 2 contains a detailed description of the proposed one-dimensional VPM. Section 3 provides a framework for two-dimensional VPM based on nonhomogeneous Poisson process. In Sect. 4, we validate the proposed models and compare them with the existing one through the extracted data set. Lastly, in Sect. 5, result interpretations and conclusions are drawn.

2 One-Dimensional Vulnerability Patching Model

When automated patching is not considered, patching the software vulnerabilities manually at regular intervals is one key aspect of secure software. Any substantial delay in patching may increase the frequency of attacks and the targets [11, 12]. Thus, the time is the major attribute that is accounted in this section for determining the intensity with which the vulnerabilities are patched. This model has defined one-dimensional stochastic process which represents the cumulative number of software failures by time t by $\{P(t), t \geq 0\}$. Here, we assumed that there is no patched vulnerability at time $t = 0$, i.e., $P(0) = 0$ with probability one. The patch failure intensity will decrease exponentially with the expected number of patched vulnerabilities. The one-dimensional NHPP is given as shown below:

$$\Pr[P(t) = n] = \frac{[\rho(t)]^n}{n!} \cdot \mathrm{e}^{-\rho(t)} \quad , \quad n = 0, 1, 2, \ldots, \tag{1}$$

Here, we also assumed that while patching the discovered vulnerabilities, some additional vulnerabilities that were not discovered are also patched. The following equation describes the one-dimensional vulnerability patching model:

$$\frac{\mathrm{d}\hat{\rho}}{\mathrm{d}t} = A \cdot (B - \hat{\rho}) + C \cdot \frac{\hat{\rho}}{B} \cdot (B - \hat{\rho}) \tag{2}$$

Solving the above equation under the initial condition $\hat{\rho}(t = 0) = 0$, we get,

$$\hat{\rho}(t) = \frac{B \cdot \left(1 - \mathrm{e}^{-(A+C) \cdot t}\right)}{1 + \frac{C}{A} \cdot \left(\mathrm{e}^{-(A+C) \cdot t}\right)} \tag{3}$$

If a large number of vulnerabilities are patched, then a time point may be reached where the rate of dependent patched vulnerabilities goes higher than the rate of unique patched vulnerabilities (i.e., $C < A$).

3 Two-Dimensional Vulnerability Patching Model

When patches are released, the success rate of patches depends on the effort applied by software users in deploying the patch and the time at which the vulnerability is patched. A study reveals that the probability of vulnerability being exploited is a function of the number of elapsed days from the date of vulnerability announcement to the date of its patch release [13]. Therefore, we can infer that the patching rate gets affected by both software users and the patching time. A new mathematical model named two-dimensional vulnerability patching model (2D_VPM) thus needs to develop.

To deal with the collective effect, we have used the Cobb–Douglas production function that represents the relationship between the dependent (output) and independent (input) variables [14]. In VPM, the patching time and software users are used as independent variables. These two variables are highly associated with unknown parameters denoted as α and β. The mathematical form of the Cobb–Douglas production function is given as

$$r \cong u^{\alpha} \cdot t^{\beta} \quad 0 \leq \alpha, \beta \leq 1 \tag{4}$$

where r collectively refers to the patching resources, u refers to the software users, and t refers to the patching time. Assuming the perfect availability of resources in operational phase, we have taken the α, β as the degree of impact to the vulnerability patching process. It may be noted that if $\alpha + \beta$ is not equal to 1, then it indicates that there are other external factors that may also affect the patching rate of vulnerabilities [15]. If r increases, the number of vulnerabilities patched ($\rho(t, u)$) will also increases. The two-dimensional model also follows nonhomogeneous Poisson process (NHPP). Our model has defined a two-dimensional stochastic process which represents the cumulative number of software failures by time t and with the usage of resources r by $\{P(t), t \geq 0, u \geq 0\}$. The two-dimensional NHPP is given as shown below

$$\Pr[P(t,u) = n] = \frac{[\rho(t,u)]^n}{n!} \cdot \mathrm{e}^{-\rho(t,u)}, \quad n = 0, 1, 2, \ldots, \tag{5}$$

We have considered the following assumptions for developing a 2D_VPM:

- More number of vulnerabilities are patched with more number of software users.
- Number of software users increases with time.

The proposed model focuses on improving software reliability such that the number of released/installed patches increases. The differential equation representing the rate of change of the cumulative number of successfully patched vulnerabilities with respect to the combined effect of time and users is given as

$$\frac{\mathrm{d}\hat{\rho}}{\mathrm{d}r} = A \cdot (B - \hat{\rho}) + C \cdot \frac{\hat{\rho}}{B} \cdot (B - \hat{\rho}) \tag{6}$$

Under the initial condition $\hat{\rho}(r = 0) = 0$ and solving the above equation, we get,

$$\hat{\rho}(r) = \hat{\rho}(u,t) = \frac{B \cdot \left(1 - e^{-(A+C)\cdot\left(u^{\alpha}\cdot t^{\beta}\right)}\right)}{1 + \frac{C}{A} \cdot \left(e^{-(A+C)\cdot\left(u^{\alpha}\cdot t^{\beta}\right)}\right)} \tag{7}$$

When $\alpha = 0$, the proposed two-dimensional VPM converts into time-dependent vulnerability patching model (that is similar to Eq. 3). If r increases, the number of patches released with respect to time and vulnerabilities, i.e., $\rho(v, t)$ increases.

4 Parameter Estimation and Comparison

In this section, we estimate the unknown parameters of the proposed one-dimensional VPM and two-dimensional VPM through statistical package for social sciences (SPSS). Nonlinear regression is performed in SPSS to model the dependent variables as a nonlinear function of model parameters.

Here, we have used the vulnerability data set of Windows 8.1 to validate the proposed model that is extracted from common vulnerability exposure (CVE) database [16]. The data set consists of a total of 232 vulnerabilities that are patched in the period of January 2016–May 2017 due to the effort spent by 297 software users. Figure 1 plots the number of patched vulnerabilities and number of software users with respect to time.

For the mentioned data set, we estimated the patching parameters A, B, C, α, β with Eqs. (3) and (7) as shown in Table 2. We performed estimations for both the proposed models independently. The idea of introducing the second dimension to the vulnerability patch modeling is to ensure the simultaneous impact of patching time and software users on the vulnerability patching rate. Therefore, the results obtained were compared for checking the significance of both models. Table 1

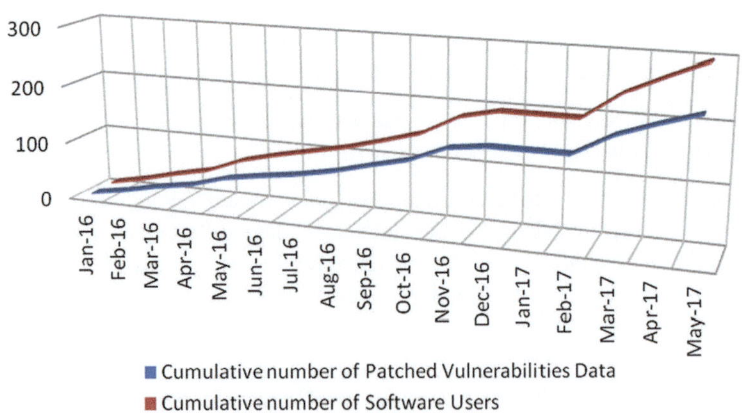

Fig. 1 Windows 8.1 Data from Jan 2016 to May 2017

Table 2 Parameter estimates of the proposed two-dimensional model

Model	B	A	C	α	β
1D_VPM	1616.46	0.048	0.329	–	–
2D_VPM	363.297	0.054	0.390	0.59	0.186

represents the obtained parameter estimation results. It gives the statistical coefficient value of the proposed models, and the value of α and β is significant for the proposed two-dimensional model as expected, indicating a high dependency on software users.

Note that we have observed that in 2D_VPM, the value of $\alpha + \beta$ is not equal to 1 that indicates that there are some unknown external factors that are affecting the patching rate. Figures 2 and 3 show the cumulative patched vulnerabilities for Windows 8.1, comparing the actual versus predicted values for the proposed 1D_VPM and 2D_VPM.

4.1 Model Comparison Criteria

1. The bias [17] is defined as the sum of difference between the estimated and actual data.

$$\text{Bias} = \frac{\sum_{i=1}^{k} (\hat{\rho}(t_i) - \rho_i)}{k} \tag{8}$$

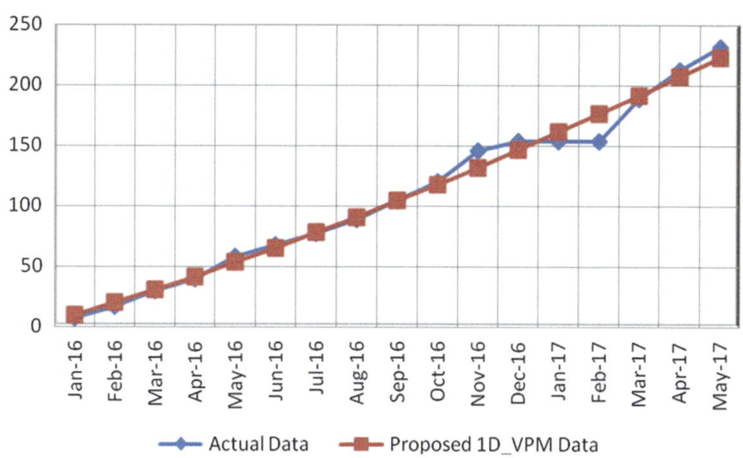

Fig. 2 Comparison of actual versus predicted data for Windows 8.1 through 1D_VPM

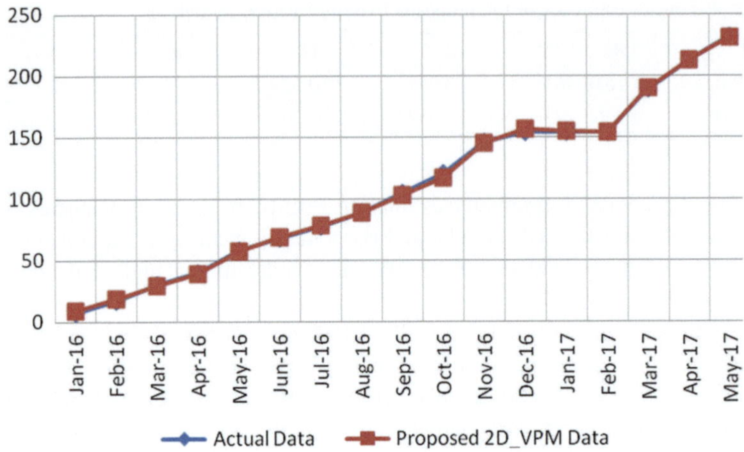

Fig. 3 Comparison of actual versus predicted data for Windows 8.1 through 2D_VPM

2. The mean square error (MSE) [4] measures the deviation between the predicted values with actual observations

$$\mathrm{MSE} = \frac{\sum_{i=1}^{k} (\rho_i - \hat{\rho}(t_i))^2}{k - n} \tag{9}$$

3. The predictive ratio risk (PRR) [4] measures the error between actual and estimated values

$$\mathrm{PRR} = \sum_{i=1}^{k} \frac{\hat{\rho}(t_i) - \rho_i}{\hat{\rho}(t_i)} \tag{10}$$

4. The standard deviation (SD) [17] measures as

$$\mathrm{SD} = \sqrt{\frac{1}{k-1} \sum_{i=1}^{k} (\rho_i - \hat{\rho}(t_i) - \mathrm{Bias})^2} \tag{11}$$

5. The root mean square prediction error (RMSPE) [17] measures the closeness of model predicts

$$\mathrm{RMSPE} = \sqrt{\mathrm{SD}^2 + \mathrm{Bias}^2} \tag{12}$$

6. R^2 (Coefficient of Determination) measures the successful fit rate in variation of data.

$$R^2 = 1 - \frac{\sum_{i=1}^{k} (\rho_i - \hat{\rho}(t_i))^2}{\sum_{i=1}^{k} \left(\rho_i - \sum_{j=1}^{k} \rho_j / k \right)^2} \tag{13}$$

where $\hat{\rho}(t_i)$ represents the estimated data, ρ_i represents the actual data, and k represents the number of data points. Higher the value of R^2 the better the fitness. On the contrary, lower the value of MSE, Bias, PRR, SD, and RMSPE, the better the fitness.

Since the value of bias, mean square error (MSE), predictive ratio risk (PRR), standard deviation (SD), and root mean square percentage error (RMSPE) for the proposed two-dimensional VPM were observed to be lower than the proposed one-dimensional (as shown in Table 3), and 2D_VPM is proved to be significantly fit with the software patching data of Windows 8.1. The results imply that while considering patch modeling for discovered vulnerabilities, software users are of utmost importance that improve the performance of patching rate.

Figure 4 represents the goodness of fit comparison between proposed one-dimensional VPM and two-dimensional VPM on the basis of criteria discussed above. As discussed, the lower value of criteria such as MSE, Bias, PRR, SD, and RMSPE represents the better fitness while the higher value of (R^2) represents the better fitness. From Table 2 and Fig. 4, the 2D_VPM is proved to have better fitness than 1D_VPM.

Table 3 Comparison of performance measures of existing and proposed models

Model	Bias	MSE	PRR	SD	RMSPE	R^2
One-dimensional model	0.163	67.799	0.2846	0.7371	0.7549	0.987
Two-dimensional model	0.142	11.414	0.2564	0.6390	0.6545	0.998

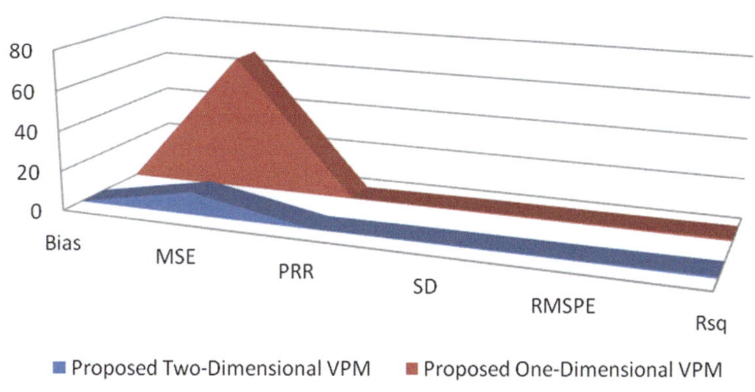

Fig. 4 Goodness of fit measures for one-dimensional and two-dimensional model

5 Conclusions

We have proposed a new time-based vulnerability patching model under the assumption that the patching of discovered vulnerabilities can also cause patching of some additional vulnerabilities without causing any patch failure. Further, we have extended the proposed one-dimensional vulnerability patching model to two-dimensional vulnerability patching model by introducing software users as new dimension. We have used Cobb–Douglas production function for showing the combined effect of software users and patching time. Moreover, the performance of all the three models is measured on the basis of goodness of fit criteria like bias, mean square error (MSE), predictive ratio risk (PRR), standard deviation (SD), root mean square percentage error (RMSPE), and coefficient of determination (R^2).

Here, we have analyzed that the performance measures of two-dimensional vulnerability patching model are significantly better than the one-dimensional model. This research proves that other than patching time, the number of software users also has substantial effect on software patches that may help in improving the prediction. In future, patch release time of discovered vulnerabilities can be optimized using the best fitted proposed two-dimensional vulnerability patching model.

References

1. Ozment, A., & Schechter, S. E. (2006, July). Milk or wine: Does software security improve with age? In *Usenix Security*.
2. Cavusoglu, H., Cavusoglu, H., & Zhang, J. (2008). Security patch management: Share the burden or share the damage? *Management Science, 54*(4), 657–670.
3. Beattie, S., Arnold, S., Cowan, C., Wagle, P., Wright, C., & Shostack, A. (2002, November). Timing the application of security patches for optimal uptime. In LISA (Vol. 2, pp. 233–242); Bilge, L., & Dumitras, T. (2012, October). Before we knew it: An empirical study of zero-day attacks in the real world. In *Proceedings of the 2012 ACM Conference on Computer and Communication Security* (pp. 833–844). ACM.
4. Cavusoglu, H., Cavusoglu, H., & Zhang, J. (2006). Economics of security patch management. In *The Fifth Workshop on the Economics of Information Security (WEIS06)*.
5. Okamura, H., Tokuzane, M., & Dohi, T. (2009, November). Optimal security patch release timing under non-homogeneous vulnerability-discovery processes. In *2009 20th International Symposium on Software Reliability Engineering* (pp. 120–128). IEEE.
6. Kansal, Y., Kumar, D., & Kapur, P. K. (2016). Vulnerability patch modeling. *International Journal of Reliability, Quality and Safety Engineering, 23*(06), 1640013.
7. Burum, S., & Holmes, G. Apple v. FBI: Privacy vs. Security? *National Social Science, 9*.
8. https://www.symantec.com/content/dam/symantec/docs/reports/istr-21-2016-en.pdf.
9. Sandelin, B. (1976). On the origin of the Cobb-Douglas production function. *Economy and History, 19*(2), 117–123.
10. Sharma, K., Garg, R., Nagpal, C. K., & Garg, R. K. (2010). Selection of optimal software reliability growth models using a distance based approach. *IEEE Transactions on Reliability, 59*(2), 266–276.
11. Arora, A., Telang, R., & Xu, H. (2004). Optimal time disclosure of software vulnerabilities. In *Conference on Information Systems and Technology*, Denver CO, October, 23–24.

12. Arora, A., Telang, R., & Xu, H. (2008). Optimal policy for software vulnerability disclosure. *Management Science, 54*(4), 642–656.
13. Schryen, G. (2011). Is open source security a myth? *Communications of the ACM, 54*(5), 130–140.
14. Rescorla, E. (2005). Is finding security holes a good idea? *IEEE Security and Privacy, 3*(1), 14–19.
15. Sachdeva, N., Kapur, P. K., & Singh, O. (2016). An innovation diffusion model for consumer durables with three parameters. *Journal of Management Analytics, 3*(3), 240–265.
16. Common Vulnerability Exposure [online]: www.cvedetails.com.
17. Arora, A., Krishnan, R., Nandkumar, A., Telang, R., & Yang, Y. (2004, May). Impact of vulnerability disclosure and patch availability-an empirical analysis. In *Third Workshop on the Economics of Information Security* (Vol. 24, pp. 1268–1287).
18. Dohi, T., & Yun, W. Y. (2006). Advanced reliability modeling II: Reliability testing and improvement. In *Proceedings of the 2nd Asian International Workshop (AIWARM 2006)*, Busan, Korea, 24–26 August 2006. World Scientific.
19. Kapur, P. K., Pham, H., Gupta, A., & Jha, P. C. (2011). *Software reliability assessment with OR applications*. London: Springer.
20. Kapur P. K., Sachdeva, N., & Khatri, S. K., (2015). Vulnerability discovery modeling. In *International Conference on Quality, Reliability, Infocom Technology and Industrial Technology Management* (pp. 34–54).

A Hybrid Intuitionistic Fuzzy and Entropy Weight Based Multi-Criteria Decision Model with TOPSIS

Nitin Sachdeva and P. K. Kapur

Abstract In a scenario where decision-makers are always faced with the challenge of selecting the right technology for their IT needs posed due to the availability of multiple advanced technologies in the market and consequences related to wrong selection, Intuitionistic Fuzzy Sets (IFSs) have demonstrated effectiveness in dealing with such vagueness and hesitancy in the decision-making process. Here in this paper, we propose a hybrid IFS and entropy weight based Multi-Criteria Decision Model (MCDM) with Technique for Order of Preference by Similarity to Ideal Solution (TOPSIS) method. The model helps measure the exactness and vagueness of each alternative over several criteria. An Intuitionistic Fuzzy Weighted Approach (IFWA) operator for aggregating individual decision-maker's opinions regarding each alternative over every criterion is employed. Additionally, Shannon's entropy method is used to measure criteria weights separately. We apply the proposed model in selection of cloud solution for managing big data projects.

Keywords Intuitionistic fuzzy set (IFS) · Intuitionistic fuzzy weighted approach (IFWA) · TOPSIS · Big data · MCDM · Cloud solution

1 Introduction

Literature provides for several MCDMs applied to varied fields of engineering, economics, social sciences, policy making, management, etc. Zadeh [21] introduced the concept of fuzzy sets, by assigning to each set element a value between 0 and 1 as single membership. Later on, this single value concept was challenged by Gau and Buehrer [5], as it failed to confirm the accuracy and, hence came the

N. Sachdeva (✉)
Institute of Management Technology, Ghaziabad, Uttar Pradesh, India
e-mail: nitin.sach@gmail.com

P. K. Kapur
Amity Center for Interdisciplinary Research, Amity University, 201313 Noida, Uttar Pradesh, India
e-mail: pkkapur1@gmail.com

P. K. Kapur et al. (eds.), *System Performance and Management Analytics*, Asset Analytics, https://doi.org/10.1007/978-981-10-7323-6_27

concept of vague sets. Bustince and Burillo [4] suggested that this notion of vague set relates to Atanassov [3] concept of Intuitionistic Fuzzy Sets (IFSs).

Intuitionistic Fuzzy Sets are often characterized by membership value and nonmembership value of elements. The whole idea of highlighting the membership and nonmembership is to manage both the ambiguity and uncertainty inherently present in almost all decision-making process majorly due to either incomplete or imprecise information. In the past two decades or so, IFSs have been applied in the field of medicines [6, 12, 13] multi-criteria decision problems [1, 8, 17, 19] and also in recognizing patterns [16, 18, 23].

Further, one of the classical MCDM approaches, TOPSIS (technique for order performance by similarity to ideal solution) has been proposed by Hwang and Yoon [7]. In TOPSIS, the most preferred alternative has the shortest distance from the positive ideal solution and farthest distance from the negative ideal solution, simultaneously [7, 22]. Some of the major advantages of using TOPSIS are the simplicity of both understanding and implementation along with good computational efficiency and the ability to measure the relative performance of each alternative in an objective mathematical form [20].

Here in this paper, we propose a hybrid intuitionistic fuzzy [3] and entropy [10] based MCDM to understand the various cloud solutions available in the market for handling big data projects in terms of their important characteristics being evaluated using the proposed approach. We then apply the famous TOPSIS method in order to generate a rank among these alternatives based on some predefined criteria. The biggest challenge faced by almost all decision-makers is to crisply express the impact of various alternatives on these criteria. To resolve this conflict of ambiguity, we make use of Intuitionistic Fuzzy Sets in this Multi-Criteria Decision-Making environment. In this process, we aggregate individual decision-maker's opinions regarding each criterion over every shortlisted alternative using an IFWA operator. We then employ Shannon's entropy method on the decision-maker's opinion on each criterion separately. Finally, deviations of these aggregations from both positive and negative ideal solution are carried out using famous TOPSIS approach to finally reach a ranking of these shortlisted alternatives.

Five major market players in the area of cloud computing used primarily because of their capability to handle big data projects are considered: HP Cloud, Microsoft Azure, Rackspace, Amazon, and Google Cloud Platform.

2 Proposed Methodology

2.1 Preliminaries

1. **Intuitionistic Fuzzy Set**

Atanassov [3] extended the classical Fuzzy Set Theory to propose Intuitionistic Fuzzy Set in order to address the fundamental problem of vagueness likely to be present in almost any decision-making process. An Intuitionistic Fuzzy Set I_{FS} in a finite set Y may be easily defined as

$$I = \{(y, \alpha_I(y), \beta_I(y))|y \in Y\}$$

where $\alpha_I(y)$ and $\beta_I(y) : Y \to [0, 1]$ are membership and nonmembership function, respectively, with the condition,

$$0 \leq \alpha_I(y) + \beta_I(y) \leq 1$$

$\forall\, I$ in Y, the third parameter is $\chi_I(y)$, called the intuitionistic fuzzy index or hesitation degree of whether y belongs to I or not

$$\chi_I(y) = 1 - \alpha_I(y) - \beta_I(y) \tag{1}$$

It is obviously seen that for every $y \in Y,\ 0 \leq \chi_I(y) \leq 1$.

Smaller value of $\chi_I(y)$ implies that the knowledge about y is more certain as compared to a large $\chi_I(y)$ value. Obviously, when $\alpha_I(y) = 1 - \beta_I(y)\ \forall\, y$, the ordinary fuzzy set concept is recovered [11]. Let C and D be IFSs of the set Y, then multiplication operator is defined as follows [3]:

$$C \otimes D = \{\alpha_C(y) \cdot \alpha_D(y), \beta_C(y) + \beta_D(y) - \chi_C(y) \cdot \chi_D(y)|y \in Y\} \tag{2}$$

2. **Entropy of IFS**

Shannon [10] proposed the entropy function in 1948 $H(p_1, p_2, p_3, \ldots, p_n) = -\sum_{i=1}^{n} p_i \log(p_i)$ as a measure of hesitation in a discrete distribution wherein, $p_i (i = 1, 2, 3, \ldots, n)$ are computed as random variable probabilities using P, probability mass function. Later, a non-probabilistic based entropy of Shannon's function on a finite universal set $Y = \{y_1, y_2, y_3, \ldots, y_n\}$ was proposed by De Luca and Termini [9] and can be given as

$$E_{\mathrm{LT}}(I) = -k \sum_{i=1}^{n} [\alpha_I(x_i) \ln \alpha_A(y_i) + (1 - \alpha_I(y_i)) \ln(1 - \alpha_I(y_i))], \quad k > 0 \tag{3}$$

Here, we make use of entropy as a measure of uncertainty in the decision-making process while selecting an appropriate cloud solution for managing big data. Entropy measurement helps us to evaluate the joint entropy when we are dealing with more than one variable. The proposed research discusses a hybrid intuitionistic based MCDM framework to help evaluate the inherited uncertainty in the decision-making process by co-incorporating entropy concept. When the probability is uniformly distributed among the variables, we have the highest uncertainty with the outcome. This means entropy is maximum in such cases. MCDM frameworks like the one proposed here incorporate the idea to measure this uncertainty in the system by making use of joint entropy.

De Luca and Termini [9] axioms were then extended by Szmidt and Kacprzyk [12, 15] to enhance their entropy measure on IFSs(Y). Lately, following equation has been proposed by Vlachos et al. [16] that kind of satisfies all the four axiomatic requirements to measure intuitionistic fuzzy entropy:

$$E_{\mathrm{LT}}^{\mathrm{IFS}}(A) = -\frac{1}{n \ln 2} \sum_{i=1}^{n} [\alpha_I(y_i) \ln \alpha_I(y_i) + \beta_I(y_i) \ln \beta_I(y_i) \\ -(1-\chi_I(y_i)) \ln(1-\chi_I(y_i)) - \chi_I(y_i) \ln 2] \tag{4}$$

It is noted that $E_{\mathrm{LT}}^{\mathrm{IFS}}(A)$ is composed of the hesitancy degree and the fuzziness degree of the IFS I.

2.2 Proposed Intuitionistic Fuzzy and Entropy-Based Decision Model with TOPSIS

Let $I = \{I_1, I_2, \ldots, I_m\}$ be the alternative set, and $Y = \{Y_1, Y_2, \ldots, Y_n\}$ be the criteria set. Now, we present the algorithm to evaluate the proposed hybrid model, Fig. 1.

Step 1 ***Decision-Makers Weights***

We make use of the linguistic terms expressed in intuitionistic fuzzy numbers to express say 'l' decision-maker's importance.

Let $D_k = (\alpha_k, \beta_k, \chi_k)$ be an intuitionistic fuzzy number for a rating of the kth decision-maker. Then, the weight of kth decision-maker can be obtained as:

$$\psi_k = \frac{\left(\alpha_k + \chi_k\left(\frac{\alpha_k}{\alpha_k+\beta_k}\right)\right)}{\sum_{k=1}^{l}\left(\alpha_k + \chi_k\left(\frac{\alpha_k}{\alpha_k+\beta_k}\right)\right)} \quad \text{Where,} \sum_{k=1}^{l} \psi_k = 1 \tag{5}$$

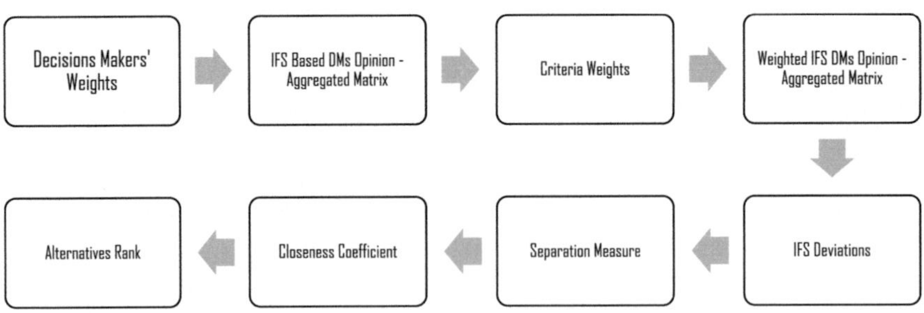

Fig. 1 Step-by-step procedure for applying the proposed hybrid model

Step 2 ***IFS-Based DMs Opinion—Aggregated Matrix***

Let $N^{(k)} = (n_{ij}^{(k)})_{mxn}$ be an intuitionistic fuzzy decision matrix of each decision-maker. $\psi = \{\psi_1, \psi_2, \psi_3, \ldots \psi_l\}$ is the weight of each decision-maker and $\sum_{k=1}^{l} \psi_k = 1, \psi_k \in [0, 1]$. In group decision-making process, all the individual decision opinions need to be fused into a group opinion to construct aggregated intuitionistic fuzzy decision matrix. In order to do that, IFWA operator proposed by Xu [19] is used. $N = (\mathrm{n}_{ir})_{mxn}$, where

$$\begin{aligned} n_{ij} &= \mathrm{IFWA}_{\psi}\left(n_{ir}^{(1)}, n_{ir}^{(2)} \ldots n_{ir}^{(l)}\right) \\ &= \psi_1 n_{ir}^{(1)} \oplus \psi_2 n_{ir}^{(2)} \oplus \cdots \oplus \psi_l n_{ir}^{(l)} \\ &= \left[1 - \prod_{k=1}^{l}\left(1 - \alpha_{ir}^{(k)}\right)^{\psi_k}, \prod_{k=1}^{l}\left(\beta_{ir}^{(k)}\right)^{\psi_k}, \prod_{k=1}^{l}\left(1 - \alpha_{ir}^{(k)}\right)^{\psi_k} - \prod_{k=1}^{l}\left(\beta_{ir}^{(k)}\right)^{\psi_k}\right] \end{aligned} \tag{6}$$

Here $n_{ir} = (\alpha_{I_i}(y_r), \beta_{I_i}(y_r), \chi_{I_i}(y_r))\ (i = 1, 2 \ldots m; r = 1, 2, 3 \ldots n)$.

The aggregated intuitionistic fuzzy decision matrix can be defined as follows:

$$N = \begin{bmatrix} (\alpha_{I_1}(y_1), \beta_{I_1}(y_1), \chi_{I_1}(y_1)) & (\alpha_{I_1}(y_2), \beta_{I_1}(y_2), \chi_{I_1}(y_2)) & \ldots & (\alpha_{I_1}(y_n), \beta_{I_1}(y_n), \chi_{I_1}(y_n)) \\ (\alpha_{I_2}(y_1), \beta_{I_2}(y_1), \chi_{I_2}(y_1)) & (\alpha_{I_2}(y_2), \beta_{I_2}(y_2), \chi_{I_2}(y_2)) & \ldots & (\alpha_{I_2}(y_n), \beta_{I_2}(y_n), \chi_{I_2}(y_n)) \\ \cdot & \cdot & & \cdot \\ \cdot & \cdot & & \cdot \\ (\alpha_{I_m}(y_1), \beta_{I_m}(y_1), \chi_{I_m}(y_1)) & (\alpha_{I_m}(y_1), \beta_{I_m}(y_2), \chi_{I_m}(y_2)) & \ldots & (\alpha_{I_m}(y_n), \beta_{I_m}(y_n), \chi_{I_m}(y_n)) \end{bmatrix}$$

$$N = \begin{bmatrix} n_{11} & n_{12} & \ldots & n_{1n} \\ n_{21} & n_{22} & \ldots & n_{2n} \\ \cdot & \cdot & & \cdot \\ \cdot & \cdot & & \cdot \\ n_{m1} & n_{m2} & \ldots & n_{mn} \end{bmatrix} \tag{7}$$

Step 3 ***Criteria Weights.***

Let $w_r^k = \left\{\alpha_r^{(k)}, \beta_r^{(k)}, \chi_r^{(k)}\right\}$ be defined as an IFS element denoting criterion Y_j by the kth decision-maker. Then using entropy method [7, 22], the criteria weights are evaluated as

$$\begin{aligned} E_{\mathrm{LT}}^{\mathrm{IFS}}(C_r) = -\frac{1}{m \ln 2}\sum_{i=1}^{m} & [\alpha_{ir}(C_r) \ln \alpha_{ir}(C_r) + \beta_{ir}(C_r) \ln \beta_{ir}(C_r) \\ & -(1 - \chi_{ir}(C_r)) \ln(1 - \chi_{ir}(C_r)) - \chi_{ir}(C_r) \ln 2] \end{aligned} \tag{8}$$

Such that, $1/(m \ \ln 2)$ is fixed with $0 \leq E_{\mathrm{LT}}^{\mathrm{IFS}}(C_j) \leq 1$ and $j = 1, 2, \ldots n$. Accordingly d_j, the degree of divergence expected on the intrinsic information provided on criterion C_j is given as

$$d_r = 1 - E_{\mathrm{LT}}^{\mathrm{IFS}}(C_r), \quad r = 1, 2, 3, \ldots n. \tag{9}$$

The value of d_r represents the inherent contrast intensity of criterion C_r, then the entropy weight of the rth criterion is

$$w_r = \frac{d_r}{\sum_{r=1}^{n} d_r} \tag{10}$$

Step 4 ***Weighted IFS DMs Opinion—Aggregated Matrix***

Now, we construct the aggregated weighted intuitionistic fuzzy decision matrix using the definition

$$\overline{Z} = W^T \otimes \overline{D} = W^T \otimes [\overline{y}_{ir}]_{m \times n} = \left[\overline{\overline{y}}_{ir}\right]$$

where,

$$W = (w_1, w_2, w_3, \ldots w_r, \ldots w_n)$$

$$\overline{\overline{y}}_{ir} = \left\langle \overline{\overline{\alpha}}_{ir}, \overline{\overline{\beta}}_{ir}, \overline{\overline{\chi}}_{ir} \right\rangle = \langle 1 - (1 - \alpha_{ir})^{w_r}, \beta_{ir}^{w_r}, 1 - (1 - (1 - \alpha_{ir})^{w_r}) - \beta_{ir}^{w_r} \rangle, \quad w_r > 0 \tag{11}$$

Step 5 ***IFS Deviations***

In general, the evaluation criteria R_1 and R_2 are benefit and loss (cost) criteria, respectively. δ^+ and δ^- denote the Intuitionistic Fuzzy Positive (IFPIS) and Fuzzy Negative Ideal Solution (IFNIS), which are defined as follows:

$$\delta^+ = \left(\overline{\overline{\alpha}}_{\delta^+ W}(C_r), \overline{\overline{\beta}}_{\delta^+ W}(C_r)\right) \quad \text{and } \delta^- = \left(\overline{\overline{\alpha}}_{\delta^- W}(C_r), \overline{\overline{\beta}}_{\delta^- W}(C_r)\right)$$

where

$$\begin{aligned}
\overline{\overline{\mu}}_{\delta^+ W}(C_r) &= \left(\left(\max_i \overline{\overline{\alpha}}_{\delta_i.W}(C_r) | r \in R_1\right), \left(\min_i \overline{\overline{\mu}}_{\delta_i.W}(C_r) | r \in R_2\right)\right) \\
\overline{\overline{\upsilon}}_{\delta^+ W}(C_r) &= \left(\left(\min_i \overline{\overline{\beta}}_{\delta_i.W}(C_r) | r \in R_1\right), \left(\max_i \overline{\overline{\alpha}}_{\delta_i.W}(C_r) | r \in R_2\right)\right) \\
\overline{\overline{\mu}}_{\delta^- W}(C_r) &= \left(\left(\min_i \overline{\overline{\alpha}}_{\delta_i.W}(C_r) | r \in R_1\right), \left(\max_i \overline{\overline{\alpha}}_{\delta_i.W}(C_r) | r \in R_2\right)\right) \\
\overline{\overline{\upsilon}}_{\delta^- W}(C_r) &= \left(\left(\max_i \overline{\overline{\beta}}_{\delta_i.W}(C_r) | r \in R_1\right), \left(\min_i \overline{\overline{\beta}}_{\delta_i.W}(C_r) | r \in R_2\right)\right)
\end{aligned} \tag{12}$$

Step 6 ***Separation Measures***

We now assess the separation between alternatives on IFS using distance measures [2, 12]. These separation measures, $d_{\text{IFS}}(\delta_i, \delta^+), d_{\text{IFS}}(\delta_i, \delta^-)$, are calculated using intuitionistic Euclidean distance [14] for each alternative from Intuitionistic Fuzzy Positive Ideal and Negative Ideal Solutions to finally get the ranks of each alternative

$$\begin{aligned} d_{\text{IFS}}(\delta_i, \delta^+) &= \sqrt{\sum\nolimits_{r=1}^{n} \left[(\alpha_{\delta_i W}(C_r) - \alpha_{\delta^+ W}(C_r))^2 + (\beta_{\delta_i W}(C_r) - \beta_{\delta^+ W}(C_r))^2 + (\chi_{\delta_i W}(C_r) - \chi_{\delta^+ W}(C_r))^2 \right]} \\ d_{\text{IFS}}(\delta_i, \delta^-) &= \sqrt{\sum\nolimits_{j=1}^{n} \left[(\alpha_{\delta_i W}(C_j) - \mu_{\delta^- W}(C_r))^2 + (\beta_{\delta_i W}(C_r) - \beta_{\delta^- W}(C_r))^2 + (\chi_{\delta_i W}(C_r) - \chi_{\delta^- W}(C_r))^2 \right]} \end{aligned} \tag{13}$$

Step 7 ***Closeness Coefficient***

Now the closeness coefficient of an alternative δ_i from its IFPIS δ_i^+ is defined as follows:

$$\text{CC}_{i^+} = \frac{d_{\text{IFS}}(\delta_{ir}, \delta_i^-)}{d_{\text{IFS}}(\delta_{ii}, \delta_i^+) + d_{\text{IFS}}(\delta_{ii}, \delta_i^-)}, \quad \text{where, } 0 \leq \text{CC}_{i^+} \leq 1, \quad i = 1, 2, 3, \ldots m. \tag{14}$$

So based on the CC_{i^+} value, the alternatives are ranked in a descending order highlighting the closeness to IFPIS. So the alternative with the highest CC value is chosen as the best.

Step 8 ***Alternatives Rank***

Based on the CC_{i^+} values, the ranking is assigned based on its decreasing order.

3 Empirical Illustration—Which Cloud Solution for Big Data Projects?

Corporates, big or small sized, today are seemingly worried about managing their data. The three most important concerns facing them include: e-governance, business continuity and security. Although the literature provides for several other criteria we tend to restrict our study to these three main criteria and assume that the rest of the criterions like monitoring system, IT capital expenditures, implementation cost, confidentiality, etc. can always be subcategorized into the three criteria chosen for this study. Using these three important criteria as the basis of our proposed hybrid methodology, we provide for an interesting framework to help decision-makers evaluate five cloud solutions currently available in the market to help them analyze big data projects. Our hybrid approach as explained in the last section is based on the intuitionistic fuzzy and entropy method wherein the idea is to aggregate decision-maker's opinions about the solutions considering the shortlisted criteria and develop a ranking methodology using TOPSIS to help managers select the most effective solution for their big data projects need. The five existing market players: Amazon, HP Cloud, Google, Rackspace, and Microsoft Azure are shortlisted and evaluated using our proposed methodology based on the three selected criteria: e-governance, business continuity, and security.

Step 1 ***Decision-Makers Weights***

The importance of each decision-maker is calculated using Eq. (5) and the linguistic terms presented in Table 1. These weights are shown in Table 2.

One of the key features of making use of the intuitionistic fuzzy based methodology is to incorporate each decision-makers weight in the decision-making process. As a matter of fact, no one expert is capable of taking an objective decision based on his experience but can truly provide useful insights into the process depending on his capabilities. IFS ensures that their individual capabilities are well thought of while making use of their judgments in the process.

Step 2 ***IFS-Based DMs Opinion—Aggregated Matrix***

Using linguistic terms presented in Table 3, DMs opinion of each alternative based on the selected criteria is recorded, Table 4 and the final aggregated matrix is shown in Table 5 obtained using Eq. (7).

Step 3 ***Criteria Weights***

Now, we compute the weights of each criterion by applying entropy method using Eqs. (8)–(10) and the results are shown in Table 6. Clearly, the inherited uncertainty in assigning weights to the criteria is handled by making use of the entropy function. To evaluate weights of each criterion, we calculate divergence using Eq. (9). By making use of these average distance measures, we then evaluate the criteria weights using Eq. (10) which includes the inherited uncertainty in deciding these weights along with average intrinsic information.

Table 1 Linguistic terms

Linguistics terms		Very important	Important	Medium	Unimportant	Very unimportant
IFNs	α	0.9	0.75	0.5	0.35	0.1
	β	0.1	0.2	0.45	0.6	0.9

Table 2 Decision-makers weights

ψ_i		
I	II	III
Very important	Medium	Important
0.406	0.238	0.356

Table 3 Linguistics terms

Linguistics terms	Extremely good (EG)	Very very good (VVG)	Very good (VG)	Good (G)	Medium good (MG)	Fair (F)	Medium bad (MB)	Bad (B)	Very bad (VB)	Very very bad (VVB)	
IFNs	α	1	0.9	0.8	0.7	0.6	0.5	0.4	0.25	0.1	0.1
	β	0	0.1	0.1	0.2	0.3	0.4	0.5	0.6	0.75	0.9

Step 4 ***Weighted IFS DMs Opinion—Aggregated Matrix***

We now construct the aggregated weighted intuitionistic fuzzy decision matrix that includes: (1) importance of each decision-maker; (2) weights assigned to each alternative by DMs; (3) aggregated decision matrix including both importance of each DM and weights assigned by them to each alternative; and (4) weights of each criterion based on inherited uncertainty using entropy. This proposed aggregated weighted IFS-entropy based decision matrix is constructed using Eq. (11) and the results are shown in Table 7:

Step 5 ***IFS Deviations***

Now, the IFPIS and IFNIS are evaluated in order to prioritize the alternatives based on their closeness to the ideal solutions. We now make use of Eq. (12) and evaluate both the positive and negative ideal solutions and the results are shown in Table 8:

Step 6 and 7 ***Separation Measures and Closeness Coefficient***

Finally, negative and positive separation measures of each alternative along with their closeness coefficients, Table 9, are calculated using Eqs. (13) and (14).

Stage 8 ***Alternatives Rank***

Final ranking based on CC_{i+}'s is: *Google>Amazon>Rackspace>HP Cloud>Microsoft Azure.*

4 Conclusion

In the last two decades, several applications of Intuitionistic Fuzzy Sets as a Multi-Criteria Decision-Making approach have been proposed in wide areas of engineering and management. In this paper, we propose a hybrid intuitionistic entropy based decision-making model with TOPSIS to evaluate and rank the most suitable cloud solution for big data projects implementation. In information theory, entropy as a measure has been used in relation to the average information available for a source. With the help of Intuitionistic Fuzzy Set theory, we first assign importance to the decision-makers and evaluate the importance of each alternative over predefined criteria. Then based on the principle of entropy, the optimal criteria weights are obtained using Shannon's entropy method. We then make use of IFWA operator to create an aggregated weighted intuitionistic fuzzy and entropy-based matrix. Finally, relatively best alternative is chosen based on the classical TOPSIS methodology. The proposed hybrid model stands apart from the existing MCDMs in literature because of its simple application procedure and how the introduction of objective entropy weight in an intuitionistic fuzzy environment with TOPSIS can help rank various alternatives.

Table 4 Alternative weights based on criteria

Criteria		e-governance (X_1)					Business continuity (X_2)					Security (X_3)				
Cloud solutions		Amazon	HP Cloud	Google	RS[a]	MS Azure	Amazon	HP Cloud	Google	RS[a]	MS Azure	Amazon	HP Cloud	Google	RS[a]	MS Azure
Decision-makers	I	G	MG	VVG	MG	F	MG	F	VG	F	MB	VG	G	VG	VG	G
	II	VG	G	VG	G	MG	G	MG	G	F	F	G	MG	VG	G	G
	III	G	F	VG	G	MG	MG	G	VG	MG	F	VG	MG	G	G	MG

[a]*RS*, Rackspace

Table 5 Intuitionistic fuzzy aggregated matrix

Stage 2		e-Gov			Business continuity			Security		
		α	β	χ	α	β	χ	α	β	χ
Aggregated IF decision matrix based on DM	Amazon	0.728	0.17	0.102	0.626	0.272	0.102	0.78	0.118	0.102
	HP Cloud	0.596	0.302	0.102	0.605	0.292	0.103	0.644	0.254	0.102
	Google	0.849	0.1	0.051	0.78	0.118	0.102	0.769	0.128	0.103
	Rackspace	0.663	0.236	0.101	0.538	0.361	0.101	0.746	0.151	0.103
	MS Azure	0.562	0.337	0.101	0.462	0.438	0.1	0.668	0.231	0.101

Table 6 Criteria weights

Criteria	e-governance	Business continuity	Security
Entropy $(E_{\text{LT}}^{\text{IFS}}(C_r))$	0.7957	0.8793	0.7285
Distance measure (d_r)	0.2043	0.1207	0.2715
Weights (w_r)	0.3425	0.2024	0.4551

Table 7 Aggregated weighted IF decision matrix

Stage 4		e-governance			Business continuity			Security		
		α	β	χ	α	β	χ	α	β	χ
Aggregated weighted IFS-entropy decision matrix	Amazon	0.3598	0.545	0.0952	0.1805	0.7683	0.0512	0.498	0.3781	0.1239
	HP Cloud	0.2669	0.6636	0.0695	0.1714	0.7795	0.0491	0.375	0.536	0.089
	Google	0.4766	0.4545	0.0689	0.264	0.6489	0.0871	0.4867	0.3924	0.1209
	Rackspace	0.311	0.6099	0.0791	0.1447	0.8137	0.0416	0.464	0.423	0.113
	MS Azure	0.2463	0.689	0.0647	0.1179	0.8461	0.036	0.3946	0.5133	0.0921

Table 8 IFPIS and IFNIS

Stage 5	e-governance			Business continuity			Security		
	α	β	χ	α	β	χ	α	β	χ
δ^+	0.201943	0.429374	0.337811	0.298996	0.451199	0.201943	0.429374	0.337811	0.298996
δ^-	0.086031	0.008562	0.205006	0.033716	0.000909	0.086031	0.008562	0.205006	0.033716

Table 9 Separation measures and closeness coefficient

Separation measures and closeness coefficient		Amazon	HP Cloud	Google	Rackspace	MS Azure
	$d_{\text{IFS}}(\delta_i, \delta^+)$	0.4494	0.6553	0.5812	0.5468	0.6717
	$d_{\text{IFS}}(\delta_i, \delta^-)$	0.2933	0.0925	0.4528	0.1836	0.0302
	CC_{i^+}	0.3949	0.1237	0.4379	0.2514	0.043

References

1. Atanassov, K., Pasi, G., & Yager, R. R. (2005). Intuitionistic fuzzy interpretations of multi-criteria multi-person and multi-measurement tool decision making. *International Journal of Systems Science, 36*(14), 859–868.
2. Atanassov, K. T. (1999). *Intuitionistic fuzzy sets*. Heidelberg: Springer.
3. Atanassov, K. (1986). Intuitionistic fuzzy sets. *Fuzzy Sets and Systems, 20,* 87–96.
4. Bustince, H., & Burillo, P. (1996). Vague sets are intuitionistic fuzzy sets. *Fuzzy Sets and Systems, 79,* 403–405.
5. Gau, W. L., & Buehrer, D. J. (1993). Vague sets. *IEEE Transactions on Systems Man and Cybernetics, 23*(2), 610–614.
6. Hashem, I. A. T., Yaqoob, I., Anuar, N. B., Mokhtar, S., Gani, A., & Khan, S. U. (2015). The rise of big data on cloud computing: Review and open research issues. *Information Systems, 47,* 98–115.
7. Hwang, C. L., & Yoon, K. (1981). *Multiple attribute decision making-methods and applications: A state-of-the-art survey*. New York: Springer.
8. Liu, H. W., & Wang, G. J. (2007). Multi-criteria decision-making methods based on intuitionistic fuzzy sets. *European Journal of Operational Research, 179,* 220–233.
9. De Luca, A., & Termini, S. (1972). A definition of a non-probabilistic entropy in the setting of fuzzy sets theory. *Information and Control, 20,* 301–312.
10. Shannon, C. E. (1948). The mathematical theory of communication. *Bell System Technical Journal, 27*(379–423), 623–656.
11. Shu, M. S., Cheng, C. H., & Chang, J. R. (2006). Using intuitionistic fuzzy sets for fault tree analysis on printed circuit board assembly. *Microelectronics Reliability, 46*(12), 2139–2148.
12. Szmidt, E., & Kacprzyk, J. (2001). Intuitionistic fuzzy sets in some medical applications. *Lecture Notes in Computer Science, 2206,* 148–151.
13. Szmidt, E., & Kacprzyk, J. (2004). A similarity measure for intuitionistic fuzzy sets and its application in supporting medical diagnostic reasoning. *Lecture Notes in Computer Science, 3070,* 388–393.
14. Szmidt, E., & Kacprzyk, J. (2000). Distances between intuitionistic fuzzy sets. *Fuzzy Sets and Systems, 114,* 505–518.
15. Szmidt, E., & Kacprzyk, J. (2001). Entropy of intuitionistic fuzzy sets. *Fuzzy Sets and Systems, 118,* 467–477.
16. Vlachos, I. K., & Sergiadis, G. D. (2007). Intuitionistic fuzzy information—Applications to pattern recognition. *Pattern Recognition Letters, 28,* 197–206.
17. Wang, P. (2009). QoS-aware web services selection with intuitionistic fuzzy set under consumer's vague perception. *Expert Systems with Applications, 36*(3), 4460–4466.
18. Wang, W. Q., & Xin, X. L. (2005). Distance measure between intuitionistic fuzzy sets. *Pattern Recognition Letters, 26,* 2063–2069.
19. Xu, Z. S. (2007). Intuitionistic fuzzy aggregation operators. *IEEE Transactions on Fuzzy Systems, 15*(6), 1179–1187.
20. Yeh, C.-H. (2002). A problem-based selection of multi-attribute decision-making methods. *International Transactions in Operational Research, 9,* 169–181.
21. Zadeh, L. A. (1965). Fuzzy set. *Information and Control, 8*(3), 338–356.
22. Zeleny, M. (1982). *Multiple criteria decision making*. New York: McGraw-Hill.
23. Zhang, C. Y., & Fu, H. Y. (2006). Similarity measures on three kinds of fuzzy sets. *Pattern Recognition Letters, 27,* 1307–1317.

The Role of Website Personality and Website User Engagement on Individual's Purchase Intention

Kokil Jain and Devnika Yadav

Abstract **Purpose**: The paper aims to understand the effect of website personality and website user engagement on an individual's purchase intention in an online purchase environment. **Design/Approach/Research Methodology**: A total of 221 valid online questionnaires were utilized to empirically test the research model using multiple regression approach. The study sample includes online shoppers who performed shopping via Internet medium. **Findings**: The study deduced that there exists an impact of website personality and website user engagement on individual's purchase intention, although the interaction differs for each of the following dimension of personality trait—solidity, enthusiasm, genuineness, sophistication, and unpleasantness. **Originality/Value**: The study has been conducted in National Capital Region of India. There have been very few researches conducted in the region where website personality has been taken as an important dimension to understand its role in purchase probability. Also, studies conducted otherwise, have not included the interaction of website personality and user engagement on purchase intention.

Keywords Website's personality · Purchase intention · Website user engagement

1 Introduction

The growing Internet era and the technologies revolving around it have opened a lot of doors for advertisers to promote their websites and create an image of their brands in the eyes of the customers. Developments in electronic technology have increased the exposure of consumers to websites and digital brands. They have become aware of the variety that is being offered to them. Today Internet is easily

K. Jain (✉) · D. Yadav
Amity International Business School, Amity University, Noida, Uttar Pradesh, India
e-mail: kjain@amity.edu

P. K. Kapur et al. (eds.), *System Performance and Management Analytics*,
Asset Analytics, https://doi.org/10.1007/978-981-10-7323-6_28

available to everyone. The growing dependency on technology has made it a necessity more than just a facility. Advertisers today have become aware of this trend and are willing to use this is as an opportunity.

E-commerce has changed the way people shop today. Without having to step out of the house, people can order for clothes, shoes, accessories, groceries, electronics goods, etc. Many websites like Amazon, Flipkart, Snapdeal, etc. have managed to create a brand out of their name. There are a few websites that have been able to build a relationship with the customer. There are many factors that lead to this for example price sensitivity, payment options, versatility of goods provided, quality of products, website's user interface, etc. The interaction of the website and user builds a relationship that could probably define the future customer brand relationship.

The users unknowingly relate to the brand, store, or the retail website with certain human personality traits. Therefore, it becomes essential for marketers to identify what personality they want their brand or website to embody so that the user can build a relationship with them based on the congruity that exists between the personality of the user and that of the website leading to a synergistic relationship. It becomes important to understand what is brand personality and website personality. For a retail website, it all ends up on how much a customer feels comfortable to purchase from the website. The paper is an attempt to understand the role of website personality on user engagement and purchase intention. The study is probably one of its kind done on Indian audience and can further be utilized to create effective websites for the brands.

2 Literature Review

The earliest research that discussed the product personality concept dates back to [27] that was done by Martineau (1957). The researches that followed focused on identifying own self [3] or the ideal-self [25] or just a few characteristics [20] using a brand. Experts view brand identity as a key approach to separate the brands' exclusiveness [6].

Many studies recommended that higher the similarity between human traits stronger is the inclination towards the brand [26]. Sites are likewise a type of ad [14, 38], which serves as a characteristic transporter of the brand. In the past, there have been number of studies done to comprehend brand identity and furthermore its linkage with site identity. There have been scales that have been produced to quantify brand identity and site identity.

2.1 Human Personality

It can be defined as the characteristic traits portrayed by people that make them carry on and respond distinctively to their environment. Thus, these attributes help advertisers and researchers to understand the consumer's behavior [24]. Purchasing choices depend on the attributes reflected by the human personality. Henceforth knowing about these can help marketers plan better personalities for their brands [6, 18].

The marketplace on the Internet cannot be felt physical and henceforth creating a website personality through simple visual impact becomes essential [6].

When consumers associate with computers, they tend to view the computers personality characteristics as "real" [28]. Hence, it becomes difficult to associate human characteristics with the virtual interface of the computer [6].

This specific research aims at understanding how human personalities would relate with the Internet as a marketplace.

2.2 Brand Personality

A brand personality is a collection of distinctive human attributes connected with a brand [1]. While creating a website personality, it becomes distinctly important to incorporate brand personality as well as investigate the contrast between the brand and website personality. Aaker [1] defined brand personality in terms of five dimensions—sincerity, excitement, competence, sophistication, and ruggedness. These dimensions were further divided into different facets which were as follows—Sincerity (down to earth, wholesome, cheerful, honest), excitement (Daring, imaginative, daring, up to date, spirited), competence (reliable, intelligent, successful), sophistication (upper class, charming), and ruggedness (tough, outdoorsy).

The main contrast between human personality and brand personality is how it is framed. While human personality attributes are inferred on the premise of an individual's behavior, physical qualities, states of mind, attitudes, and demographic characteristics [30], the view of brand identity characteristics is shaped on the basis of direct or indirect contact the consumers have with the brand [31].

Researchers have tried to relate a wide variety of marketing factors to various theories of personality. The factors considered include product usage [7, 41], decision behavior [19, 42], purchase behavior [11, 12, 21], brand loyalty [4], innovative buying behavior [5, 10], response to advertising and design features [17, 43], and product acceptance and rejection [9] as cited in Chen and Rodgers [6].

Famous personalities like celebrities, politicians, athletes, etc. likewise add to the image of the brand in the conscious of the buyer. Purchasers likewise relate one's self with the brand [13]. These help in shaping the overall identity of the brand and thus a distinctive personality. The marketers tap on this behavior for having an impact on the psyches of the consumer.

2.3 Website Personality

Recent studies suggest that website design should not restrict itself as just a medium for interface; rather it should have deeper and more engaging connections through website personality characterization, e.g., Leen et al. [22]. According to Ailawadi and Keller [2], the importance of creating a unique and strong personality cannot be undermined as it serves as major differentiating parameter creating a strong competitive edge in marketplace where offerings are almost similar. A similar strategy will work in an online retail environment as well where the market is slowly treading towards saturation. The strength of relationship between website personality and involvement of the customer was further established through the research carried out by Shobeiri et al. [37].

d'Astous and Levesque [8] established through their research that just like the brand personality, a store also expresses a human personality trait. They argued that brand personality dimensions cannot be fully applied to the store personality. They further argued on the importance of the quality of interactions done by salesperson in defining store personality. They suggested that store personality also has a negative dimension that requires separate attention from the overall brand personality construct.

A retail website also has its own kind of interactions with the consumer. It provides special offers and discounts, recommendations according to the consumers' previous purchases, handles complaints and grievances, helps out with the solution of one's problem, and offers various payment options and much more. All these are very similar to what the offline stores offer to the customer. However, in an online store, the personality is considered for a virtual store which is not physically present.

There were special scales that were developed by researchers to measure the personality of the website. Chen and Rodgers [6] developed a website personality scale by investigating the presence of human and brand personality attributes as well as information characteristics in over 100 websites. They tried to understand the impact of the personality traits on relationship, intention, satisfaction, comfort, and value of the customers with the E-Retailers website. The factors they used were intelligent, fun, organized, candid, and sincere.

According to Shobeiri et al. [37], there exists a relationship between website personality and two variables of e-retailing websites, site involvement and site attitude. Their research was based on the factors that determined the consumer attitude toward the site. While enthusiasm has a positive effect, unpleasantness had a negative influence on the customer's attitude towards the site. Some other dimensions they considered influenced only site attitude (genuineness), site involvement (solidity), or neither of the two (sophistication). Poddar et al. [32] adapted the concept of brand/store personality to Internet marketing by investigating the effects of website personality on perceived Web site quality and consumer purchase intention.

2.4 Website's User Engagement

The studies done by O'Brien and Toms [29] created and assessed a scale to quantify consumer's engagement in online shopping situations. Usability and involvement were seen as essential parts of the way show that the connections between aesthetics, novelty, and focused attention with the result variable, endurability. By and large, the six variables recognized were altogether interconnected, exhibiting that there is a need to the outline a process to consider—the entire consumer experience as opposed to—a single dimension [33].

Tractinsky [40], Tractinsky et al. [39], and also Schenkman and Jönsson [36], have also shown how the aesthetic qualities of interfaces impacted the consumers' judgments on the usage towards the website. According to them, the various components of experience must be examined simultaneously and identified with each other, and presume that engagement, a nature of consumers' involvement with innovation, is a multidimensional construct. As indicated by O'Brien and Toms [29], engagement is a nature of user's involvement with innovation which has a multidimensional construct.

These dimensions of website's user engagement will be utilized as a part of the study and will attempt to comprehend the ramifications of the dimensions of website personality on the dimensions of website's user engagement and in turn on purchase intention of the consumer.

2.5 *Purchase Intention*

The intent to buy certain products refers to as purchase intention. This has been conceptualized very well by Yoo and Donthu [44]. Despite the fact that websites qualities have been distinguished [15] and used by various marketing researchers [14], a considerable number of the attributes are similar. Websites qualities help to decide how people see and process webpage content [35] and fill in as determinants of site adequacy and buying goals [6].

3 Model Discussion and Hypothesis Development

3.1 *Website Personality, Website User Engagement and Purchase Intention*

In line with the above discussion, it can be said that a website with a strong and appealing personality will have a positive impact on the user's engagement toward the website and purchase intention of the user. The paper aims to investigate the user's opinion about the website personality and its effect on the user's engagement and purchase intention. Also, the research model proposes to study the relationship between the users' engagement and their purchase intention.

Poddar et al. [32] adapted the concept of brand/store personality to Internet marketing by investigating the effects of website personality on perceived website quality and consumer purchase intention. The model was tested on apparel websites using data from multiple sources. In the present research, similar model has been utilized to understand the relationship between website personality and purchase intention. The research will draw a relationship between five personality dimensions—enthusiasm, solidity, genuineness, sophistication and unpleasantness, and purchase intention. But, the novelty of this research lies in the fact that website personality interactions have also been studied independently on website user engagement.

Thus, following hypotheses are proposed.

H1: Website Personality affects Website User Engagement

$H1_a$: The perceived solidity of the website has an impact on website user engagement

$H1_b$: The perceived enthusiasm of the website has an impact on website user engagement

$H1_c$: The perceived genuineness of the website has an impact on website user engagement

$H1_d$: The perceived sophistication of the website has an impact on website user engagement

$H1_e$: The perceived unpleasantness of the website has an impact on website user engagement.

H2: Website Personality has an impact on the purchase intention of the user

$H2_a$: The perceived solidity of the website has a positive effect on purchase intention of the user.

$H2_b$: The perceived enthusiasm of the website has a positive effect on the purchase intention of the user.

$H3_c$: The perceived genuineness of the website has a positive effect on the purchase intention of the user.

$H3_d$: The perceived sophistication of the website has a positive effect on the purchase intention of the user.

$H3_e$: The perceived unpleasantness of the website has a negative effect on the purchase intention of the user.

3.2 Website User Engagement and Purchase Intention

In line with the discussion in the previous section, the website's user engagement also has been researched to have an impact on the individual's purchase behavior. The research done by O'Brien and Toms [29] developed and evaluated the reliability and validity of a scale to measure user engagement in online shopping environments. Perceived usability and felt involvement were integral components of the path model that mediated the relationships between Aesthetics, novelty, and focused attention with the outcome variable, endurability. Overall, the six factors identified were all interconnected, demonstrating that there is a need during the design process to consider the whole user experience rather than a single dimension [33].

The dimensions of website's user engagement that were identified by O'Brien and Toms [29], were: Perceived usability, aesthetics, focused attention, felt involvement, novelty, and endurability.

The present research aims to explore the relationship that the dimensions of website user engagement have on the individual's intention to purchase from the website. Thus, the following hypotheses are proposed

H3: Website User Engagement has an impact on the purchase intention of the user

$H3_a$: The perceived focused attention of the user towards the website has a positive effect on the purchase intention of the user.
$H3_b$: The perceived usability of the user toward the website has a positive effect on the purchase intention of the user.
$H3_c$: The perceived aesthetics of the website has a positive effect on the purchase intention of the user.
$H3_d$: The perceived endurability of the user towards the website has a positive effect on the purchase intention of the user.
$H3_e$: The perceived novelty of the website by the user has a positive effect on the purchase intention of the user.
$H3_f$: The perceived involvement of the user with website has a positive effect on the purchase intention of the user.

Figure 1 summarizes the conceptual model.

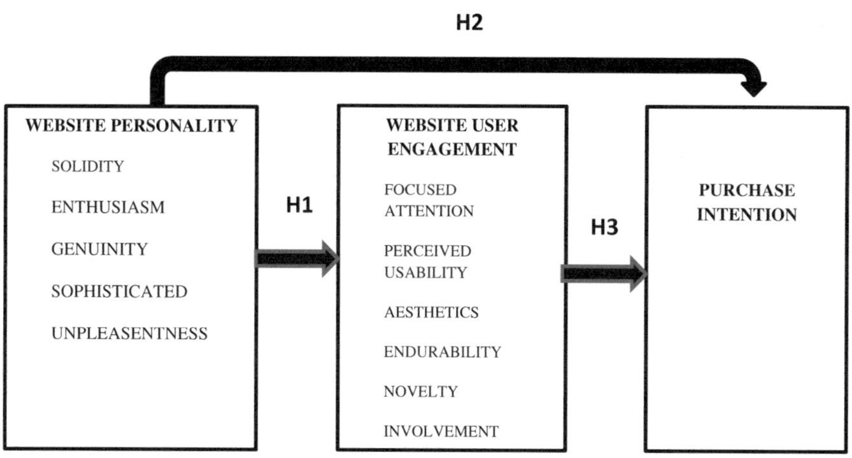

Fig. 1 Conceptual model

4 Research Methodology

In order to understand the personality of the e-retail website and how these personalities affect the purchase intention and the web user engagement, a questionnaire was developed with the help of the literature review. Data were collected using the survey Monkey online application. A total of 250 questionnaires were sent, out of which 221 were included in the final calculation of results as the rest were incompletely filled. The survey was conducted in and near Amity University, Noida. The survey consisted of student, employees, and a few nonworking people. Majority were students since they are active shoppers and web users [23, 44]. This survey was sent to them through a web link using the WhatsApp application. This method was most appropriate because of its convenience as well as it directly reached its target audience. These were the people that used e-commerce the most. Each of the participants in this survey was made to contemplate on their recent purchases and based on that answer the survey appropriately. The participants were allowed to choose their own e-retailer website so that they could relate more towards the questions. The responses that were incomplete were eliminated. To measure website personality, 11-item personality scale developed by Poddar et al. [32] was used.

5 Hypothesis Test and Discussion

The framed hypotheses were tested using SPSS 21. Stepwise regression was conducted to test the relationship between dimensions of website personality and purchase intention and website user engagement and purchase intention. Pearson correlation was done to test the relationship between website personality and website user engagement.

5.1 Website Personality and Website User Engagement

The results of stepwise regression on dimensions of website personality and website user engagement presented a statistically significant model, $F\ (4, 216) = 102.373$, $p < 0.05$ (Refer Appendix, Table 1). The R^2 value of the model is 0.652 (Refer Appendix, Table 2).

The dimensions of website personality which defined website user engagement as deduced from the results were solidity ($ß = 0.247$, $p < 0.05$), genuineness ($ß = 0.237$, $p < 0.05$), sophisticated ($ß = 0.288$, $p < 0.05$), and unpleasantness ($ß = 0.346$, $p < 0.05$) (Refer Appendix, Table 3). Thus, we accept hypothesis $H1_b$ and $H1_{c.}$

The results failed to produce any significant relationship between enthusiasm dimension of website personality and website user engagement.

Thus, Hypotheses $H1_b$ stands rejected.

The results majorly support the role of website personality on user engagement thus stressing the role of unique personality creation for the website so as to increase user engagement which finally increases the probability of purchase. The results also clearly identify the aspects of website personality which will specifically have an impact on user engagement.

5.2 *Website Personality and Purchase Intention*

The results of stepwise regression on dimensions of website personality and purchase intention presented a statistically significant model, F (2, 218) = 102.373, $p < 0.0005$ (Refer Appendix, Table 4). The R^2 value of the model is 0.484 (Refer Appendix, Table 5).

The dimensions of website personality which defined purchase intention as presented in the results were genuineness ($ß = 0.389$, $p < 0.05$) and enthusiasm ($ß = 0.373$, $p < 0.05$) (Refer Appendix, Table 6). Thus, we accept hypothesis $H1_b$ and $H1_{c.}$

The results failed to produce any significant relationship between other three dimensions of website personality, i.e., solidity, sophistication and unpleasantness, and purchase intention.

Thus, Hypotheses $H2_a$, $H2_d$, and $H3_e$ stand rejected.

It can be further concluded that although website personality traits—solidity, genuineness, sophistication, and unpleasantness affect website user engagement, but when it comes to purchase intention, enthusiasm and genuineness are the main predictors.

The results identifying enthusiasm as one of the main predictors of purchase intention shows that the fun element of the site is an important predictor in measuring whether the user will purchase from the website or not. Previous researches have also shown that exciting and fun websites have a far greater influence on the customer's attitude toward the site in comparison to the other dimensions [37, 44].

Genuineness emerged as a factor influencing purchase intention. This shows that it is important for customers to be able to trust the website they are purchasing from or the intent to purchase from. This factor is also significant because the products that one intends to purchase are not physically present in order to decide the quality of the product. Hence, the customer puts his/her faith in the website and makes the purchase. The website therefore should develop a relationship with the customer that is based on honesty and trust.

5.3 *Website User Engagement and Purchase Intention*

The results of stepwise regression on dimensions of website user engagement and purchase intention presented a statistically significant model, $F\ (2, 218) = 206.984$, $p < 0.0005$ (Refer Appendix, Table 7). The R^2 value of the model is 0.655 (Refer Appendix, Table 8).

The dimensions of website personality which defined purchase intention as presented in the results were endurability ($ß = 0.595$, $p < 0.05$) and novelty ($ß = 0.457$, $p < 0.05$), (Refer Appendix, Table 9). Thus, we accept hypothesis $H2_d$ and $H2_{e.}$

The results failed to produce any significant relationship between other four dimensions of website user engagement (i.e., focused attention, perceived usability, aesthetics, and involvement) and purchase intention.

Thus, Hypotheses $H3_a$, $H3_{b,}$ $H3_c$, and $H3_f$ stands rejected.

Thus, it can be concluded that endurability which includes providing a rewarding and successful shopping experience to the consumer is one of the major predictor intention to purchase. Also, novelty which induces curiosity and presents individual with a unique experience results in the final purchase.

6 Marketing Implications and Limitations

The Indian consumer is getting more comfortable with online shopping and therefore it becomes imperative for marketers to understand in depth the factors affecting their attitude and purchase intention towards a particular website. As the results of the research suggest that genuineness and enthusiasm are important website personality traits that drive purchase, marketers should have design and experience elements which help to build the desired personality traits. Similarly, the research helps to understand dimensions of user engagement which leads to individuals purchasing from the website.

A major insight for marketers from this research is that website dimensions which effect website user engagement and purchase intention are not perfectly similar. For example, sophistication of the website might result in enhanced user engagement, but would not result in purchase. Therefore, marketers need to clearly identify traits governing both these important factors and thus accordingly design their websites.

The research provides a better understanding of the underlying dimensions defining interaction of an individual with the website and therefore gives major understanding of the important points to be considered while designing effective websites.

The limitation of this study is the small size of the sample, i.e., 221 and also the respondents are from NCR region. These two limitations restrict the generalization of the results to entire consumer base involved in online shopping in India. Therefore, future studies should test the findings on a wider audience. Democratic factors like age and gender can also be included in the study to understand their interaction with the website personality.

Annexure

See Tables 1, 2, 3, 4, 5, 6, 7, 8 and 9.

Table 1 ANOVA results—website personality and website user engagement

ANOVA[a]						
Model		Sum of squares	*df*	Mean square	*F*	Sig.
4	Regression	101.153	4	25.288	101.222	0.000[b]
	Residual	53.963	216	0.250		
	Total	155.116	220			

[a]Dependent variable: website user engagement
[b]Predictors: (constant), genuineness, unpleasantness, sophistication, solidity

Table 2 Model summary—website personality and website user engagement

Model	*R*	*R* square	Adjusted *R* square	Std. error of the estimate	Change statistics	Durbin Watson
					R square change	
3	0.808	0.652	0.646	0.499830	0.031	2.057

Predictors: (constant), genuineness, unpleasantness, sophistication, solidity
Dependent variable: website user engagement

Table 3 Results of stepwise regression between website personality and website user engagement

Factors	Unstandardized coefficients		Standardized coefficients	*T*	Sig.
	B	Std. error	Beta		
(Constant)	0.726	0.205		3.546	0.000
Genuineness	0.162	0.044	0.237	3.696	0.000
Unpleasantness	0.224	0.027	0.346	8.203	0.000
Sophistication	0.221	0.044	0.288	5.075	0.000
Solidity	0.182	0.041	0.247	4.389	0.000

Table 4 ANOVA results—website personality and purchase intention

ANOVA						
Model		Sum of squares	*df*	Mean square	*F*	Sig.
2	Regression	168.327	2	84.163	102.373	0.000[a]
	Residual	179.224	218	0.822		
	Total	347.550	220			

Dependent variable: purchase intention
[a]Predictors: (constant) genuineness, enthusiasm

Table 5 Model summary—website personality and purchase intention

Model	*R*	*R* square	Adjusted *R* square	Std. error of the estimate	Durbin-Watson
2	0.696[a]	0.484	0.480	0.90671	2.134

Dependent variable: purchase intention
[a]Predictors: (constant) genuineness, enthusiasm

Table 6 Results of stepwise regression between website personality and purchase intention

Factors	Unstandardized coefficients		Standardized coefficients	*t*	Sig.
	B	Std. error	Beta		
Genuineness	0.389	0.081	0.380	4.799	0.000
Enthusiasm	0.373	0.083	0.356	4.495	0.000

Table 7 ANOVA for website user engagement and purchase intention

ANOVA[a]						
Model		Sum of squares	*df*	Mean square	*F*	Sig.
2	Regression	227.662	2	113.831	206.984	0.000[b]
	Residual	119.889	218	0.550		
	Total	347.550	220			

[a]Dependent variable: purchase intention
[b]Predictors: (constant), endurability, novelty

Table 8 Model summary for website user engagement and purchase intention

Model summary[a]					
Model	*R*	*R* square	Adjusted *R* square	Std. error of the estimate	Durbin-Watson
2	0.809[b]	0.655	0.652	0.74159	1.958

[a]Predictors: (constant), ENDURABILITY_INDEX, NOVELITY_INDEX
[b]Dependent variable: PURCHASE_INTENTION_INDEX

Table 9 Results of stepwise regression between website user engagement and purchase intention

Factors	Unstandardized coefficients		Standardized coefficients	*t*	Sig.
	B	Std. error	Beta		
(Constant)	0.246	0.259		0.951	0.343
Endurability	0.595	0.073	0.455	8.155	0.000
Novelty	0.427	0.056	0.422	7.560	0.000

References

1. Aaker, J. L. (1997). Dimensions of brand personality. *Journal of Marketing Research, 34*(3), 347–356.
2. Ailawadi, K. L., & Keller, K. L. (2004). Understanding retail branding: Conceptual… priorities. *Journal of Retailing, 80*(4), 331–342.
3. Belk, R. (1988). Possessions and the extended self. *Journal of Consumer Research, 15*(2), 139–168.
4. Brody, R. P, & Cunningham, S. M. (1968). Personality variables and the consumer decision process. *Journal of Marketing research,* 50–57. http://dx.doi.org/10.2307/3149793.
5. Bruce, G. D., & Witt, R. E. (1970). Personality correlates of innovative buying behavior. *JMR, Journal of Marketing Research (pre-1986), 7*(000002), 259.
6. Chen, Q., & Rodgers, S. (2006). Development of an instrument to measure web site personality. *Journal of Interactive Advertising, 7*(1), 4–46. https://doi.org/10.1080/15252019.2006.10722124.
7. Cohen, J. B. (1967). An interpersonal orientation to the study of consumer behavior. *Journal of Marketing Research, 4*(August), 270–278.
8. d'stous, A., & Levesque, M. (2003). A scale for measuring store personality. *Psychology & Marketing, 20*(5), 455–469.
9. Dolich, I. J. (1969). Congruence relationships between selfimages and product brands. *Journal of Marketing Research, 6*(1), 80–84.
10. Donnelly, J. H. (1970). Social character and acceptance of new products. *Journal of Marketing Research, 7*(1), 111–113.
11. Evans, F. B. (1959). Psychological and objective factors in the prediction of brand choice Ford versus Chevrolet. *The Journal of Business, 32*(4), 340–369.
12. Evans, F. B. (1961). The brand image myth. *Business Horizons, 4*(3), 19–28.
13. Fournier, S. (1994). A consumer-brand relationship framework for strategic brand management. *Doctoral dissertation*. University of Florida).
14. Griffith, D. A., & Chen, Q. (2004). The influence of virtual direct experience (VDE) on on-line ad message effectiveness. *Journal of Advertising, 33*(1), 55–68.
15. Ha, L., & James, E. L. (1998). Interactivity reexamined: A baseline analysis of early business web sites. *Journal of Broadcasting & Electronic Media, 42*(4), 457–474.
16. Halliday, J. (1996). Chrysler brings out brand personalities with '97 ads. *Advertising Age, 67*(40), 3–4.
17. Holbrook, M. B. (1986). Aims, concepts, and methods for the representation of individual differences in esthetic responses to design features. *Journal of consumer research, 13*(3), 337–347.
18. Horton, R. L. (1979). Some relationships between personality and consumer decision making. *Journal of Marketing Research, 16*(2), 233.
19. Kernan, J. B. (1968). Choice criteria, decision behavior, and personality. *Journal of Marketing Research,* 155–164.
20. Kleine, R. E., Kleine, S. S., & Kernan, J. B. (1993). Mundane consumption and the self: A social-identity perspective. *Journal of consumer psychology, 2*(3), 209–235.
21. Koponen, A. (1960). Personality characteristics of purchasers. *Journal of Advertising Research.*
22. Leen, J. Y., Ramayah, T., & Omar, A. (2010). The impact of website personality on consumers' initial trust towards online retailing websites. *World Academy of Science, Engineering and Technology, 66,* 820–825.
23. Lester, D. H., Forman, A. M., & Loyd, D. (2006). Internet shopping and buying behavior of college students. *Services Marketing Quarterly, 27*(2), 123–138.
24. Liebert, R. M., & Liebert, L. L. (1998). *Liebert & Spiegler's personality: Strategies and issues*. Thomson Brooks/Cole Publishing Co.

25. Malhotra, N. K. (1988). Self concept and product choice: An integrated perspective. *Journal of Economic Psychology, 9*(1), 1–28.
26. Malhotra, N. K. (1981). A scale to measure self-concepts, person concepts, and product concepts. *Journal of marketing research,* 456–464.
27. Martineau, P. (1957). *Motivation in Advertising*. New York: McGraw-Hill.
28. Moon, Y., & Nass, C. (1996). How "real" are computer personalities? Psychological responses to personality types in human-computer interaction. *Communication research, 23* (6), 651–674.
29. O'Brien, H. L., & Toms, E. G. (2010). The development and evaluation of a survey to measure user engagement. *Journal of the American Society for Information Science and Technology, 61*(1), 50–69.
30. Park, B. (1986). A method for studying the development of impressions of real people. *Journal of Personality and Social Psychology, 51*(5), 907.
31. Plummer, J. T. (1985, February). Brand personality: A strategic concept for multinational advertising. In *Marketing Educators' Conference* (pp. 1–31). New York, NY: Young & Rubicam.
32. Poddar, A., Donthu, N., & Wei, Y. (2009). Web site customer orientations, web site quality, and purchase intentions: The role of web site personality. *Journal of Business Research, 62* (4), 441–450.
33. Quesenbery, W. (2003). *The five dimensions of usability* (Vol. 20, pp. 89–90). Mahwah, NJ: Lawrence Erlbaum Associates.
34. Richard, M. O., & Chandra, R. (2005). A model of consumer web navigational behavior: Conceptual development and application. *Journal of Business Research, 58*(8), 1019–1029.
35. Rodgers, S., & Thorson, E. (2000). The interactive advertising model: How users perceive and process online ads. *Journal of Interactive Advertising, 1*(1), 41–60.
36. Schenkman, B. N., & Jönsson, F. U. (2000). Aesthetics and preferences of web pages. *Behaviour & Information Technology, 19*(5), 367–377.
37. Shobeiri, S., Mazaheri, E., & Laroche, M. (2015). How would the E-retailer's website personality impact customers' attitudes toward the site? *Journal of Marketing Theory and Practice, 23*(4), 388–401.
38. Singh, S. N., & Dalal, N. P. (1999). Web home pages as advertisements. *Communications of the ACM, 42*(8), 91–98.
39. Tractinsky, N., Katz, A. S., & Ikar, D. (2000). What is beautiful is usable. *Interacting with Computers, 13*(2), 127–145.
40. Tractinsky, N. (1997, March). Aesthetics and apparent usability: Empirically assessing cultural and methodological issues. In *Proceedings of the ACM SIGCHI Conference on Human Factors in Computing Systems* (pp. 115–122). ACM.
41. Tucker, W. T., & Painter, J. J. (1961). Personality and product use. *Journal of Applied Psychology, 45*(5), 325.
42. Westfall, R. (1962). Psychological factors in predicting product choice. *The Journal of Marketing,* 34–40.
43. Wright, P. (1975). Factors affecting cognitive resistance to advertising. *Journal of Consumer Research, 2*(1), 1–9.
44. Yoo, B., & Donthu, N. (2001). Developing a scale to measure the perceived quality of an Internet shopping site (SITEQUAL). *Quarterly Journal of Electronic Commerce, 2*(1), 31–45.

Impact of Social Media on Society—Analysis and Interpretation

Gurinder Singh, Loveleen Gaur and Kumari Anshu

Abstract Since ages, forms of media and technology have endured drastic modification referring to transformation in time, necessities, upgradation of technology, using comfort within one's means, availability, etc. Media aids in disseminating evidences, sensitizing, and instructing people. Social media is rendered to be the succeeding groundbreaking upheaval in the field of human communication. The research paper studies the influence of social media on the Habits and conducts of the community/public. Study is conducted to check the significance of social media on lives of various sections of populaces, causes for the advancements in social media, professional prospects accessible with this progression in social media, etc. Social media has transformed the standards of understanding, learning, interface, media habit, and usage for individual adults as well as the teenagers. Utilization of social media by businesses and working professionals for advertising, communicating, and networking is also emphasized in this paper. The prospects of latest communication trends have sown its seeds in the form of current social media disruption. With the help of this paper, we purpose to understand and decode the shifting forms of communication by the usage of social media.

Keywords Social media · Networking · Social lives · Digital marketing

1 Introduction

In today's world, knowledge is supreme. Social media is a platform which provides ideas, awareness, information, and knowledge of the latest trends to people. Nowadays, social media acts as a significant instrument in persuading our principles, believes, attitude, behaviour, lifestyle, and our holistic viewpoint towards the

---

G. Singh · L. Gaur · K. Anshu (✉)
Amity University, Noida, UP, India
e-mail: anshu_c4@yahoo.co.in

L. Gaur
e-mail: lgaur@amity.edu

P. K. Kapur et al. (eds.), *System Performance and Management Analytics*,
Asset Analytics, https://doi.org/10.1007/978-981-10-7323-6_29

world. Social media is a novel stage that draws people from all around the world to interchange their thoughts, associate with, share and rally for a reason, search out for recommendations, and offer advices and assistance.

Usage of social media has gone up significantly. In India, smartphone infiltration has increased immensely. It is projected to reach 520 million by the year 2020, and this will rank India among the largest smartphone using economies in the world. With the advent of increased usage of smartphone, the broadband infiltration is also bound to surge from 14 to 40% by the year 2020. The usage of Internet in smartphone was at 1.73 EB per month (nearly 69% of entire smartphone Internet use), and this figure is expected to grow 10 times by 2019 and a CAGR of 60%. It is also expected that Internet usage traffic on Tablets will also raise 20 times from 2014 till 2019, and a CAGR of 83%, to reach 3.2 EB per month [1].

Social media is a remarkable publicizing tool too. In lieu of the incredible consumer base growth in social media, businesses have started exploiting this medium for merchandise marketing through promotions, where they endorse brands, talk over about the attributes, and generate cognizance among people [2]. Although social media has conveyed countless positives, allowing us to unite virtually with our friends and families across the sphere, letting us disrupt borders and cultural barricades. But at the same time, social media comes with a cost tagged to it. It shows an undesirable effect on day-to-day lives, as the amalgamation of seclusion and universal spread has weathered our ethos. Social media is taking away our belief, conviction, hope, and security which we used to have in each other [3]. The Internet has always been a disruptive power which affects the dissemination and consumption network for major media segments. Statistics from the surveys have revealed that in past 4 years, there has been an enthralling upswing in the time that individuals devote on social media in countries like America and the UK [4], [5]. This is shown in (Fig. 1).

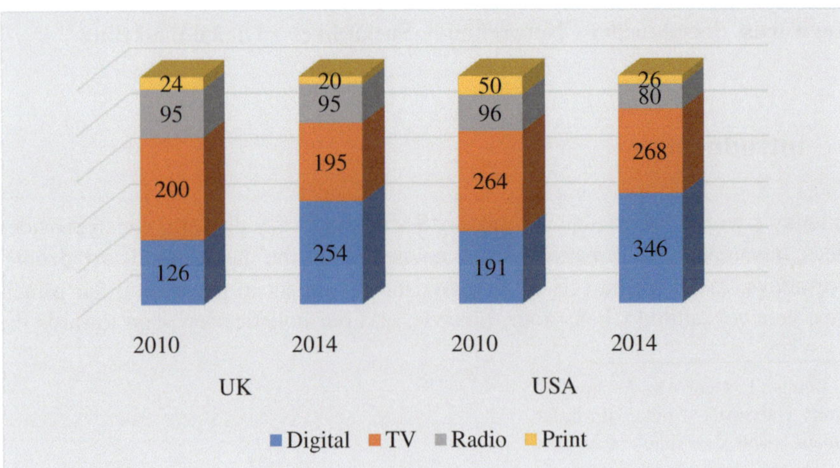

Fig. 1 Real time spent on different media sources in minutes per day for an average adult (minutes per day). *Source* eMarketer

2 Literature Review

There are bearings of social media on different sections of the society, youngsters, juveniles, and families. These sites are used to pool resources on class assignments with their friends and mates and cherish the expansion of the arena of their creativity, imagination, and pioneer thinking [6]. But at the same time, it also has its gloomy face in the form of cyber harassment, setting, Facebook dejections, blackmails, stalking, etc. [7].

The subject on the worth and usefulness of social networking and the reason the teenagers are using them is studied. The research elucidates on a methodology which will further assist in striking an equilibrium between social media and their studies [8].

According to the study, the children of different age groups and in various parts of the country devote ample amount of time on social media. The parents are also beginning to generate an amiability for the technology along with their kids at an early stage of life [9].

This paper focused on knowing if the pupils were reading right kind of digital content, there is a principal problem in the changeover between old-style textbooks and digital content. It was found that students do not buy books anymore [10]. So, an appropriate device should be in place to check if the pupils are reading these digital subjects or not.

In the paper, efforts are made to discover the repercussions on domestic culture adaptation process occurring due to social media. The conclusion was also enthused that the people sustain and flourish their associations and connections with the help of social media [11].

According to the paper, the role of Internet in social life was found. It was concluded that Internet is exclusive and nonpareil, has quality alteration abilities as a channel, and also possesses the capacity to generate a connection and association with same likings, principles, integrities, beliefs, and values that could inspire confidence and assertiveness [12].

The discussion paper suggested that media has a critical role in education and administrative side of the story is required for connecting education policy and communication policy [13]. An appropriately formulated set of plans are needed to instruct and coach parents, child care, and preschools for wholesome growth of kids by application of media and technology.

This white paper deliberates the influence of digital media. They established that transformation is relentless in digital media. Innovative channels are evolving and getting attached to already prevailing formats [14].

An investigation to scrutinize the composite and multifaceted nature of social media is done in this paper. The main concentration was on the evolution of services in the field of social media, and on the study of their consequences [15].

According to study [16], customers prefer to share music, technological related, and funny contents on social media platforms.

3 Objectives of Study

1. To study the emerging trends of social media.
2. To analyze the influence of social media on the lives of people.
3. To assess the factors influencing the usage of social media.
4. To explore the emerging promotional opportunities due to social media.

4 Research Methodology

The study has been deliberated to have a combination of secondary and primary research. Secondary research is carried out through the study of research papers. Quantitative study is carried out through a structured survey of the respondents to understand the impact of social media on their lives. Interviews were carried out with the help of structured questionnaire of respondents which majorly consisted of Likert and rank order scale based questions on their day-to-day habit and usage design or experience with social media.

4.1 Hypothesis

1. Ho1: Social media has impact on the personal lives of the people.
2. Ho2: Social media is a bane for students.
3. Ho3: Social media is effective for advertising.
4. Ho4: Social media is the principal source of engagement during leisure.

4.2 Analysis and Interpretations

This research paper implicit a principal study done on 168 respondents. Responses were collected from 180 respondents out of which only 168 qualified for further analysis. This number consists of male gender and female gender in the ratio of 4:3 (Table 1). Further, the data collected from sample have 43% students and 57% working members, engaged either in services or business. These are the people among whom social media is very popular and are mostly in the age bracket of 22–25 years.

From the analysis it is inferred that as a medium for accessing social media, mobiles are preferred as compared to computers (Fig. 2).

Table 1 Descriptive/frequency distribution

Variable		Frequency	Percent
Age group	18–21 years	12	7.1
	22–25 years	96	57.1
	26–29 years	43	25.6
	42–45 years	17	10.2
Occupation	Service	53	31.5
	Business	45	26.8
	Student	70	41.7
Gender	Male	96	57.1
	Female	72	42.9

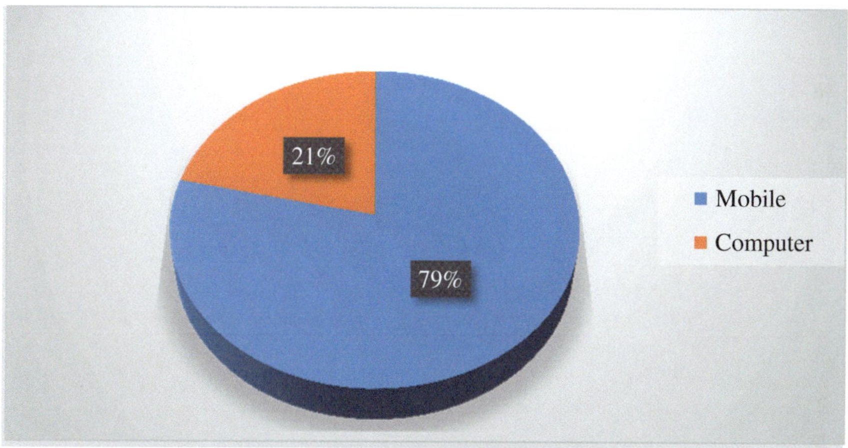

Fig. 2 Medium preferred for accessing social media. (*Source* Author's interpretation from data analysis)

4.3 Testing of Hypothesis

Ho1: Social media has impact on the personal lives of the people.

From the investigation of the information gathered from the respondents utilizing multiple regression, a model summary is acquired (Table 2), where the estimation of adjusted R square is >0.40, i.e., online networking is considered to affect the existence of individuals and over 40% respondents in the sample are impacted by web-based social networking. The estimation of adjusted R square is 0.448. The significance level is <0.05 and F value is 46.26. Thus, we can infer from the analysis that social media has an impact on the individual existences of the general population. Consequently, we accept the null hypothesis.

Table 2 Summary of analysis for social media impact on the personal lives of the people

Model	Unstandardized coefficients		Standardized coefficients	t	Sig.
	B	Std. error	Beta		
(Constant)	−0.593	0.382		−1.552	0.123
Gender	0.335	0.13	0.173	2.571	0.011
Occupation	0.596	0.069	0.518	8.659	0
Age	0.434	0.048	0.611	9.091	0
R square	0.458				
Adjusted R square	0.448				
F	46.262				0.000

1. Dependent variable: impact of social media on personal lives
2. Predictors: (constant), age, occupation, gender

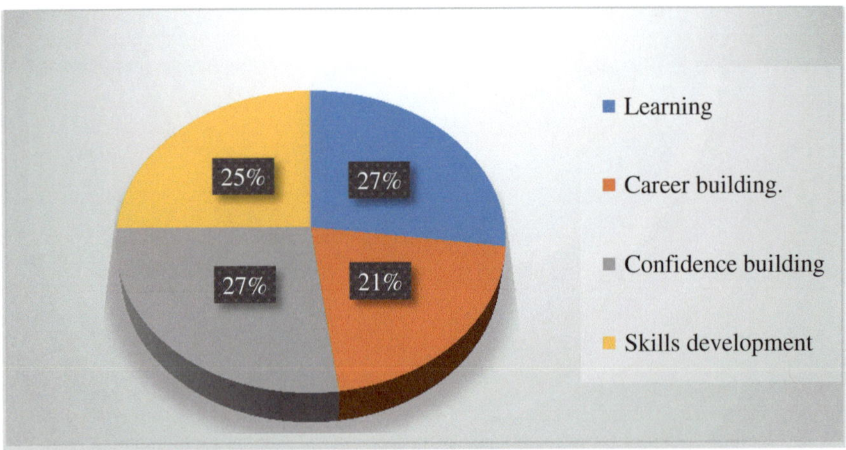

Fig. 3 Benefits of social media. (*Source* Author's interpretation from data analysis)

$$
\begin{aligned}
Y &= a + b_1X_1 + b_2X_2 + b_3X_3 \\
Y &= -0.593 + 0.335(\text{Gender}) + 0.596(\text{Occupation}) + 0.434(\text{Age})
\end{aligned}
\tag{1}
$$

Analysis shows that social media greatly impacts the personal lives of the people. It is also concluded from the research work that the smartphones are the most preferred device for accessing social media. Other devices that were highlighted were mobile with Internet and tablet. Among the various social media technologies used for accessing social media, Twitter and social media sharing are the most popular among the respondents. If we try to find the major activities that are performed by people on their smartphones then gaming, talking, and social networking appear to be the most performed activities.This is shown in (Fig. 3).

Ho2: Social media is a bane for Students

From the investigation of the information gathered from the respondents utilizing multiple regression, a model summary is acquired (Table 3), where the estimation of adjusted R square is <0.40, i.e., online networking is considered as bane for students by less than 40% of the respondent in the sample. The estimation of adjusted R square is 0.130. The significance level is also <0.05 and F value is 9.34. Along these lines, we can decipher from the examination that social media is not considered as bane for students. Henceforth, we reject the null hypothesis.

$$\begin{aligned} Y &= a + b_1X_1 + b_2X_2 + b_3X_3 \\ Y &= 2.54 + (-)0.59(\text{Gender}) + (-)0.12(\text{Occupation}) + (-)0.045(\text{Age}) \end{aligned} \tag{2}$$

It is observed from the research that respondents are using the social media sites for their benefits and going to educational sites for getting information required for the studies.

Research also indicated that there is a shift in the mindset of the people. Television which was previously considered as source of information and entertainment has now been swapped by Internet. This is shown in (Fig. 4).

Ho3: Social media is effective for advertising

From the investigation of the information gathered from the respondents utilizing multiple regression, a model outline is acquired (Table 4), where the estimation of adjusted R square is >0.40, i.e., online networking is viewed as effective for publicizing by over 40% of the respondents in the sample. The estimation of adjusted R square is 0.545. The significance level is also <0.05 and F value is 67.75. Thus, we can interpret from the analysis that social media is a powerful wellspring of promoting. Subsequently, we accept the null hypothesis.

Table 3 Summary of analysis for social media as bane for students

Model	Unstandardized coefficients		Standardized coefficients	*t*	Sig.
	B	Std. error	Beta		
(Constant)	2.547	0.35		7.268	0
Gender	−0.589	0.12	−0.416	−4.924	0
Occupation	−0.121	0.063	−0.144	−1.913	0.057
Age	−0.045	0.044	−0.087	−1.033	0.303
R square	0.146				
Adjusted *R* square	0.13				
F	9.339				0.000

1. Dependent variable: Internet is a bane for students
2. Predictors: (constant), age, occupation, gender

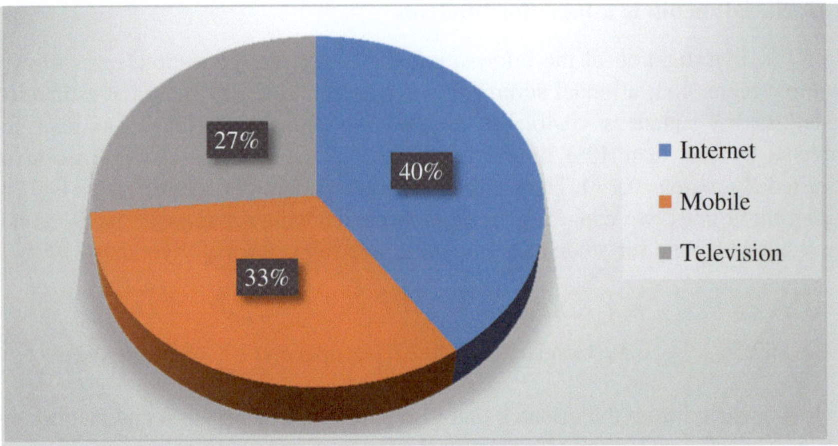

Fig. 4 Preference of learning tools with educational value. (*Source* Author's interpretation from data analysis)

$$
\begin{aligned}
Y &= a + b_1X_1 + b_2X_2 + b_3X_3 \\
Y &= -0.153 + 1.088(\text{Gender}) + 0.117(\text{Occupation}) + 0.497(\text{Age})
\end{aligned} \tag{3}
$$

It is found from the research that slowly and steadily respondents are becoming more receptive towards social media advertising. They are finding these social media advertisements useful. Among the several means of getting interface with the commercials, respondents favored e-mails and social media the most.

Ho4: Social media is the principal source of engagement during leisure.

From the examination of the information gathered from the respondents utilizing multiple regression, a model outline is acquired (Table 5), where the estimation of adjusted R square is >0.40, i.e., social media is considered as the principal source of

Table 4 Summary of analysis for social media effectiveness for advertising

Model	Unstandardized coefficients		Standardized coefficients	t	Sig.
	B	Std. error	Beta		
(Constant)	−0.153	0.297		−0.516	0.607
Gender	1.088	0.101	0.656	10.735	0
Occupation	0.117	0.054	0.119	2.195	0.03
Age	0.497	0.037	0.816	13.381	0
R square	0.553				
Adjusted R square	0.545				
F	67.752				0.000

1. Dependent variable: effectiveness of social media for advertising
2. Predictors: (constant), age, occupation, gender

Table 5 Summary of analysis for social media as major source of engagement during leisure

Model	Unstandardized coefficients		Standardized coefficients	t	Sig.
	B	Std. error	Beta		
(Constant)	1.699	0.159		10.695	0
Gender	−0.435	0.054	−0.524	−8.02	0
Occupation	0.207	0.029	0.421	7.245	0
Age	−0.086	0.02	−0.284	−4.345	0
R square	0.489				
Adjusted R square	0.481				
F	52.399				0.000

1. Dependent variable: visiting social media sites of interest enjoy doing most at leisure
2. Predictors: (constant), age, occupation, gender

engagement during leisure time by more than 40% of the respondents in the sample. The value of adjusted R square is 0.480. The significance level is also <0.05 and F value is 52.4. In this way, we can decipher from the examination that social media is the essential wellspring of engagement amid recreation. Thus, we accept the null hypothesis.

$$\begin{aligned} Y &= a + b_1X_1 + b_2X_2 + b_3X_3 \\ Y &= -1.699 + (-)0.435(\text{Gender}) + 0.207(\text{Occupation}) + (-)0.086(\text{Age}) \end{aligned} \quad (4)$$

It is found from the research that respondents are now more often using social media and enjoying performing activities during their leisure. Respondents are now getting habitual of the social media. Surfing the Internet or visiting websites of interest, social networking, creating graphic arts, etc. on the computer (e.g., digital photos, blogs, websites, digital art, etc.), and playing games on a mobile device or phones are some of the activities enjoyed doing most during leisure. At the same time, respondents are concerned about the issues like speed and privacy. This is shown in (Fig. 5). It is also observed that slowly and steadily traditional means have been replaced by social media which has brought far away people together connecting them and keeping them in close touch with families. It has become the quickest source of sharing pictures, music, and any professional or personal information.

5 Finding

It became well known from the results of the research that social media has become an essential part of our day-to-day life. People are using it widely for social networking, relating to new people, making friends, and are getting used to it. It has collapsed the communication obstructions and has created an open communication

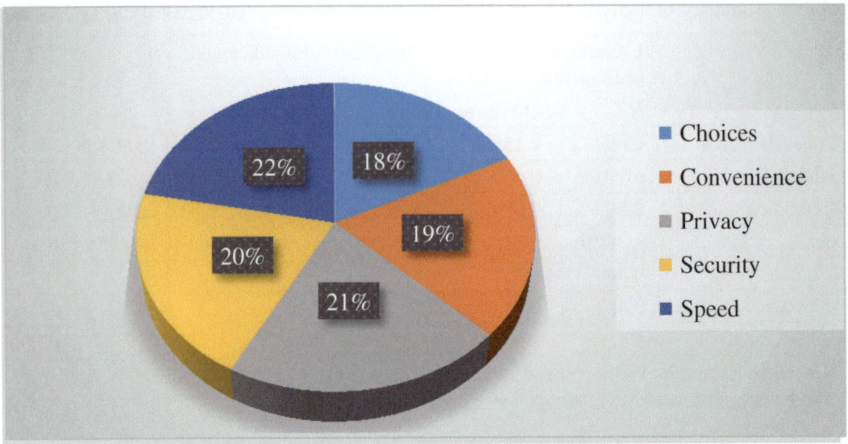

Fig. 5 Area of concern for social media. (*Source* Author's interpretation from data analysis)

channel, bringing far-off people together connecting and keeping them in close touch with families. It has become the quickest source of sharing pictures, music, and any professional or personal information.

Social media convenes an unobstructed flow of information to build upon the information and knowledge bank. It is used for the benefit of the students, for accessing educational, entertainment, and health and wellness sites for getting relevant information.

Slowly but surely people are developing more receptiveness for social media promotion. They think that the social media commercials are beneficial, and are aiding both the professionals and the individuals in generating product cognizance.

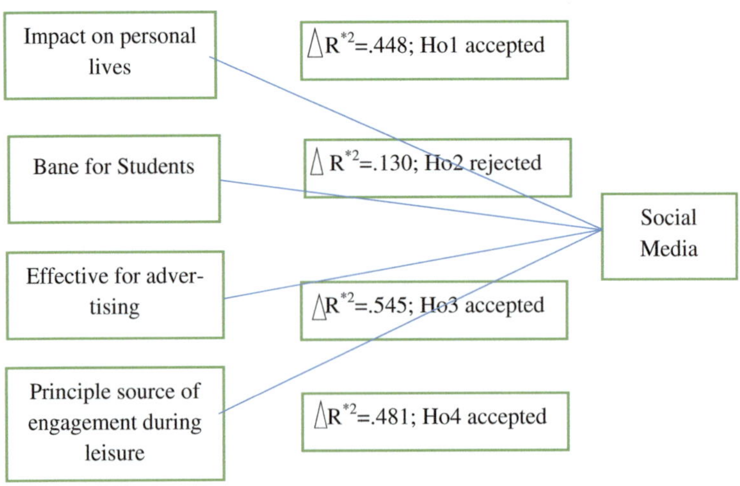

Fig. 6 Framework with findings of hypothesis testing

But as the saying goes everything comes at a cost and so is the case with social media also. Social media is very addictive and has filled in for the human fellowship, their physical, mental and emotional support and replaced the credence, hope, security with virtual, computer-generated connection [3]. The framewrk for hypothesis testing is shown in (Fig. 6).

6 Conclusion

The application of social media aims at improvising the people's insight, degree of communication, and constructing a robust and healthy affiliation amongst people and between people and businesses. It has marked a prominent impression on our society and has impacted people from different profiles. Social media sites have developed as a means of keeping oneself entertained and engaged during leisureliness but also a crucial means for gaining knowledge and insight, and attaining information. Eventually, each social media activity is to be designed and customized to redirect the target audience for the commercial setup and productive information and connectivity for individuals. Social media is also a significant element of publicity crusades, disseminating cognizance, marketing which is commendably used by proficient enterprises and new business setups to construct a stage for them. These advances in technology have enabled several businesses to outreach masses in very short span of time and deliver the messages to persuade them to make consumption choices.

References

1. Cisco. (2016). Cisco VNI *Mobile Forecast (2015–2020)—Cisco*. http://www.cisco.com/c/en/us/solutions/collateral/service-provider/visual-networking-index-vni/mobile-white-paper-c11-520862.html.
2. Newman, D. (2016). *Social media is no longer a marketing channel, it's a customer experience channel*. https://www.forbes.com/sites/danielnewman/2016/01/12/social-media-is-no-longer-a-marketing-channel-its-a-customer-experience-channel/#c68aa7963a57.
3. Schurgin O'Keeffe, G., & Clarke-Pearson, K. (2011). The impact of social media on children, adolescents, and families, council on communications and media. *From the American Academy of Pediatrics Clinical Report, 127*(4).
4. eMarketer. (2014). *UK Consumers Spend over 9 Hours per Day Consuming Media*. https://www.emarketer.com/Article/UK-Consumers-Spend-over-9-Hours-per-Day-Consuming-Media/1011314.
5. eMarketer. (2014). *Mobile Continues to Steal Share of US Adults' Daily Time Spent with Media*. https://www.emarketer.com/Article/Mobile-Continues-Steal-Share-of-US-Adults-Daily-Time-Spent-with-Media/1010782.
6. Schurgin O'Keeffe, G., & Clarke-Pearson, K. (2011). Clinical report—The impact of social media on children, adolescents, and families. *Official Journal of American Academy of Pediatrics*.

7. Is the Internet hurting children? (n.d.). Retrieved December 19, 2016, from CNN. http://edition.cnn.com/2012/05/21/opinion/clinton-steyer-internet-kids/.
8. Qingya Wang, W. C. (2011). *The effects of social media on college students*. http://scholarsarchive.jwu.edu/mba_student.
9. Gutnick, A. L., & Robb, M. (2008). *Always connected: The new digital media habits of young children*. New York: The Joan Ganz Cooney Center at Sesame Workshop.
10. Feldstein, A. (n.d.). *Digital delivery of course content: Is Johnny Reading?* Virginia State University. http://web.mit.edu/comm-forum/mit7/papers/Feldstein_MIT7.pdf.
11. Sawyer, R. (2011). *The impact of new social media on intercultural adoption*. http://digitalcommons.uri.edu/srhonorsprog.
12. McKenna, J. A. (2004). *The internet and the social life*. www.uvm.edu/pdodds/files/papers/others/2004/bargh2004a.pdf.
13. Patricia Edgar, D. D. (2008). *Discussion paper: Television, digital media and children's learning*. Victorian Curriculum and Assessment Authority.
14. Jonny Bentwood, E. (2007). *Distributed influence: Quantifying the impact of social media*. https://technobabble2dot0.wordpress.com/2008/01/16/white-paper-distributed-influence-quantifying-the-impact-of-social-media/.
15. Trottier, D. (2012). *Social media as surveillance: Rethinking visibility in a*. Ashgate Publishing.
16. Erdogmusa, I. E., & Cicek, M. (2016). The impact of social media marketing on brand loyalty. *Procedia—Social and Behavioral Sciences,* 1353–1360.

Conceptual Framework of How Rewards Facilitate Business Operations in Knowledge-Intensive Organizations

Shweta Shrivastava and Shikha Kapoor

Abstract Knowledge workers contribute to operations of their employers through their skills and expertise to solve complex business problems. This makes then indispensable and crucial to the success of such firms.

Following the framework developed by Tsui et al. (Academy of Management 40(5):1089–1121, [12]), this paper addresses the debate regarding the effectiveness of intrinsic and extrinsic rewards in providing a climate of satisfaction in knowledge-intensive organizations. It creates propositions about the how the perception of knowledge workers changes in response to changes in the intrinsic and extrinsic components of rewards. Keeping employment relationships as backdrop and inherent work-related needs of knowledge workers into consideration, this paper makes propositions regarding the possible impact of changes in intrinsic and extrinsic rewards on 'α' or the perceived value of employment in unbalanced relationships. Knowledge workers are known to be more driven by characteristics of their work than extrinsic aspects of rewards. Therefore, the paper suggests that the share of intrinsic rewards must be either more or (at least) equal to that of extrinsic rewards.

This paper adds to the long drawn debate between the effectiveness of intrinsic and extrinsic rewards. It draws attention to the need for organizations to focus on the existing employment relationship and the work-related needs of knowledge workers while taking reward decisions.

Keywords Rewards · Intrinsic · Extrinsic · Employment relationships Overinvestment · Underinvestment · Knowledge workers

S. Shrivastava
Faculty of Management Studies, University of Delhi, Delhi, India
e-mail: shwetashrivastava.17@fms.edu

S. Kapoor (✉)
Amity International Business School, Amity University, Noida, India
e-mail: skapoor2@amity.edu

P. K. Kapur et al. (eds.), *System Performance and Management Analytics*, Asset Analytics, https://doi.org/10.1007/978-981-10-7323-6_30

1 Introduction

Organizations today have realized that their employees are the biggest asset that they possess. It is only with the performance, intellect and skills of their employees that organizations can keep their business operations afloat and can combat the competitive work environment. Therefore, management of organizations has started to put in considerable effort to hire, develop and retain the best talent available for as long as possible. They develop policies and practices that are conducive for employees to put in their maximum effort towards business operations [1].

The employment relationship between organizations and employees has changed since the Great Recession in the first decade of the twenty-first century. That was the time when organizations had to resort to headcount management to remain competitive, to gain flexibility and essentially, to remain afloat [2]. This new relationship has altered the dynamics between employers and employees and is said to have an impact on various human resource practices as well [3]. The need to take a close look at such practices gets more pronounced when it comes to managing ‘knowledge workers’ who have emerged as a key economic resource today [4]. Characteristics of the new relationship make it a challenging task for organizations to balance their own conflicting priorities with the those of these knowledge workers, especially since traditional management systems have been found to be ineffective in retaining them [5].

The new employment relationship is visibly characterized by the absence of a *mutual commitment* between employer and employees, unlike what existed earlier [2, 6]. Organizations have replaced the assurance of a secured job by an agreement to provide a challenging job, a pre-decided compensation package and learning opportunities to employees [3]. From employees’ perspective, the notion of a lifelong employment with one organization has been replaced by a more flexible and agile relationship. Employees do not always visualize a long-term career with their employer and continue working so long as the value of their rewards is commensurate with their contributions [7].

Offered Inducements	Expected Contributions: Low/Narrow	Expected Contributions: High/Broad
Low/Narrow	1) Quasi-spot contract	2) Underinvestment
High/Broad	3) Overinvestment	4) Mutual investment

Fig. 1 Four-quadrant model of EOR (Reproduced from Tsui et al. [12])

Apart from this, what has emerged as a pressing need for both the employers and employees in this new equation, is the need for equity in employment. A surge in the adoption of rewards practices such as 'pay for performance' and 'variable pay' indicated the desire of organizations to increase focus on critical goals and to reduce the risk of disproportionate and undeserved bonuses [8]. Employees, too, wish to equally, fairly and adequately compensated for their efforts. A study conducted on metal casting workers in 2007 to examine how employers could persuade workers to increase their productivity revealed that employees' cooperation towards organizational activities was short lived, unless their extra efforts were matched by extra rewards from employers [9]. This highlights the need for employees to be engaged in equitable employment. The notion that employment is an employer's prerogative stands challenged today and employees form an extremely integral part of the employment relationship [10]. If their needs are not met, employees can then take steps such as voicing dissatisfaction, reducing performance and adjusting inputs to match the low value derived or look for employment elsewhere [11].

Therefore, in this new age relationship, rewards play a crucial role in achieving a balanced relationship and are of paramount importance to both employers and employees. The objective of this paper is to discuss how skilled knowledge workers perceive rewards and how they respond when organizations alter their reward offerings to establish equitable relationships. Following the four-quadrant EOR framework developed by Tsui et al. [12], this paper hypothesizes about the impact of changes in the intrinsic and extrinsic components of rewards on knowledge workers' perception of inequitable employment relationships. It does so while taking into consideration their inherent work-related needs and the differing nature of reward components. It proposes a framework for changes in the employees' perception of inequitable employee–organization relationship (EOR). We begin by visiting the four-quadrant matrix developed by Tsui et al. [12] to explore various employment approaches.

2 Review of Literature

2.1 Employee–Organization Relationships

The term 'employee–organization relationship' (EOR) signifies the employer's perspective of balance between contributions expected from employees (such as commitment, performance, etc.) and the inducements (created through a bundle of human resource practices) to affect those contributions [12]. Most of the research on EOR is based on the social exchange theory and the inducement-contribution theory, where the former involves recurring, mutual obligation to exchange benefits

[13]. The four-quadrant model developed by Tsui, Pearce, Porter and Tripoli in 1997 draws upon the inducement-contribution model and views employment as a relationship based on exchange of inducements provided by organization in return for contributions from employees, with reciprocity being its underlying principle (March and Simon 1958).

The four quadrants refer to the four approaches that organizations can adopt for rewarding their employees; an approach of overinvestment, underinvestment, quasi-contract or mutual investment (Fig. 1: Four-quadrant model of EOR).

In an underinvestment approach, contributions exceed inducements, which makes the approach unfit in the new employment relationship. Providing inadequate rewards is disadvantageous to organizations since knowledgeable employees have the potential to engage in alternative employment [7] and thus may lead to loss of skilled manpower. It can also impact an employee's motivation and may encourage him to take steps such as voicing dissatisfaction, reducing performance and adjusting inputs to match the low value derived or look for employment elsewhere [11].

In an overinvestment relationship, inducements offered to employees exceed contributions expected from them and as discussed earlier, are unfavourable for organizations. However, this also does not make for a favourable proposition for organizations as disproportionately high rewards can lead to creation of a low-performance culture in the organization where poor performers receive substantial returns without the expectation of performance (Shaw, et al. 2009). It can also lead to retention of low performers, loss of high performing talent and to a negative impact on the organization's financial agility [14] Interestingly, even though it may seem like a favourable condition for employees, in the long term, it can cause harm to their employability and work attitude as they get accustomed to being rewarded highly for mediocre performance.

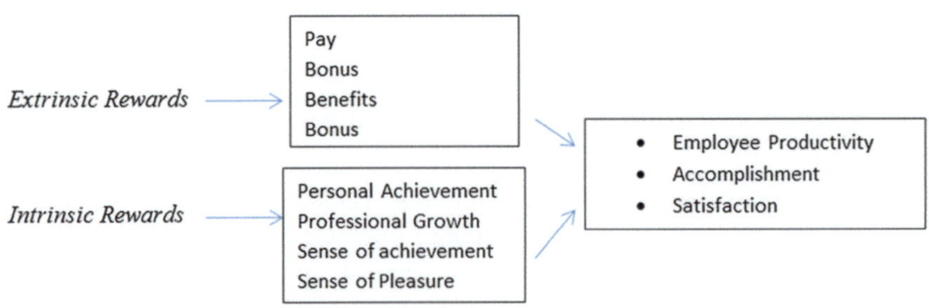

Fig. 2 Extrinsic and intrinsic rewards

A quasi-spot contract relationship is short term or temporary in nature and is just like an economic exchange. Mutual investment approach entails high contributions from employees in exchange for high inducements from employer. This approach has been said to be most effective in creating long-term relationships as it is perfectly equitable and both entities are equally invested [12].

The inducement and contribution process is a cyclic process where employees need to contribute to the organization sufficiently to get inducements from the organization, which in turn needs to be attractive enough for employees to be motivated enough to contribute [15]. Thus, it is required that organizations understand the nature rewards, the needs and characteristics of their skilled manpower and, explore how reward decisions will impact them both the organization and these employees.

2.2 *Rewards*

Decisions related to rewards are amongst the most important decisions for organizations as it allows them to elicit necessary role behaviours from their employees. Rewards are one of the most critical influences on the quality and effectiveness of the human capital [16]. Rewards comprise of various elements and, primarily, can be tangible or intangible, extrinsic or intrinsic. Tangible and intangible rewards are qualitatively different as the former are more identifiable, demonstrable and measurable as compared to latter which emanate from psychological, internal and attitudinal components [17]. Intrinsic rewards are said to be directly associated with doing a task and are intangible [18]. Hackman and Oldham proposed task identity, task significance, skill variety, autonomy and feedback as dimensions associated with doing a task that constitute intrinsic rewards or 'job characteristics' [19]. The new age workforce today seeks meaningful work, opportunities to learn and expand existing skills and the autonomy to do accomplish tasks on their own [20]. Extrinsic rewards, such as pay, benefits, etc. are external to the work itself, are tangible [21] and are controlled by external entities [20] such as the management, supervisors and other stakeholders. These rewards have often been criticized as it is believed by many that they reduce intrinsic motivation towards the task itself [22]. Social rewards indicate the workplace interactions that an employee experiences such as working relationships with superiors, peers, *subordinates,* etc. [23] (Fig. 2).

For rewards to act as an inducement for employees to compensate the organization for beneficial treatment, it is important that what is offered to employees is valued by them [24]. The next section discusses knowledge workers and the characteristics of rewards that are valued by them.

2.3 Knowledge Workers and Work-Related Needs

The definition of a knowledge worker remains ambiguous and has been criticized for the lack of 'theoretical and methodological rigour' [25]. Even though the work of knowledge workers has been found to be undefined and unstructured, it is said to consist of a few characteristics that demarcate it as a separate category. Such jobs refer to what Daniel Pink refers to as 'heuristics' or jobs which do not work on set algorithms and require novel solutions as opposed to 'algorithmic', which follow a set of established instructions [26].

Knowledge workers include those who engage in complex, analytical and abstract work. It includes professions where one is required to deal with and manipulate symbols, concepts and information and is more 'unstructured and organizationally contingent' [25] Such workers form a majority of the workforce in knowledge-intensive firms where most of the work done is intellectual in nature and products/services are produced using the knowledge of the personnel as a major resource [27]. Knowledge workers are said to possess the ability to apply their advanced education and skills to identify and solve organizational problems [28]. This may include professions such as legal services, accounting, engineering, researchers, software designers, etc.

Emanating from the characteristics of work performed by knowledge workers, complex, unstructured and abstract, are the needs of such employees from reward offered to them. For instance, they are said to value the importance of their skills more than the formal hierarchy in the organization [29] and are more committed to their profession than to their organization. Therefore, intrinsic or job-related rewards have been found to be impactful for their motivation. Owing to the uncertainty around their work, knowledge workers need to decide how they wish to schedule, arrange and perform their work. Hence, autonomy is of great importance [28] Driven primarily by their professions, challenging work is also believed to impact the intrinsic motivation of knowledge workers [30]. Social networks or social relationships of knowledge workers with their peers are important for them to engage in collaborative activities and enhance each other's skills. A supportive and collaborative work culture where knowledge sharing and creation is facilitated is important from the point of view of knowledge workers [31].

Knowledge workers have been said to differ from service-oriented employees in their need for work characteristics such as involvement in decision-making, autonomy, skill variety, task significance, etc. Owing to such differences, their perception of inequitable relationships can also be expected to be different. Therefore, it will be of value to understand how the context of the employee–organization relationship impact their intellectually driven employees.

The forthcoming section further builds on the theory of inequitable employment relationships (overinvestment and underinvestment) using characteristics of rewards and links it with knowledge workers' perception of employment.

3 Attaining Equity in Overinvestment and Underinvestment Approaches in Knowledge-Intensive Firms

Both overinvestment and underinvestment employment relationships are unbalanced relationships as the former is skewed towards employees and the latter is skewed towards employers [10]. Changes to rewards in these two relationships to make the inducements and contributions equitable require substantial thought on the part of organizations. It will entail deciding ***which element of rewards*** needs to be altered by ***how much***. These decisions are important as different types of rewards have different influences on employees in the short and the long term. The forthcoming analysis has been developed to explore the answer to the question of 'which element' regarding knowledge workers. It analyses the expected impact on the perception of inequitable relationships in the case of knowledge workers, who have their preferences and fundamental work-related needs. Two time frames have been explored in the paper since, with time, the impact of changes may get altered—either magnified or diluted. It hypothesizes the impact of alterations in quantum of intrinsic and extrinsic rewards in over and underinvestment relationships assuming that:

- Contribution of employee towards the organization remains constant
- Rewards consist of extrinsic (pay and benefits) and intrinsic (job characteristics) kinds only; social rewards have not been included
- Intrinsic and extrinsic rewards are provided in equal proportion prior to changes in the reward combination
- One-time change *either* to extrinsic *or* intrinsic rewards (the other component remains constant) are made only to an extent that establishes equity between overall inducements and contributions. In other words, an overinvestment relationship does not get converted to underinvestment relationship and vice versa.
- Before the change is made, employees do not differentiate between intrinsic or extrinsic reward
- Short term and long term refer to a period of six months and one year or more, respectively (Table 1).

α—*'Perceived' value*

α refers to a 'perceived' value created in the mind of an employee when the inducements offered to him are higher than the contributions expected from him. This value assumes the direction of the aforesaid relationship between inducements

Table 1 Perceived value in over and under investment approaches

Approach	Expected contribution	Offered inducements	Perceived value (α)
Over-investment	Low	High	Positive ($\alpha 1$)
Underinvestment	High	Low	Negative ($\alpha 2$)

and contributions. In an overinvestment scenario, employee's contributions are lower than inducements offered to him and hence α is positive ($\alpha 1$). A negative α value ($\alpha 2$) is created in an underinvestment approach where an employee's contribution is higher than the inducements offered to him. In a purely equitable relationship, α will equal 0. For an employment relationship to sustain in the long term, it needs to be equitable. $\alpha 2$ (negative α) implies that employees do not perceive the employment relation as equitable or beneficial, which is the case in an underinvestment relationship. Therefore, this relationship is unlikely to sustain for long. $\alpha 2$ is positive, despite being positive and in favour of employees, is an unfavourable proposition for employers in the long term, as discussed earlier. To eliminate inequity of either kind, organizations can choose to reduce the quantum of either intrinsic or the extrinsic component of overall inducements. Owing to differing nature of these two components, such reduction is likely to lead to different impacts on the perceived $\alpha 1$ and $\alpha 2$. The impact is also expected to vary as per time frame in consideration.[1]

In the section below, we hypothesize the impact of a one-time alteration in intrinsic and extrinsic components of rewards on $\alpha 1$ and $\alpha 2$, both in the short and long term. This will eventually determine whether the employee chooses to leave or stay with the organization.

3.1 Overinvestment

Table 2 depicts the proposed impacts of changes in rewards program in both under and overinvestment approaches. Although employees favour an imbalance in their favour (March and Simon 1958), inducements in an over-investment relationship should be reduced to achieve equity.

3.1.1 Reduction in Intrinsic Rewards

In this favourable employment relationship, employees feel well rewarded with both intrinsic and extrinsic components. Since intrinsic rewards are intangible and less demonstrable, their absence is likely to take a while to impact. Therefore, when intrinsic rewards are reduced, employees may not feel a real or immediate change in the short term.

Intrinsic needs of employees correspond with self-actualization (Maslow's Need Hierarchy Theory), need for growth (Alderfer's ERG Theory) and need for motivation (Theory X and Theory Y) [32]. This need becomes more heightened in the

[1]Impact in the short term and the long term refers to the increase or decrease in α in comparison to the initial value of α; ***before*** the change in reward combination.

Table 2 Reward alterations in overinvestment approach

Rewards reduced	Impact on perceived value ($\alpha 1$)	
	Short term	Long term
Intrinsic	Marginal reduction	Significant reduction
Extrinsic	Significant reduction	Marginal reduction

case of knowledge workers who are mindful of their skills and value the intrinsic aspects of their jobs. Therefore, in the long term, these rewards become conspicuous through their absence and employees are likely to realize the loss of intrinsic rewards which they no longer receive. This realization makes $\alpha 1$ significantly reduce in the long term as knowledge workers perceive a significant deterioration in their work. The following proposition has been offered below therefore:

Proposition 1a In an overinvestment approach, a decrease in intrinsic rewards is likely to lead to a marginal decrease in α in the short term and a significant decrease in α in the long term.

3.1.2 Reduction in Extrinsic Rewards

Financial rewards are a core reward offered by employers and have a significant impact on the job attitudes of workers [33]. Some researchers have been of the view that that providing extrinsic rewards leads to creation of an external locus of control and an external source of motivation as well [34]. In the current context, employees who were significantly endowed with extrinsic rewards till now lose sight of the intrinsic rewards that they continue to receive and thus resent this reduction in extrinsic component. This leads to an immediate and significant decrease in $\alpha 1$ when extrinsic rewards are reduced.

In the long term, a reduction in extrinsic reward is likely to make such employees regain the concept of why they are working and redirect their focus to the intrinsic happiness they continue to derive from their work [35]. Extrinsic rewards have often believed to reduce intrinsic motivation and therefore, this decrease is likely to rightly shift the employees' focus away from extrinsic rewards towards the high quantum of intrinsic rewards they receive. This, therefore, leads to little or just a marginal decrease in $\alpha 1$.

Proposition 1b In an overinvestment approach, a decrease in extrinsic rewards is likely to lead to a significant decrease in $\alpha 1$ in the short term and a marginal decrease in $\alpha 1$ in the long term.

3.2 *Underinvestment*

Rewards underline the relationship that the organization wishes to establish with its employees. In an underinvestment relationship, it becomes clear to employees that the organization wants to be the relationship to be advantageous to itself rather than to its employees. This realization sets the foundation for perception of any kind of changes in the rewards offering (Table 3).

3.2.1 Increase in Intrinsic Rewards

Increase in intrinsic rewards along coupled with expectation of high contribution is expected to significantly increase α as it may signal to employees that they are important and indispensable to the organization [10]. Even though intrinsic rewards have little bearing on employees' decisions of continuance [36], since intellectually driven employees value their work and skills, such an increase in intrinsic rewards is likely to increase $\alpha 2$ significantly in the long run as well.

Proposition 2a In an underinvestment approach, an increase in intrinsic rewards is likely to create a significant increase in $\alpha 2$ in the short term as well as the long run.

3.2.2 Increase in Extrinsic Rewards

An increase in extrinsic rewards is a highly observable step taken by an organization, making the relationship more equitable in the employee's mind. A likely reason for this is that changes in intrinsic rewards have an immediate effect on an employee's perception of employer's intentions. This is, therefore, likely to significantly and immediately improve $\alpha 2$ and to discourage employees from terminating employment, in the short term.

However, extrinsic rewards have been criticized for their ability to 'foster short-term thinking' and due to their 'addictive' nature [26]. In the current scenario, in the long term, employees are likely to get accustomed to the high extrinsic rewards. Since knowledge workers have a high need for achievement, in the long term their focus shifts back to the low value of intrinsic rewards that is being derived out of the employment. The impact of such increase in $\alpha 2$ does not hold for very long and is later overshadowed by the absence of intrinsic rewards.

Table 3 Reward alterations in underinvestment approach

Rewards reduced	Impact on perceived value ($\alpha 1$)	
	Short term	Long term
Intrinsic	Significant increase	Significant increase
Extrinsic	Significant increase	Marginal increase

Proposition 2b In an underinvestment relationship, increasing extrinsic rewards leads to a significant increase to $\alpha 2$ in the short term but only a marginal increase in the long run.

4 Implications and Conclusion

The debate regarding the effectiveness of extrinsic and intrinsic components of rewards has been going on for long. While few researchers have highlighted the harmful effects of extrinsic rewards on the intrinsic desire of workers, the claim has also been refuted in various studies [37]. A practical view of the matter is that both components of rewards serve varying needs of employees and are thus important. The paper supports the view that employees have multifaceted needs from rewards and it is unlikely for them to have one extreme driving need [38]. In other words, needs of employees from rewards cannot be bucketed wholly into one category and thus need to be viewed in parts and in totality, as a unified model of inducements. A complex web of factors such as the nature of job, the preference and paying capacity of the employer organization, etc., also play a role in the determination of reward elements, their quantum, the mode of payment, etc. This paper adds to the discussion by adding another factor in the equation; that of the of the existing employment relationship which sets the foundation of such perception.

Few thoughts emerge from the discussion. First is that skilled employees value intrinsic attributes of their jobs such as autonomy, feedback, etc. and their absence is likely to make them more dissatisfied with their job than that of extrinsic rewards. Therefore, the share of intrinsic rewards must be either more or at least equal to that of extrinsic rewards. Second, reward decisions in organizations must revolve around existing employment relationships along with crucial aspect of needs and characteristics of the workforce. These two aspects determine how the elements of rewards are observed and perceived by employees. Third, there can be differences between the perception of employers and employees regarding HR practices and it is the perception of employees towards adequacy and appropriateness of rewards that affects the work attitudes and not the objective state of rewards [39, 40] Therefore, appropriate communication of rewards and the related decisions is also extremely important. This will help by making employees perceive organization's intentions better.

Addressing the debate of rewards with specific reference to certain sections of workforce (knowledge workers in this case) is likely to get better results and understanding, especially since the business ecosystem is changing rapidly. The paper provides propositions which can be further used as directions to study intrinsic and extrinsic rewards in employment relationships. Inclusion of social rewards in research studies will be beneficial to understand if they have a bearing on the morale and perception of employees.

Employees' contribution to organizational operations, profits and revenue is of extreme importance. Therefore, the importance of different types of rewards for employees and their indirect impact on employees' mindset should be acknowledged by organizations.

References

1. Kapoor, S. (2015). *Human Resource Management (Text and Cases)* (pp. 364–365).
2. Tsui, A. S., & Wu, J. B. (2005). The new employment relationship versus the mutual investment approach: Implications for human resource management. *Human Resource Management,* 115–121.
3. Roehling, M. V., et al. (2000). The nature of the new employment relationship: A content analysis of the practitioner and academic literatures. *Human Resource Management,* 305–320.
4. Drucker, P. (2011). *Management challenges for the 21st Edition. Classic Drucker Collection.* New York: Routledge.
5. Lowendahl, B. (2000). *Strategic management of professional service firms.* Copenhagen: Copenhagen School Press.
6. Stewart, T. A. (1998). Grey flannel suit? *Fortune, 137*(5), 76–82.
7. Bremen, J. M., & Davenport, T. O. (2013). Treat your employees as consumers. *Workspan* (May, pp. 27–33).
8. Hay Group. (2010). *The changing face of rewards.* Philadelphia: Hay Group.
9. Imberman, W. (2012). Motivating employees: What works and what doesn't. *Foundry Management and Technology* (November, pp. 23–26).
10. Yan, Z. A., et al. (2008). How do I trust thee? The employee-organization relationship, supervisory support, and middle manager trust in the organization. *Human Resource Management, 47*(1), 111–132.
11. Adams, S. J. (1963). Toward an understanding of inequity. *The Journal of Abnormal and Social Psychology, 67*(5), 422–436.
12. Tsui, A. S., et al. (1997). ALternative approaches to the employee-organization relationship: Does investment in employees pay off? *Academy of Management, 40*(5), 1089–1121.
13. Coyle-Shapiro, J., & Shore, L. M. (2007). The employee–organization relationship: Where do we go from here? *Human Resource Management Review, 17*(2), 166–179.
14. Berggren, E., & Bernshteyn, R. (2007). Organizational transparency drives company performance. *Journal of Management Development, 26*(5), 411–417.
15. Coyle-Shapiro, et al. (2007) The employee-organization relationship: where do we go from here? *Human Resource Management Review, 17*(2), 166–179. ISSN 1053-4822.
16. Gupta, N., & Shaw, J. D. (2014). Employee compensation: The neglected area of research. *Human Resource Management Review, 24*(4), 1.
17. Butler, S., & Charles, M. (1999). *Adoption & Fostering* (pp. 48–58).
18. Mottaz, C. J. (1985). The relative importance of intrinsic and external rewards as determinants of work satisfaction. *The Sociological Quarterly,* 165–385.
19. Hackman, R. J., & Oldham, G. R. (1976). Motivation through the design of work: test of a theory. *Organizational Behaviour and Human Performance* (pp. 250–279).
20. Thomas, K. (2009). *Ivey Business Journal.* [Online] Available at: http://iveybusinessjournal.com/publication/the-four-intrinsic-rewards-that-drive-employee-engagement/.
21. Malhotra, N., et al. (2007). Linking rewards to commitment: an empirical investigation of four UK call centres. *The International Journal of Human Resource Management,* 2095–2128.

22. Giancola, F. L. (2014). Should HR Professionals devote more time to intrinsic rewards. *Compensation and Benefits Review, 46*(1), 25–31.
23. Williamson, I. O., et al. (2009). The interactive effect of collectivism and organizational rewards on affective organizational commitment. *Cross Cultural Management,* 28–43.
24. Eisenberger, R. et al. (2001). Reciprocation of perceived organizational support. *Journal of Applied Psychology, 86*(1), 42–51.
25. Scarbrough, H. (1999). Knowledge as work: Conflicts in the management of knowledge workers. *Technology Analysis & Strategic Management, 11,* 5–16.
26. Pink, D. H. (2009). *Drive: The surprising truth about what motivates us.* New York: Riverhead Hardcover.
27. Alvesson, M. (2000). Social identity and the problem of loyalty in knowledge-intensive companies. *Journal of Management Studies, 37*(8), 1101–1123.
28. Newell, S., Robertson, M., Scarbrough & Swan, J. (2002). *Managing knowledge work.* Basingstoke and New York: Palgrave Macmillan.
29. Blackler, et al. (1993). Editorial introduction: Knowledge workers and contemporary organizations. *Journal of Management Studies, 30*(6), 851–862.
30. Teece, D. J. (2003). Expert talent and the design of (professional services) firms. *Industrial and Corporate Change, 12*(4), 895–916.
31. Iles, P., Yolles, M., & Altman, Y. (2001). HRM and knowledge management: Responding to the challenge. *Research and Practice in Human Resource Management, 9*(1), 3–33.
32. Wiersma, U. J. (1992). The effects of extrinsic rewards in intrinsic motivation: A meta-analysis. *Journal of Occupational and Organizational Psychology, 65*(2), 101–114.
33. Gkorezis, P., & Petridou, E. (2012). The effect of extrinsic rewards on public and private sector employees' psychological empowerment: A comparative approach. *The International Journal of Human Resource Management, 23*(17), 3596–3612.
34. deCharms, R. (1968). *Personal causation: The internal affective determinants of behaviour.* New York: Academic Press.
35. Festinger, L. (1967). *The effect of compensation on cognitive process.* New York: McKinsey.
36. O'Driscoll, M. P., & Randall, D. M. (1999). Perceived organizational support, satisfaction with rewards, and employee job involvement and organizational commitment. *Applied Psychology: An International Review, 48*(2), 197–209.
37. Ledford, G. E., et al. (2013). *Negative effects of extrinsic rewards on intrinsic motivation: More smoke than fire.* New York: WorldatWork.
38. Reif, W. (1975). Intrinsic vs externa rewards - Resolving the controversy. *Human Resource Management, 14*(2), 1–10.
39. Edgar, F., & Geare, A. (2005). HRM practice and employee attitudes: Different measures—different results. *Personnel Review, 34*(5), 534–549.
40. Hackman, R. J., & Oldham, G. R. (1976). Motivation through the design of work: Test of a theory. *Organizational Behaviour and Human Performance,* 250–279.
41. Jacobsen & Thorsvik. (2002). *Overview of Incentive System* (p. 303).

Tunnel QRA: Present and Future Perspectives

Jajati K. Jena, Ajit Kumar Verma, Uday Kumar and Srividya Ajit

Abstract With the vision of faster in-land transportation of humans and goods, long tunnels with increasing engineering complexities are being designed, constructed and operated. Such complexities arise due to terrain (network of small tunnels) and requirement of multiple entries and exits (network of traffics leading to non-homogenous behaviour). Increased complexities of such tunnels throw unique challenges for performing QRA for such tunnels, which gets compounded due to handful number of experiments performed in real tunnels, as they are costly and dangerous. A combined approach of CFD modelling of scaled down tunnels could be a relatively less resource intensive solution, nevertheless, associated with its increased uncertainties due to introduction of scaling multiplication factors. Further, with the advent of smart system designs and cheap computational cost, a smart tunnel which manages its own traffic of both dangerous goods carriers and other passenger vehicles based on continuously updated dynamic risk estimate, is not far from reality.

Keywords Quantitative risk assessment (QRA) · Tunnels · Fire dynamics
Risk analysis · F-N · EV · QRAM · Froude scaling · Smart tunnel
Dynamic risk of tunnel

J. K. Jena
Cyient Limited, Hyderabad, India

J. K. Jena
Lulea University of Technology, Lulea, Sweden

A. K. Verma (✉)
Western Norway University of Applied Sciences, Haugesund, Norway
e-mail: AjitKumar.Verma@hvl.no

Uday Kumar
Operation and Maintenance Engineering, Lulea University of Technology, Lulea, Sweden

S. Ajit
Department of Civil Engineering, Lulea University of Technology, Lulea, Sweden

P. K. Kapur et al. (eds.), *System Performance and Management Analytics*,
Asset Analytics, https://doi.org/10.1007/978-981-10-7323-6_31

1 Introduction

Tunnels are constructed as a cost-effective and time reducing measure to join two geographical locations separated by a hill or mountain. Tunnels are bore through these mountains and hills for easing surface transportation without a long route that circumscribes them. This not only reduces the cost of transportation but also cuts down the fossil foil induced CO_2 production that has a significant impact on global warming. One important example is the *Gotthard Base Tunnel*, 57.09 km long railway base tunnel through the Alps in Switzerland. It is also the first flat route through the Alps or any other major mountain range, with a maximum height of 549 m (1801 ft) above sea level, corresponding to that of Berne. Whereas, tunnels are bore through earth crust for subsurface transportation. One such example is the famous subsea *Channel Tunnel* (50.45 km) between France and England. A subset of subsurface tunnels is the metro railway transportation systems of various big cities of the world, which are primarily aimed to provide rapid transportation of daily goers. Among the road tunnels, the *Lærdal Tunnel* (24.51 km), in Norway is the longest road tunnel in the world succeeding the Swiss *Gotthard Road Tunnel.*

As necessity for faster transportation increases, a good number of long tunnels are under construction or at advanced level of planning. Tables 1 and 2 provide the list of important upcoming transportation tunnels globally (source Wikipedia).

In a global scale, many long tunnels are being constructed with multiple entry and exit; with a network of tunnels making these tunnels a complex engineering system. The design and construction of long tunnels throws unique challenges like determining the exact path of the tunnel; maintaining structural integrity during construction while using heavy equipment for boring huge diameter paths; ensuring structural integrity of the tunnel during a 10,000 year return period seismic event; designing a ventilation system for providing healthy air for travellers; providing a safe evacuation path during an accident (especially during fire and explosion); laying out a fire fighting strategy or an emergency operating procedure for firefighting and rescue operation by firemen during a large size fire arising due to vehicular collisions; designing a highly reliable firefighting system, etc.

Due to the advancement of hardware technology and computational capability, the challenges have been analysed, effective solutions have been found and implemented during construction and operational phase of tunnels. However, due to inherent space constraint inside a tunnel, increased traffic and the necessity to transport hazardous commodities using Heavy Goods Vehicles (HGVs), the risk arising in the tunnel due to fire and explosion caused by vehicular accidents though have been brought to ALARP region, but have not come down to significant low level.

Table 1 Important upcoming tunnels under construction

Name	Location, Country	Length (km)	Expected year of completion
Railway tunnels			
Brenner Base Tunnel	Stubai Alps, Austria—Italy	55.00	2026
Semmering Base Tunnel	Lower Austria/Styria, Austria	26.00	2026
Koralm Tunnel	Austria Koralpe, Austria	32.90	2022
Follo Line	Oslo, Norway	19.50	2021
Ceneri Base Tunnel	Lepontine Alps, Switzerland	15.40	2021
Musil Tunnel	Wonju-Jecheon (Jungang Line), South Korea	25.08	2018
Metro tunnels			
Mass Rapid Transit (Singapore): Thomson-East Coast Line	Singapore	42.00	2024
Paris Metro Line 15	Paris Petite Couronne, France	75.00	2030
Third Interchange Contour	Moscow Metro, Russia	58.30	2019
Road tunnels			
Förbifart Stockholm	Stockholm, Sweden	16.50	2025
Ryfast	Stavanger-Strand, Norway	14.30	2019

Table 2 Important upcoming tunnels at advanced planning stage

Name	Location, Country	Length (km)	Expected year of completion
Railway tunnels			
Bohai Strait tunnel	Bohai Strait, China	123.00	2023
Mont d'Ambin Base Tunnel	Cottian Alps, France—Italy	57.00	2023
Fehmarn Belt Fixed Link	Germany–Denmark	17.60	2024
Gulf of Finland Tunnel	Helsinki, Finland-Tallinn, Estonia	100.00	–
Barrandov Tunnel	Prague—Beroun	24.70	–
Metro tunnels			
Athens Metro Line	Athens, Greece	33.00	2023
Road tunnels			
Rogfast	Stavanger, Norway	25.00	2023
Fehmarn Belt Fixed Link	Germany–Denmark	17.60	2024
Agua Negra Tunnel	Chile–Argentina	14.00	–

2 Safety Features of Road and Train Tunnels

Safety requirements [1, 2], safety criteria [3], safety systems, safe operating and maintenance practices are devised and adopted to safeguard human life and the tunnel from structural damage. To ensure continuous adherence of safety standards and practices and to compare the level of safety among important European Tunnels, yearly tests are conducted by ADAC in conjunction with EuroTap [3]. Important tunnels throughout Europe are tested each year and benchmarked based on a group of assessment criteria [3] that are devised on the foundation of EU Directive 2004/54/EC [1]. The safety systems can be grouped into following categories based on Annexure-I [1] (Table 3).

3 Major Tunnels Accidents

In spite of adherence to both design and operational safety standards and practices, numerous accidents have occurred in the past, sometimes with fatal consequences. Table 4 provides the comprehensive list of significant fatal road tunnel accidents [4].

Table 3 EU Directive 2004/54/EC minimum safety requirements on safety systems and their primary objective

Safety system		Primary objective
Lighting	• Normal lighting • Safety lighting • Evacuation lighting	Crucial for safe evacuation and firefighting
Ventilation	• Mechanical ventilation • Special provisions for (semi-) transverse ventilation	Prevention of back-layering and control of fires; safe breathing during accidental condition; and toxic gas control
Emergency stations	At least every 150 m	Safe evacuation
Water supply	At least every 250 m	Firefighting
Road signs	Mandatory for all tunnels	Safe driving and safe emergency evacuation
Control and monitoring	• Visual monitoring • Automatic incident detection and/or fire detection • Control centre	Effective monitoring Emergency handling
Equipment to close the tunnel	• Traffic signals before the Entrances • Traffic signals inside the tunnel at least every 1000 m	Emergency measure to prevent entry into the tunnel during accidents

Table 4 Road fire accidents involving fire caused casualties, 1978–2017

Accident	Country	Year	Cause	Deaths
Sierre Tunnel	Switzerland	2012	Coach crash onto the wall	28
Sasago Tunnel	Japan	2012	Concrete ceiling collapse	9
Big Dig Tunnel	USA	2006	Concrete ceiling collapse	1
Fløyfjell Tunnel	Norway	2003	Car fire due to accident	1
St. Gotthard tunnel	Switzerland	2001	Two HGV collision	11
Gleinalm tunnel	Austria	2000	Two car collision	5
Rotsethhorn tunnel	Norway		Collision	2
Tauern tunnel	Austria	1999	HGV collision	12(4*)
IsoladelleFemmine	Italy	1996	LPG tanker collision	4
Pfänder tunnel	Austria	1995	Car–truck collision	1
Huguenot tunnel	South Africa	1994	Bus fire	1
Serra a Ripoli tunnel	Italy	1993	Car collision involving HGV	4
Gumefens tunnel	Switzerland	1987	Collision involving HGV	2
L'arme tunnel	France	1986	Car collision with trailer	3
Pecorile tunnel	Italy	1983	Collision involving fish lorry	9
Salang tunnel	Afghanistan	1982	Gas tanker explosion	176+
Caldecott tunnel	USA	1982	Car collision involving tanker	7
Sakai tunnel	Japan	1980	Truck collision	5
Kajiwara tunnel	Japan	1980	Truck collision	1
Nihonzaka tunnel	Japan	1979	Collision	7
Velsen tunnel	Netherlands	1978	Collision involving HGV	5

Source [4] * = Fire induced

If an accident does happen in a tunnel, the severity of injuries sustained is significantly higher than on open stretches of motorways. In a tunnel, the risk of being killed in a traffic accident is higher as on open stretches of motorways. Traffic safety is significantly higher in tunnels with unidirectional traffic than in tunnels with bidirectional traffic. In tunnels with bidirectional traffic, the probability of being killed in a traffic accident is expected to be higher as in tunnels with unidirectional traffic [5].

4 Risk Assessment of Tunnels

Risk assessment is a systematic scientific investigation process to analyse and assess potential incidents or accidents, thereby identifying the weak areas in the engineered system; estimate the damage to human life and assets; and find the scope of possible improvement of safety to minimize the occurrence probability of unwanted damaging consequence. Tunnels are expected to cause societal risk, the risk to a group of people those use it, rather than individual risk, the probability that a person among local population and up to a certain distance from the tunnel dies due to accidental scenarios in the tunnel. Primarily, this is because the individual risk is several orders of magnitude less that of the societal risk. The entire risk assessment process for an engineered system can be divided into three parts [6].

i. **Risk analysis**: Involves the analysis of possible dangerous scenarios in and around the engineered facility due to triggering of various internal and external hazards. Depending upon the cost–benefit, either a qualitative or a quantitative or both can be performed. However, quantitative risk analysis is performed for tunnels.
ii. **Risk evaluation**: The estimated risk values are compared against the relevant regulatory standards for their acceptability.
iii. **Safety management**: If the estimated risk is found to be unacceptable, additional safety measures in terms of safety system or procedure as safety barrier are proposed.

5 General Approach to Tunnel QRA

Quantitative Risk Assessment (QRA) is a systematic risk analysis method to quantify the risk associated with an engineering facility (or activity) due to potential hazards producing undesirable consequence, creating risk to humans and other assets. A numeric estimate of risk has several advantages as they are unambiguous, and can be used as absolute criteria or relative criteria to compare the risk levels of more than tunnels. Therefore, QRA improves confidence in risk-informed decision-making process.

Traditionally, hazards are characterized by their occurrence frequency or return period. Risk is defined as the product of hazard occurrence frequency and the probability of undesirable consequence. Mathematically,

$$\text{Risk} = \text{Hazard occurrence frequency} \times \text{Probability of undesirable consequence}$$

where, risk as well as hazard occurrence frequency are expressed in per annum. Probability is a numeric fraction in the closed interval [0, 1].

Risk analysis considers all possible hazardous scenarios (human induced and natural causes) and their possible prevention, protection and mitigations measures to ensure that the estimated risk is within acceptable limit based on governing regulations. Risk analysis considers all realistic protection and mitigation features and doesn't credit any very high cost protection or mitigation feature or heroic actions. If risk analysis outcome shows a risk value (say fatality rate) higher than the tolerable limit, it recommends additional protection and mitigation feature to be introduced in the tunnel interms of design feature or operational procedure. Risk analysis needs to be updated with the newly introduced feature to demonstrate that the estimated risk is within the tolerable limit as per local regulation.

For road tunnels, the hazards could be natural like earthquake, accidents causing fire or explosion arising from hazardous substances carried by various vehicles. The frequency of hazard occurrence can be estimated by a statistical analysis of traffic data of each hazard carrier type. The consequence of each hazard type is estimated based on the consequential scenario analysis that is likely to occur inside the specific tunnel. The expected consequences are fatalities, severe injuries, minor injuries, destruction/damage to the tunnel structure and environmental damage, etc. The scenarios and their impact zones are determined either by performing static (empirical or semi-empirical) calculations or dynamics simulation of the prevailing scenario using suitable computational fluid dynamics (CFD) software.

Risk is expressed in two metrics Fatalities per Year (F/N) and/or Expected Value (EV)

i. Fatalities per Year (F/N): Expected fatalities per year due to a particular hazard and for a particular system, (say a tunnel) preferably expressed in terms of Fatalities per Year (F/N or simply FN). The outcome of QRA is presented as an FN graph, which is a log-log scale graph with the cumulative frequencies (F) of incidents (in Y-axis) involving N or more units of damage (in X-axis).
ii. Expected Value (EV): The long-term average number of statistically expected fatalities per year due to all significant hazards inside a tunnel. Mathematically,

$$EV = \int_{1}^{\infty} F(N)\mathrm{d}N = \sum_{i=1}^{\infty} F(Ni).Ni$$

Internationally, the norms for acceptable and tolerable risk (or fatality rate F/N) arising from an engineered facility are 1 death in 100 years and 1 death in 10 years, respectively. The region above the tolerable limit is termed as non-acceptable region and the region below the acceptable limit is called acceptable region. The in-between is called the ALARP (***As Low As Reasonably Practicable***) region, where the foreseen benefits are higher than the projected cost estimates. A typical QRA process for tunnels is presented as a flowchart in Fig. 1.

Following are some of the widely used standards providing guideline for design and operation of road tunnels. Guideline for the risk analysis is an integral part of these standards/guidance documents.

- **European Directive 2004/54/EC**: Directive 2004/54/EC of The European Parliament and of The Council of 29 April 2004 on minimum safety requirements for tunnels in the trans-European road network.
- **NFPA 502**: Standard for Road Tunnels, Bridges and Other Limited Access Highways 2008 Edition from the USA (Chapter-7).
- **BD 78/99**: Design Manual for Roads and Bridges from the United Kingdom.
- **ASTRA 19004**: Risikoanalysefür Tunnel der Nationalstrassen (2014 V1.01)-Switzerland
- **Fire Safety Guidelines for Road Tunnels**: Australasian Fire Authorities Council (AFAC), Australia.

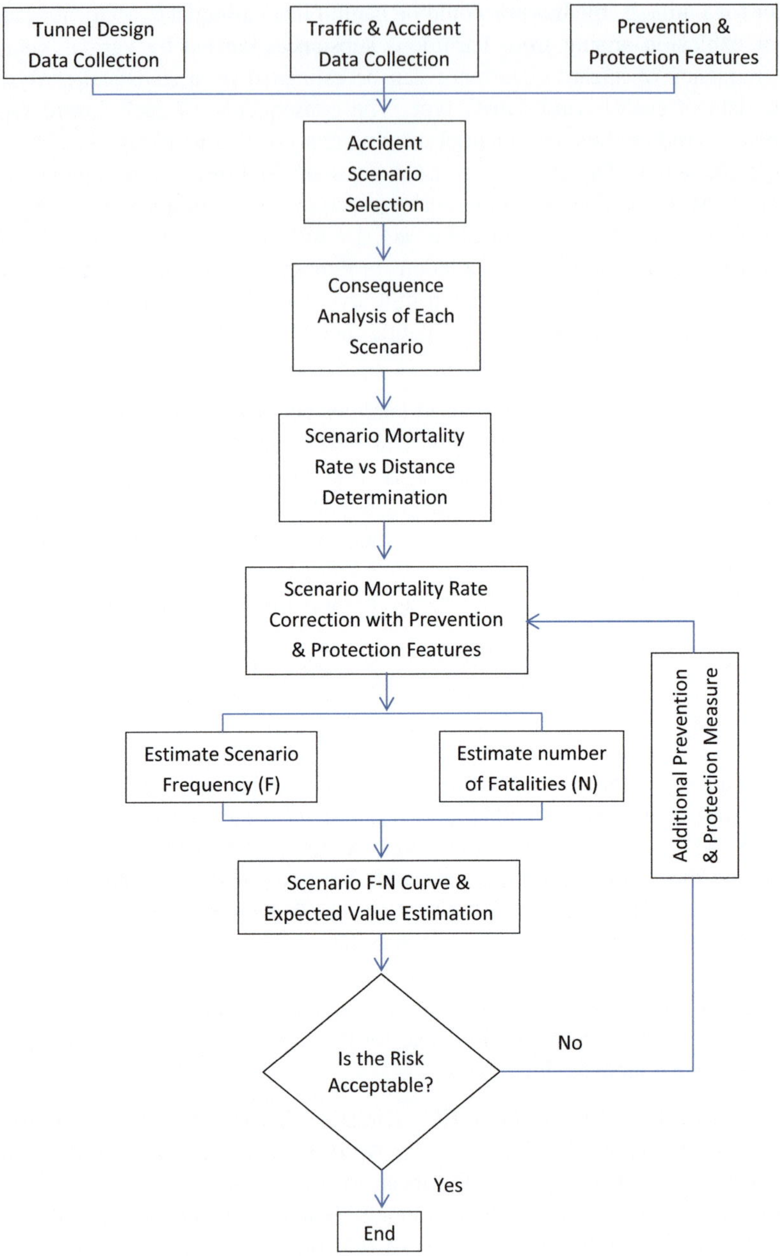

Fig. 1 A typical QRA process for a tunnel

Above-mentioned standards though prescriptive about tunnel design but provide generic guidance on risk analysis. For example: per European Directive 2004/54/EC - Article -13:

A risk analysis is an analysis of risks for a given tunnel, taking into account all design factors and traffic conditions that affect safety, notably traffic characteristics and type, tunnel length and tunnel geometry, and forecast number of heavy goods vehicles per day as well.

In Europe, based on the EU Directive 2004/54/EC, following QRA models were developed:

- OECD/PIARC DG-QRAM model—First and used by many European countries
- Austrian tunnel risk model (TURISMO)
- Dutch scenario analysis for road tunnels
- Dutch TUNPRLM model
- French specific hazard investigation
- German risk analysis for road tunnels
- Czech risk analysis for road tunnels
- Italian risk Analysis method (IRAM),—A true Risk Model

Most of the QRA models use PIARC developed QRAM to estimate the risk from transportation of Dangerous Goods (DGs). The QRAM takes into account:

- Accident frequencies (derived from historical datasets)
- Physical consequences of incidents within tunnel(s) and along the open routes
- Escape and sheltering effects
- Effects of hazards (such as heat and smoke) on people

PIARC QRAM identifies 13 accident scenarios (refer Table 5) which are representative of key dangerous goods groupings. The hazard impacts of these DGs on humans and vehicles are in the form of fire causing burning, high pressure (shock wave) damaging physical bodies or toxic gases harming human beings and structures those react chemically.

Estimation of fatalities during an accidental scenario is a multistep process. First, consequence (fire spread, toxic gas dispersion, etc.) area variation with time is estimated using consequence modelling tools. Next, the number of people (N_s) within the impact area is estimated which depend upon tunnel traffic characteristics (e.g. volume, occupancy). Further, escape probability (P_{es}) through provided escape route is determined to estimate the number of people N_s $(1 - P_{es})$ finally exposed within the impact area. Depending upon the cause (heat, toxic gas, etc.) of death specific fatality models are used to estimate fatality rate (f). Meng et al. [7], discusses the some of the standard fatality rate estimation models and used while performing the QRA of a Singapore urban non-homogenous tunnel. Gist of such model is tabulated in Table 6:

A representative F-N curve of a tunnel QRA depicting aggregated risk level encompassing all analysed scenarios is shown in Fig. 2. For any tunnel, the quantified admissible risk (residual risk) is accompanied by a set of preventive and

Table 5 PIARC QRAM hazard scenarios and their impact

Hazard	QRAM scenarios	Hazard impact
1. Heavy Goods Vehicle (No DGs)	20 MW fire	Fire and burning
2. Heavy Goods Vehicle (No DGs)	100 MW fire	Fire and burning
3. LPG cylinders	Boiling Liquid Expanding Vapour Explosion (BLEVE)	Fire and explosion generating high-pressure shock waves
4. LPG in bulk	Boiling Liquid Expanding Vapour Explosion (BLEVE)	Fire and explosion generating high-pressure shock waves
5. LPG in bulk	Vapour Cloud Explosion (VCE)	Fire and explosion generating high-pressure shock waves
6. LPG in bulk	Torch fire	Fire and explosion generating high-pressure shock waves
7. LPG in bulk	Pool fire	Huge fire and burning
8. Motor spirit in bulk	Pool fire	Huge fire and burning
9. Motor spirit in bulk	Vapour Cloud Explosion (VCE)	Fire and explosion generating high-pressure shock waves
10. Acrolein in cylinders	Toxic release	Toxic liquid and gas
11. Chlorine in bulk	Toxic release	Toxic gas
12. Ammonia in bulk	Toxic release	Toxic gas
13. Liquid CO_2 in bulk	Boiling Liquid Expanding Vapour Explosion (BLEVE)	Fire and explosion generating high-pressure shock waves

protective measures (or risk reduction measures) that form the basis of Standard Operating Procedure (SOP) and Emergency Operating Procedure (EOP) respectively.

6 Challenges in Tunnel QRA

6.1 Complex Tunnel Fire Dynamics

Fire is 'the' most risk potential hazard in tunnels. The dynamics of fire inside is complex as it is governed by many factors. Some important ones are as follows: (a) tunnel geometry (e.g. diameter, curvature, length) (b) tunnel topology (e.g. inclined) (c) fire load (e.g. Styrofoam, hydrocarbon, wood) (d) fire type (e.g. pool, jet or vehicular) (e) ventilation inside the tunnel (e.g. longitudinal, transverse) (f) external wind (g) presence of additional fire hazards during an accidental scenario (e.g. other vehicles) (h) presence of physical obstacles inside the tunnel (e.g. other vehicles) (i) type of fire suppression system (e.g. mist, spray) (j) time of actuation of fire suppression system.

Table 6 Fatality rate models used in tunnel QRA [7]

Cause of Fatality	Model	Developer
Fire heat	$F_D(t, T) = t/\exp[5.1849 - 0.0273T]$ Where, t = Exposure time (min) T = Temperature (°C)	Purser [8]
Explosion	A tabular relationship among distance from explosion site, overpressure caused and the fatality rate for the equivalence of 1000 kg TNT	Anet [12]
High CO concentration generated due to fire	$F_{CO} = 8.2925E - 04 * \frac{X_{CO}}{D} * t$ where t = Exposure time (min) D = %COHb at incapacitation (30%) X_{CO} = CO concentration	Persson [9] Nelson and Log [10] Beard [11]
High CO_2 concentration generated due to fire	$F_{CO_2} = t/\exp[6.1523 - 0.5189X_{CO_2}]$ where t = Exposure time (min) $X_{CO_2} = CO_2$ concentration	Persson [9] Nelson and Log [10] Beard [11]
Low O_2 concentration due to fire	$F_{O_2} = t/\exp[8.13 - 0.54(20.9 - X_{O_2})]$ where t = Exposure time (min) $X_{O_2} = O_2$ concentration	Persson [9] Nelson and Log [10] Beard [11]
Toxic gases such as Acrolein, Chlorine, Ammonia etc.	$Pr = a + b\ln C^n t$ where Pr = Probit value [13] t = Exposure time (min) a, b, and n are constants that have different values for different toxic gases C = Concentration of toxic gases (mg/m^3)	Wheeler et al. [14]

Complexities of the tunnel fire dynamics are compounded further because of very limited number of benchmark tunnel experiments have been performed till date as they are dangerous and costly. Some of the important outcomes of the performed as reported in [4] are as follows: (a) Fires in tunnels produce thick smoke layers which advance faster than walking speed. (b) Smoke stratification is destroyed by longitudinal ventilation. (c) Smoke stratification is destroyed if sprinklers are used. (d) Smoke does not remain stratified even in naturally ventilated tunnels. (e) There is a significant reduction in visibility before the onset of debilitating heat. (f) Naturally and semi-transversely ventilated (pool) fires burn slower than those in the open air. (g) Burning rate is enhanced by longitudinal ventilation. This is more evident in vehicle fires. (h) The burning rate of some pool fires is reduced by increased longitudinal ventilation. (i) Vehicle fires exhibit a 'fast' rate of fire development. (j) The heat release rate of car fires in tunnels may be significantly larger than in the open air. (k) High temperatures (often > 1000 °C) are common in tunnel fires. (l) High temperatures are only evident in the immediate vicinity of car fires. In most instances, it is possible to get close to a car fire. (m) High temperatures may cause explosive 'spalling' of the tunnel lining.

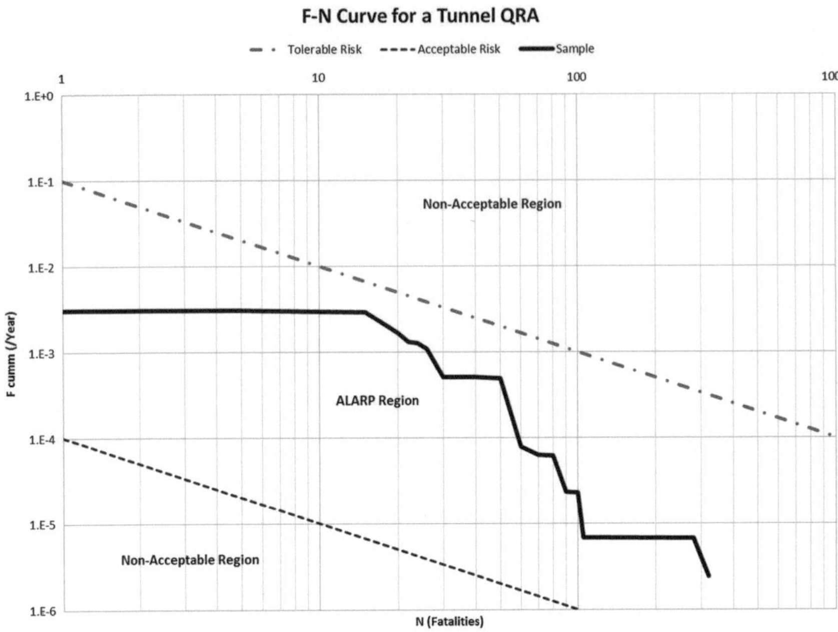

Fig. 2 A representative F-N curve of a tunnel QRA showing the aggregated risk arising from all modelled scenarios

However, the lack of numerous benchmarking experiments can be circumvented principally in following two ways:

(a) By Computational Fluid Dynamics (CFD) simulations of tunnel fires using fire dynamics simulators such as Fire Dynamics Simulator (FDS). Fire Dynamics Simulator (FDS), developed by NIST & VTT (https://www.nist.gov/publications/fire-dynamics-simulator-users-guide-sixth-edition), models subsonic large combustion problems by solving low Mach number combustion equations coupled with COM and COE equations for a given domain. This is commonly known as low Mach number (≤ 0.3 Mach) Large Eddy Solutions (LES). FDS simulation for numerous fire accidents in tunnels has been performed satisfactorily.
In spite of the development of the tremendous fast computation facility, CFD modelling can be used for fire scenarios in tunnels only. The limitation arises from the fact the numerical solutions of Navier–Stokes Equation for all possible speeds is untenable with the best computational facility as of now. Thus, different software based on semi-empirical methods issued to model other scenarios like BLEEVE, VCE, etc., while performing Tunnel QRA.

(b) By performing scaled down tunnel experiments. Scaled down experiments are basically a convergence of following three kinds of similarities as demanded by the scaling laws of physics:

i. **Geometrical similarity**: All linear dimensions must have the same scale ratio
ii. **Kinematic similarity**: The flow and model(s) will have geometrically similar motions in model and full scale
iii. **Dynamic similarity**: Ratios between different forces in full scale must be the same in model scale.

Based on Buckingham Π Theorem [15], James G. Quintiere in the paper 'Scaling applications in fire research' [16], derived 14 dimensionless scaling parameters, which can be applied in designing experiments in fire research. Since, low-speed fire (diffusion fire) propagation involves primarily gravitational and inertial force, the conservation of Froude Number (Fn) between real fire test (full scale) and reduced scale (model) tunnel fire experiment will produce identical effect as surface waves are gravity-driven implies equality in Fn will ensure that wave resistance and other wave forces are correctly scaled. If scaling parameter $\lambda = L_m/L_f$: where, f and m stand for full scale and model, various multiplication factors in term of the scaling parameter λ derived to obtain the physical quantities in the scaled down model. Following table provides the Froude scaling physical parameters [17] (Table 7).

Ingason et al. [17] have reported to perform such scaled down model test of Runehamar tunnel tests [18]. Some of the reported important results are as follows: (a) the fire growth rate is very sensitive to the longitudinal ventilation velocity [18, 19]. (b) the flame length increases linearly with the heat release rate, and is insensitive to the ventilation velocity [18, 19]. (c) the back-layering length is independent of the heat release rate for a large tunnel fire [18, 19].

In spite of significant progress made in the area of tunnel fire real and model experiments, many issues are still subject of in-depth research. A few important ones are noted below:

i. Fire dynamics in an inclined tunnel with longitudinal and transverse ventilation.
ii. Determining effect of water mist and water spray system for fire suppression in case of various types of fire scenarios.

Table 7 Multiplication factors based on Froude scaling

Physical parameter	Unit	Multiplication factor
Length	(m)	λ
Mass	(kg)	λ^3
Force	(N)	λ^3
Moment	(Nm)	λ^4
Velocity	(m/s)	$\lambda^{1/2}$
Heat Release Rate (HRR)	(kW)	$\lambda^{5/2}$
Time	(s)	$\lambda^{1/2}$
Pressure	(N/m^2)	λ
Temperature	(K)	1
Acceleration	(m/s^2)	1

iii. Drawing correlation between tunnel dimensions (frontal height and area), fire size and spread length for various types of fire with different types of ventilation system.
iv. Developing the empirical relationships for tunnel fire dynamics by combined use of CFD modelling and Froude scaled down models.

6.2 Effect of Traffic and Evacuation in Very Long Tunnels for Various Types of Fires

In present day, very long road tunnels (e.g. Lærdal Tunnel, Swiss Gotthard Road Tunnel) with bidirectional traffic are built for speedy transportation. During accidental scenarios, the spread of traffic (size and number of vehicles) is expected to largely affect the fire dynamics in such long tunnels and consequently impacting the evacuation of trapped human beings under different fire types (e.g. jet, BLEEVE, VCE etc.). Modelling of such scenario is important and a matter of research as it may significantly impact the risk level estimated in a tunnel QRA.

6.3 Uncertainty Analysis in Various Tunnel QRA Parameters

The traffic inside a tunnel including normal as well as hazardous goods carriers is very dynamic. Even under the controlled environmental condition inside a tunnel, consequence of an accident could be drastically different at different time of a day, in the presence of different sets of vehicles adjacent to the accident. Therefore, determining the worst case or even the best estimate scenario involves a huge uncertainty bound.

Similarly, uncertainty involved in the mitigating actions performed by human operators based on emergency operating procedure in the control room and uncertainty in the act of firefighting is a tough challenge in tunnel QRA.

7 Future Tunnel Designs

7.1 Smart Tunnel

A robust tunnel design that reduces the possibility of any accident is better than having a robust accident management strategy. Smart tunnels using present day intelligent system doing automatic traffic management with focus on the presence of hazardous goods carrying vehicles inside the tunnel or tunnel system is likely to be essential features of future tunnels. A hazardous goods carrying vehicle can have a predefined electromagnetic signal or a pre-designated Global Positioning System (GPS) which will be detected by the smart tunnel traffic management system (TTSM). This system will find the best possible or most intelligent safe passage condition for the goods vehicle by taking the dynamic traffic situation as input. With the continuous input of traffic situation, TTSM can operate the installed tunnel Traffic Signalling System (TTSS) for the smooth passage of the dangerous goods carrying vehicle with least possible interaction with other vehicles inside the tunnel.

7.2 *Dynamic Risk Monitoring*

Pre-emptive actuation of safety systems to prevent accident-prone situation will not only save precious human lives but also prevent humongous amount of imminent financial loss due to accidents inside the tunnels. Dynamic risk estimation inside a tunnel based on live traffic can be performed by suitable identification and categorization of various vehicle types based on their fire hazard potential with the modern day high-speed computation capability. Based on the risk level, preventive measures can be initiated by actuating safe traffic regulating system (a combination of obstacles, alarms, specific driving instruction over public annunciation system) that streamlines the vehicular movement inside the tunnel may become the norm of future tunnels.

8 Conclusions

Tunnels are complex engineering systems from risk analysis point. The biggest risk to tunnels is fire. The fire dynamics inside the tunnel depends of number of factors that are intrinsic to tunnels (viz. length, ventilation, size, fire type, etc.) and some extrinsic factors like ambient temperature, traffic density during an accident etc.). Since fire and explosion related experiments in tunnels are dangerous and costly, only a handful numbers of experiments have been performed to understand the fire dynamics. Alternate methods such as CFD modelling of low Mach speed (<0.3 Mach) accidental fire scenarios and scaled down model tests of tunnels based on Froude Scaling are used to understand the fire dynamics. In spite of significant progress made in the area of tunnel fire real and model experiments, many issues are still subject of in-depth research. A few important ones are noted below:

i. Fire dynamics in an inclined tunnel with longitudinal and transverse ventilation.
ii. Determining effect of water mist and water spray system for fire suppression in case of various types of fire scenarios.

iii. Drawing correlation between tunnel dimensions (frontal height and area), fire size and spread length for various types of fire with different types of ventilation system.
iv. Developing the empirical relationships for tunnel fire dynamics by combined use of CFD modelling and Froude scaled down models.

Similarly, a good amount of research is required to capture the effect of traffic and evacuation in present day-long tunnels for various fire types. In future, with the availability of high speed computational facility and increased application of artificial intelligence, smart tunnels that regulate the movement of traffic in the presence of hazardous goods vehicles may become operational. Similarly, with suitable GPS tagging of live traffic and categorization of vehicle types based on their fire hazard potential, dynamic risk can be estimated. This risk value can be used to manage the traffic flow inside a tunnel.

References

1. Directive 2004/54/EC dated 29 April 2004 of European Parliament and of the council on minimum safety requirements for tunnels in the Trans-European Road Network.
2. NFPA® 502, 2011 Standard for Road Tunnels, Bridges, and Other Limited Access Highways.
3. EuroTAP (European Tunnel Assessment Programm) tunnel 2/2008, P. 46–47.
4. Fire size in tunnels, PhD thesis, Richard Oswald Carvel, 2004.
5. Comparative analysis of safety in tunnel, Nussbaumer Cornelia, Austrian Road Safety Board.
6. Risk evaluation for road tunnels, Zulauf Christoph et al, 6th International Conference 'Tunnel Safety and Ventilation' 2012, Graz.
7. Meng, Q., Qu, X., Wang, X., Yuanita, V., & Wong, S. C. (2011). Quantitative risk assessment modeling for nonhomogeneous urban road tunnels. *Risk Analysis, 31*(3). https://doi.org/10.1111/j.1539-6924.2010.01503.x.
8. Purser, A. P. (1988). Toxicity assessment of the combustion products. In P. J. DiNenno (Ed.), *The SFPE Handbook of Fire Protection Engineering* (1st ed., pp. 206–276). Quincy, MA: Society of Fire Protection Engineers.
9. Persson, M. (2002). *Quantitative risk analysis procedure for the fire evacuation of a road tunnel-an illustrative example*. Sweden: Lund University, Report 5096.
10. Nilsen, A. R., & Log, T. (2009). Results from three models compared to full-scale tunnel fires tests. *Fire Safety Journal, 44,* 33–49.
11. Beard, A. N. (2009). Fire safety in tunnels. *Fire Safety Journal, 44*(1), 276–278.
12. Anet, B., Binggeli, E. (1998). *Air blast phenomena due to nuclear and conventional explosions*. CH: Gruppe Rütung, Technical Report.
13. Finney, D. J. (1980). *Probit analysis* (3rd ed.). Cambridge, UK: Cambridge University Press.
14. Weger, D., Pietersen, C. M., & Reuzel, P. G. J. (1991). Consequences of exposure to toxic gasses following industrial disasters. *Journal of Loss Prevention in the Process Industries, 4*(4), 272–276.
15. Buckingham, E. (1914). On physically similar systems; illustrations of the use of dimensional equations. *Physical Review, 4*(4), 345–376. https://doi.org/10.1103/PhysRev.4.345.
16. Quintiere, J. G. (1989). Scaling applications in fire research. *Fire Safety Journal*. https://doi.org/10.1016/0379-7112(89)90045-3.

17. Ingason, H., Li, Y. Z. (2010). Model scale tunnel fire tests with longitudinal ventilation. *Fire Safety Journal*, *45*(6–8), 371–384, https://doi.org/10.1016/j.firesaf.2010.07.004.
18. Ingason, H., Li, Y. Z., & Lönnermark, A. (2015). Runehamar tunnel fire tests. *Fire Safety Journal, 71,* 134–149. https://doi.org/10.1016/j.firesaf.2014.11.015.
19. Li, Y. Z., Lei, B., Ingason, H. Study of critical velocity and back layering length in longitudinally ventilated tunnel fires. *Fire Safety Journal*. https://doi.org/10.1016/j.firesaf.2010.07.003.

Software Vulnerability Prioritization: A Comparative Study Using TOPSIS and VIKOR Techniques

Ruchi Sharma, Ritu Sibal and Sangeeta Sabharwal

Abstract The ever-mounting existence of security vulnerabilities in a software is an inevitable challenge for organizations. Additionally, developers have to operate within limited budgets while meeting the deadlines. So they need to prioritize their vulnerability responses. In this paper, we propose an approach for vulnerability response prioritization using "closeness to the ideal" approach. We used TOPSIS and VIKOR method in this study. Both of these techniques employ an aggregating function to achieve the ranking of desired alternatives. VIKOR method determines a compromise solution on the basis of measure of closeness to a single ideal solution while TOPSIS method determines a feasible solution while taking into account the shortest distance from the positive ideal solution and the maximum distance from negative ideal solution. Both these methods share some significant similarities and differences. A comparative analysis of these two methods is done by applying them on real-life software vulnerability datasets for achieving vulnerability prioritization.

Keyword Vulnerability · Priortization · Ranking · TOPSIS · VIKOR Comparison

1 Introduction

With our growing dependence on software products and the gradually increasing number of software vulnerabilities, efficient mitigation strategies have to be developed. To maximize reliability and security of a software, the testing team has

---

R. Sharma (✉) · R. Sibal · S. Sabharwal
Department of Computer Engineering, Netaji Subhas Institute of Technology, Delhi, India
e-mail: rs.sharma184@gmail.com

R. Sibal
e-mail: ritusib@hotmail.com

S. Sabharwal
e-mail: ssab63@gmail.com

P. K. Kapur et al. (eds.), *System Performance and Management Analytics*, Asset Analytics, https://doi.org/10.1007/978-981-10-7323-6_32

to take care of the order in which the vulnerabilities are fixed. This order is important especially when the organization has to meet certain deadlines and they do not have sufficient time to fix all the detected vulnerabilities. It also helps to design suitable patches.

The security team must identify and assess vulnerabilities across disparate hardware and software platforms. They need to prioritize these vulnerabilities and remediate those that pose the greatest risk. A large number of organizations, companies, and researchers have given rating systems to rank and prioritize them. Vendors have been using their own methods for scoring software vulnerabilities, usually without detailing their criteria or processes. In the last few years, the frequency of "zero-day attack" has enhanced radically which further emphasized the need for prioritizing the process of vulnerability fixation. So practically, there is always a dearth of time for deployment of updates and patching to shield the weakness of the software systems creating opportunities for the black hat bearers. Hence, in order to handle the enormous amount of different vulnerabilities, security managers must prioritize and rank the vulnerabilities depending on their possible negative impact on software systems.

In the current literature, prioritization methods are categorized into two broad groups, viz. quantitative and qualitative. Qualitative systems suggest a rating approach to define the severity of software vulnerabilities. While quantitative scoring systems associate a numerical score with each vulnerability. In general, qualitative methods prioritize the vulnerabilities by dividing them into various severity levels. On the other hand, quantitative methods generally associate a mathematical score with the vulnerability. The most commonly used and the first open quantitative prioritization method is Common Vulnerability Scoring System (CVSS) [1–3]. The National Vulnerability Database which is the U.S. government's repository of standards based vulnerability management data also uses CVSS scoring for vulnerability categorization [4]. Over the time, it has emerged as a standard for quantitative vulnerability assessment [1–4]. In addition to these two systems, hybrid rating system also exists. They give a score for severity of each vulnerability and also assigns a rating to them by combining the quantitative and qualitative methods. Vulnerability Rating and Scoring System (VRSS) is one such hybrid vulnerability rating system [5, 6]. Later, Weighted Impact Vulnerability Scoring System (WIVSS) was proposed. This system uses the factors used in CVSS approach while assigning weights to each impact metric instead of treating them equally. It uses the same six factors that CVSS uses but it considers different weights for the impact metrics [7]. Another quantitative scoring system named Potential Value Loss (PVL) introduced seven pointers which indicates the severity of a vulnerability to compute the vulnerability score [8]. Researchers have also emphasized the use of temporal attributes and context information for severity scoring [9–11]. In recent studies, multi-criteria decision-making techniques have been employed for ranking software vulnerabilities while considering the relative importance of vulnerabilities instead of treating them as independent entities [12]. In recent past, many researchers have used "TOPSIS" (The Technique for Order of Preference by Similarity to Ideal Solution) and "VIKOR" (VIsekriterijumska Optimizacija

KOm- promisno Resenje) methods for a plethora of decision-making problems [13, 14]. These two methods helps to rank on the basis of different criteria while considering their relative importance. The TOPSIS method determines the solution by giving the shortest distance from the ideal solution and with the greatest distance from the negative ideal solution [13]. The VIKOR method determines ranking based on the particular measure of "closeness" to an ideal solution giving a compromise solution. The compromise solution is a feasible solution that is "closest" to the ideal solution. Here, compromise suggests an agreement established by mutual concessions [14]. Both these methods share some significant similarities and differences.

2 Methodology

In this section, we will describe the two techniques used for comparative study of vulnerability prioritization. Both these methods take into consideration the distance from an ideal solution to achieve ranking. The first method is the TOPSIS method which uses positive and negative ideal solutions to give the final ranking. VIKOR method, on the other hand, takes in account a single ideal solution approach.

2.1 *TOPSIS Method*

TOPSIS method was proposed by Hwang and Yoon in 1981 [13]. This method was first introduced with an idea to offer an alternative for elimination and choice expressing reality III method. It was later applied by Zeleny suggesting its various applications [15]. The basic principle suggests that suitable alternatives are present at a minimum distance from positive ideal solution and maximum distance from negative ideal solution [13]. So, it is appropriate for the process of decision-making to achieve maximum profits at minimum risk. A positive ideal solution maximizes the benefit criteria and minimizes the cost criteria, whereas a negative ideal solution maximizes the cost criteria and minimizes the benefit criteria. TOPSIS method is expressed in a succession of six steps as follows:

Step 1: Prepare a decision matrix such that rows consists of the various alternatives and columns contain the criteria for ranking.

$$D = \begin{pmatrix} & C_1 & C_2 & \cdots & C_j & \cdots & C_n \\ A_1 & x_{11} & x_{12} & \cdots & x_{1j} & \cdots & x_{1n} \\ A_2 & x_{21} & x_{22} & \cdots & x_{2j} & \cdots & x_{2n} \\ \vdots & & & & & & \\ A_i & x_{i1} & x_{i2} & \cdots & x_{ij} & \cdots & x_{in} \\ \vdots & & & & & & \\ A_m & x_{m1} & x_{m2} & \cdots & x_{mj} & \cdots & x_{mn} \end{pmatrix}$$

Here, A_i is the ith alternative and C_j is for jth criteria. x_{ij} is the numerical value corresponding to ith alternative against jth criteria.

Step 2: Calculate the normalized decision matrix. The normalized value r_{ij} is calculated as follows:

$$r_{ij} = \frac{x_{ij}}{\sqrt{\sum_{i-1}^{m} x_{ij}^2}} i = 1,2,\ldots,m \quad \text{and} \quad j = 1,2,\ldots,n. \tag{1}$$

Step 3: Calculate the weighted normalized decision matrix. The weighted normalized value v_{ij} is calculated as follows:

$$v_{ij} = r_{ij} \times w_j \; i = 1,2,\ldots,m \quad \text{and} \quad j = 1,2,\ldots,n. \tag{2}$$

where w_j is the weight of the jth criterion or attribute and $\sum_{j=1}^{n} w_j = 1$.

Step 4: Determine the ideal (A^*) and negative ideal (A^-) solutions.

$$A^* = \left\{ \left(\max_i v_{ij} \middle| j \in C_b \right), \left(\min_i v_{ij} \middle| j \in C_c \right) \right\} = \left\{ v_j^* \middle| j = 1,2,\ldots,n \right\} \tag{3}$$

$$A^- = \left\{ \left(\min_i v_{ij} \middle| j \in C_b \right), \left(\max_i v_{ij} \middle| j \in C_c \right) \right\} = \left\{ v_j^- \middle| j = 1,2,\ldots,n \right\} \tag{4}$$

Step 5: Calculate the separation measures using the m-dimensional Euclidean distance. The separation measures of each alternative from the positive ideal solution and the negative ideal solution, respectively, are as follows:

$$S_i^* = \sqrt{\sum_{j=1}^{m} \left(v_{ij} - v_j^* \right)^2}, \quad j = 1,2,\ldots,n \tag{5}$$

$$S_i^- = \sqrt{\sum_{j=1}^{m} \left(v_{ij} - v_j^- \right)^2}, \quad j = 1,2,\ldots,n \tag{6}$$

Step 6: Calculate the relative closeness to the ideal solution. The relative closeness of the alternative A_i with respect to A^* is defined as follows:

$$RC_i^* = \frac{S_i^-}{S_i^* + S_i^-}, \quad i = 1,2,\ldots,m \tag{7}$$

After the relative closeness values are obtained, rank in the order of preference.

2.2 VIKOR Method

This method works on the basis of the particular measure of closeness to the positive ideal solution. It gives a compromise solution that is the closest to the ideal solution, where compromise means an agreement established by mutual concessions. VIKOR method has following four steps as given by Opricovic and Tzeng in 2004 [14].

Step 1: Determine the best and worst values, which are known as positive ideal and negative ideal solutions.

$$\begin{aligned} & f_j^* = \max_i f_{ij} \quad f_j^- = \min_i f_{ij} \\ & \text{if the } i\text{th function represent cost then} \\ & f_j^* = \min_i f_{ij} \quad f_j^- = \max_i f_{ij} \\ & \text{where } f_{ij} = \text{value of } i\text{th alternative for } j\text{th criteria} \end{aligned} \tag{8}$$

Step 2: Calculate the values of A_i and R_i using following equations:

$$A_i = \sum_{j=1}^{n} \left(\frac{w_j \left(f_j^* - f_{ij} \right)}{f_j^* - f_j^-} \right) \tag{9}$$

$$R_i = \max_j \left(\frac{w_j \left(f_j^* - f_{ij} \right)}{f_j^* - f_j^-} \right) \tag{10}$$

Here, A_i is the maximum group of utility of the majority of alternative i; R_i is a minimum of individual regret of the opponent of alternative i, w_j is the weight of the criteria, which expresses the experts' opinion regarding relative importance of the criteria.

Step 3: Calculate the following values:

$$\begin{aligned} & A^* = \min_i A_i;\ A^- = \max_i A_i;\ R^* = \min_i R_i;\ R^- = \max_i R_i \\ & Q_i = v \frac{(A_i - A^*)}{(A^- - A^*)} + (1 - v) \frac{(R_i - R^*)}{(R^- - R^*)} \end{aligned} \tag{11}$$

v is introduced as a weight for the strategy of maximum group utility, whereas $(1 - v)$ is weight of the individual regret. The solution obtained by $\min_i A_i$ is with a maximum group utility and the solution obtained by $\min_i R_i$ is with a minimum individual regret of the opponent. The value of v is taken as 0.5 however it can be taken from 0 to 1.

Step 4: Rank the alternatives, sorting by the values of S, R and Q in decreasing order.

The results are three ranking lists. Propose a compromise solution the alternative A_1 which is the best ranked by the measure Q (minimum), if the following two conditions are satisfied.

(1) Acceptable advantage: $Q[A_2] - Q[A_1] \geq DQ$, where $DQ = 1/(M - 1)$ and A_2 is the alternative with second position in the ranking list by Q.
(2) Acceptable stability in decision-making: The alternative A_1 must also be the best ranked by S or/and R. This compromise solution is stable within a decision-making process, which could be the strategy of maximum group utility (when $v > 0.5$ is needed), or "by consensus" (v is approximately 0.5) or with veto ($v < 0.5$). If one of the above conditions is not satisfied, then a set of compromise solutions is proposed which is given as below:

- Alternative A_1 and A_2 if only condition 2 is not satisfied, or Alternatives A_1, A_2,…,A_M if the condition 1 is not satisfied. A_M is determined by the relation $Q[A_M] - Q[A_1] < DQ$ for maximum M; the positions of these alternatives are "in closeness".

3 Numerical Illustration

In this section, we present the numerical illustration of both the techniques discussed in the previous section. The dataset used in this study is extracted from CVE Details, which is a well-known data source of security vulnerabilities [16, 4]. Dataset consists of 13 different types of vulnerabilities with 3 different categories namely low (C1), medium (C2) and high (3) based on the severity levels of vulnerabilities as per the Common Vulnerability Scoring System (CVSS) values. It contains the count of vulnerabilities detected over a period of 19 years (1999–2017).

3.1 Ranking Vulnerabilities Based on TOPSIS Technique

In this study, the initial decision matrix is obtained by dividing the number of vulnerabilities in each category by the aggregate sum of each column so as to obtain the values within a range of 0–1. This initial matrix is shown in Table 1. The initial weightages of the three criteria are also mentioned. These weights are the criteria weights obtained in [12] using the analytic hierarchy process (AHP).

After the initial decision matrix is obtained, it is normalized using Eq. (1) which includes dividing each value by the square root of the sum of the squares of each column as shown in Table 2.

After the normalization is achieved, the criteria weights are multiplied column wise so as to obtain the weighted normalized decision matrix in Table 3 (Eq. 2).

Table 1 Initial decision matrix along with criteria weights

	C1 (1–3.9)	C2 (4–6.9)	C3 (7–10)
Weights (using AHP) [12]	0.065	0.199	0.735
DOS	0.2331	0.219426	0.147787
Code execution	0.0238	0.125736	0.375682
Overflow	0.0440	0.084076	0.165168
Memory corruption	0.0043	0.019841	0.060116
Sql injection	0.004279	0.026321	0.103676
XSS	0.304968	0.204722	0.003241
Directory traversal	0.02068	0.049837	0.016438
http response splitting	0.002615	0.00271	0.000275
Bypass something	0.081768	0.056765	0.035567
Gain information	0.246256	0.128776	0.010507
Gain privileges	0.027573	0.036924	0.051141
CSRF	0.004992	0.029266	0.001375
File inclusion	0.001664	0.015599	0.029027

Table 2 Normalized decision matrix

	C1	C2	C3
Weights	0.065	0.199	0.735
DOS	0.498890614	0.589005	0.3227
Code execution	0.050855465	0.337513	0.820319
Overflow	0.094081541	0.225685	0.360652
Memory corruption	0.009154839	0.053259	0.131266
Sql injection	0.009154839	0.070653	0.226381
XSS	0.652473263	0.549535	0.007077
Directory traversal	0.044244469	0.133777	0.035893
http response splitting	0.005594743	0.007274	0.0006
Bypass something	0.174941088	0.152374	0.077662
Gain information	0.52686005	0.345674	0.022943
Gain privileges	0.058991912	0.099115	0.111669
CSRF	0.010680289	0.078559	0.003002
File inclusion	0.003560096	0.041872	0.063382

Then, the positive ideal solution $(A)^*$ and negative ideal solution $(\bar{A})$ are obtained using Eqs. (3) and (4), respectively. The values are as shown in Table 4.

The prime motive behind evaluating these two ideal solutions is to find the distance between the ideal solutions to that of individual values to facilitate ranking. Hence, in the next step, separation measures are calculated from both the negative and positive ideal solutions using Eqs. (5) and (6) as shown in Tables 5 and 6.

Table 3 Weighted normalized decision matrix

DOS	0.032623	0.117459	0.237245
Code execution	0.003325	0.067307	0.603091
Overflow	0.006152	0.045006	0.265147
Memory corruption	0.000599	0.010621	0.096506
Sql injection	0.000599	0.01409	0.166434
XSS	0.042666	0.109588	0.005202
Directory traversal	0.002893	0.026678	0.026389
http response splitting	0.000366	0.001451	0.000441
Bypass something	0.01144	0.030386	0.057096
Gain information	0.034452	0.068934	0.016867
Gain privileges	0.003858	0.019766	0.082098
CSRF	0.000698	0.015666	0.002207
File inclusion	0.000233	0.00835	0.046598

Table 4 Positive and negative ideal solutions

A′	Max	0.04267	0.1175	0.6031
A*	Min	0.00023	0.00145	0.00044

Table 5 Separation measure from negative ideal solution

DOS	0.001049121	0.013457901	0.056075895	0.070582917	0.265674457
Code execution	9.56475E−06	0.004337029	0.363186166	0.36753276	0.606244802
Overflow	3.50387E−05	0.001897047	0.070068941	0.072001027	0.26833007
Memory corruption	1.33811E−07	8.40918E−05	0.009228383	0.009312608	0.096501857
Sql injection	1.33811E−07	0.000159742	0.027553543	0.027713419	0.166473478
XSS	0.001800566	0.011693663	2.26639E−05	0.013516893	0.116262175
Directory traversal	7.07763E−06	0.000636422	0.000673256	0.001316755	0.036287125
http response splitting	1.76941E−08	0	0	1.76941E−08	0.000133019
Bypass something	0.000125594	0.000837282	0.003209792	0.004172667	0.064596187
Gain information	0.00117095	0.004553995	0.000269809	0.005994754	0.077425795
Gain privileges	1.3139E−05	0.000335443	0.006667764	0.007016346	0.083763629
CSRF	2.16752E−07	0.000202084	3.11714E−06	0.000205418	0.014332421
File inclusion	0	4.76058E−05	0.002130408	0.002178014	0.046669198

Table 6 Separation measure from positive ideal solution

DOS	0.00010086	0	0.133843145	0.133944006
Code execution	0.00154767	0.00251523	0	0.004062896
Overflow	0.00133325	0.00524945	0.114206083	0.120788785
Memory corruption	0.00176966	0.011414365	0.256628177	0.269812198
Sql injection	0.00176966	0.010685205	0.190669066	0.203123927
XSS	0	6.19518E-05	0.357470811	0.357532763
Directory traversal	0.00158187	0.008241152	0.332585322	0.342408342
http response splitting	0.00178929	0.013457901	0.363186166	0.378433361
Bypass something	0.00097508	0.007581597	0.298109738	0.306666412
Gain information	6.7469E-05	0.002354666	0.343657911	0.346080046
Gain privileges	0.00150608	0.009543942	0.271433614	0.282483641
CSRF	0.00176127	0.010361724	0.361061276	0.373184272
File inclusion	0.00180057	0.011904664	0.309684396	0.323389626

The final value of the relative closeness is obtained by using Eq. (7) which uses the values of separation from positive and negative ideal solutions. Finally, vulnerabilities are ranked based on the relative closeness value. Closer is the value to 1 better is the rank. Based on this criteria, all the vulnerabilities are ranked given in Table 7.

Figure 1 shows the effective share of each vulnerability after the evaluation of relative closeness.

Table 7 Relative closeness with ideal solution and ranking

Name	Relative closeness	Rank
DOS	0.420598531	3
Code execution	0.904862367	1
Overflow	0.435687844	2
Memory corruption	0.156675047	6
Sql injection	0.26973849	4
XSS	0.162785985	5
Directory traversal	0.058391644	11
http response splitting	0.000216185	13
Bypass something	0.104461883	9
Gain information	0.116305332	8
Gain privileges	0.136144455	7
CSRF	0.022923785	12
File inclusion	0.075842605	10

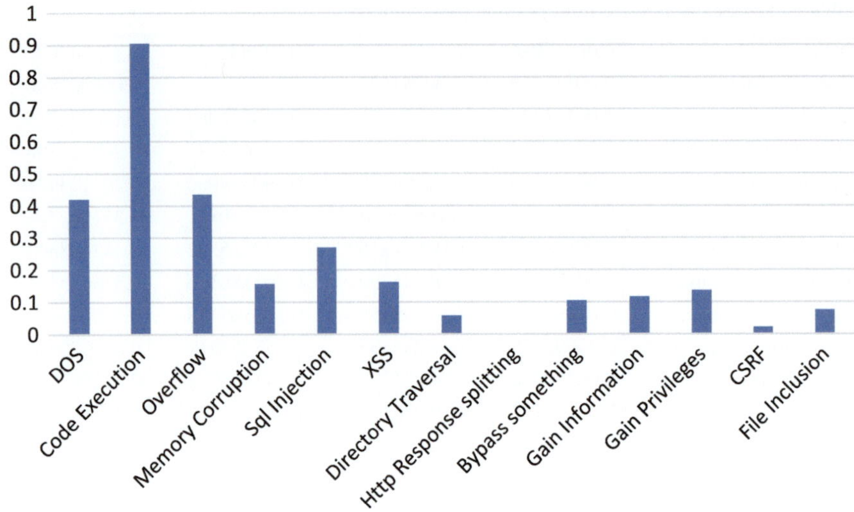

Fig. 1 Different vulnerability types and corresponding effective shares using TOPSIS

3.2 Ranking Vulnerabilities Based on VIKOR Technique

Using Eq. (9) for vulnerability type 1, we calculate the A_1 which is given below

$$A_1 = 0.063591\left(\frac{1283 - 981}{1283 - 7}\right) + 0.199419\left(\frac{9312 - 9312}{9312 - 115}\right) + 0.73519\left(\frac{19129 - 7525}{19129 - 14}\right) = 0.461783$$

Similarly $A_2, A_3 \ldots A_{13}$ can be calculated. The values of all A_i are given in Table 8.

Now using Eq. (10), we will calculate the R_i value. R_i value is given in the last column of Table 9.

Using the $A_i \& R_i$ values from Tables 8 and 9, we found the A^-, A^*, R^- and R^* values which are charted in Table 10.

Now using Eq. (11) and the values from Table 10, we can obtain the Q_i values.

$$Q_1 = 0.5\left(\frac{0.461783 - 0.146873}{0.999795 - 0.146873}\right) + (1 - 0.5)\left(\frac{0.446306 - 0.086212}{0.73519 - 0.086212}\right)$$

Table 8 A_i values

1	2	3	4	5	6	7
0.461783	0.146837	0.593084	0.866481	0.775211	0.742913	0.92088

8	9	10	11	12	13
0.999795	0.863874	0.811224	0.863315	0.972692	931832

Table 9 R_i values

R_i value (C1)	R_i value (C2)	R_i value (C3)	R_i value Max (C1, C2, C3)
0.015477	0	0.446306	0.446306
0.060625	0.086212	0	0.086212
0.056269	0.124547	0.412268	0.412268
0.064827	0.183655	0.617998	0.617998
0.064827	0.177693	0.532691	0.532691
0	0.01353	0.729382	0.729382
0.061291	0.156053	0.703536	0.703536
0.065186	0.199419	0.73519	0.73519
0.048121	0.149678	0.666075	0.666075
0.012658	0.083415	0.715152	0.715152
0.059805	0.167935	0.635575	0.635575
0.064674	0.174982	0.733036	0.733036
0.065391	0.187558	0.678882	0.678882

Table 10 Max, min values

A-	0.999795	R*	0.086212
A*	0.146837	R-	0.73519

Table 11 Q_i values of vulnerability types

1	2	3	4	5	6	7
0.462052	0	0.512796	0.831562	0.712336	0.844942	0.929353

8	9	10	11	12	13
1	0.867074	0.874022	0.843249	0.982453	0.916779

Similarly $Q_2, Q_3 \ldots Q_{13}$ can be calculated. Values of Q_i are given in Table 11. Based on the values obtained for A_i, R_i and Q_i, we have ranked the vulnerabilities given in Table 12. Lesser is the values of A_i, R_i and Q_i more is the severity of vulnerabilities.

Since we have 13 alternatives therefore DQ value is (1/(13 − 1)) = 0.08333 and $Q\ [A_2] - Q\ [A_1] = 0.462051725$ $0.462051725 \geq 0.083333$ which shows that condition 1 is satisfied. Also from the Table 12 we see that Code execution is ranked 1 by all the measures A_i, R_i and Q_i. This shows that condition 2 is satisfied. Hence, the last column of Table 12 gives the ranking of vulnerabilities based on VIKOR technique.

Figure 2 shows the effective share of each vulnerability on the basis of VIKOR method.

Table 12 Ranking based on A_i, R_i and Q_i

		Ranking		Ranking		Ranking
	A_i		R_{ij}		Q_i	
DOS	0.461783	2	0.44630629	3	0.462051725	2
Code execution	0.146837	1	0.0862118	1	0	1
Overflow	0.593084	3	0.41226794	2	0.51279551	3
Memory corruption	0.866481	9	0.61799806	5	0.831562246	5
Sql injection	0.775211	5	0.53269064	4	0.712335639	4
XSS	0.742913	4	0.72938233	11	0.844942252	7
Directory traversal	0.92088	10	0.70353625	9	0.929353262	11
http response splitting	0.999795	13	0.73519	13	1	13
Bypass something	0.863874	8	0.66607483	7	0.86707437	8
Gain information	0.811224	6	0.7151516	10	0.874022308	9
Gain privileges	0.863315	7	0.63557493	6	0.84324858	6
CSRF	0.972692	12	0.73303616	12	0.982452882	12
File inclusion	0.931832	11	0.67888248	8	0.916778653	10

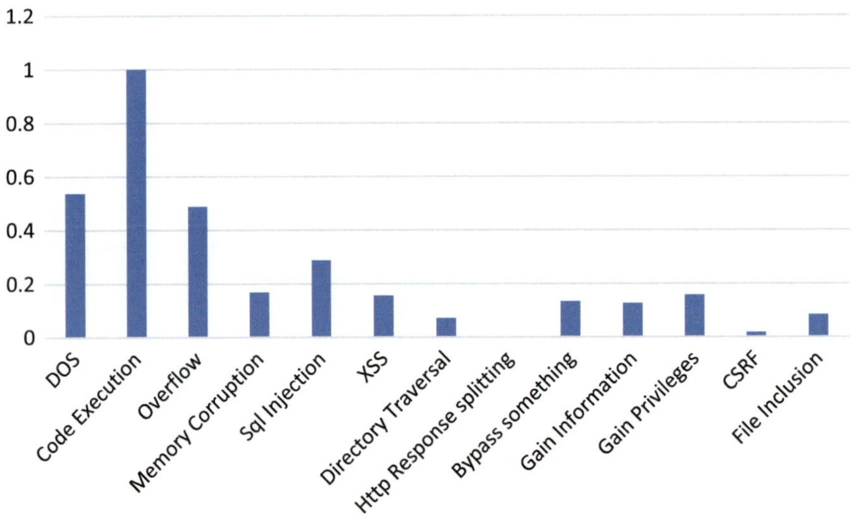

Fig. 2 Different vulnerability types and corresponding effective shares using VIKOR

4 Result Analysis

The results obtained with both the techniques used in this study gave comparable results with slight variations in the final ranking of vulnerabilities. Both the methods ranked code execution vulnerabilities to be the most critical type of vulnerability

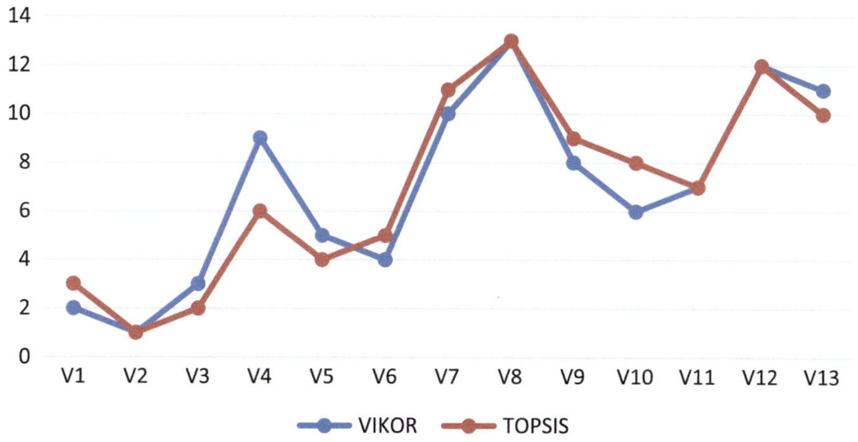

Fig. 3 Ranking of vulnerabilities by TOPSIS & VIKOR

and Http response splitting to be the least critical of all. Gain privileges and CSRF are also given same ranks in both the methods and are ranked at 7th and 12th positions, respectively. Denial of service is ranked at number 2 with the compromise solution approach while it is placed at number 3 using TOPSIS. Similarly, overflow is at rank 3 in VIKOR and at rank 2 with TOPSIS. Vulnerabilities due to memory corruption are ranked at position 9 and 6 for VIKOR and TOPSIS, respectively. Sql injection vulnerability type is ranked at number 5 using VIKOR and 4 with TOPSIS. XSS, Directory traversal, file inclusion, bypass something, gain information, and gain privilege are at positions 4, 10, 11, 8, 6, and 7 using VIKOR and 5, 11,10, 9, 8, 7 with TOPSIS, respectively.

Figure 3 shows that almost all the rankings given by these methods are comparable except for the memory corruption (V4) and gain information (V10) which shows a rank difference of 2 or more places using both the methods. Symmetrical structures are obtained with four points in the graph viz V1, V3, V5, and V6. It shows that ranks for these vulnerabilities have been swapped in both the methods. Overlapping points can be seen at V2, V8, V11, and V12 suggesting equal ranks for these vulnerabilities.

5 Conclusion and Future Research Direction

Both the methods used in this study work on the principle of "closeness to the ideal" solution. Although the steps involved in both the methods are different, the rankings obtained are quite close to each other which suggest that no particular method has advantage over the other. With respect to the number of calculations involved, VIKOR can be preferred over TOPSIS to minimize the calculations.

In the current study, we have not considered the uncertainty in the data. In future research, we can use the fuzzy or intuitionistic fuzzy approach combined together with the above techniques to get more improved ranking methods.

References

1. Schiffman, M., & Cisco, C. I. A. G. (2005, June). A complete guide to the common vulnerability scoring system (cvss). In *Forum Incident Response and Security Teams* (http://www.first.org/).
2. Mell, P., Scarfone, K., & Romanosky, S. (2007, June). A complete guide to the common vulnerability scoring system version 2.0. In *Published by FIRST-Forum of Incident Response and Security Teams, Vol. 1, p. 23.*
3. Mell, P., Scarfone, K., & Romanosky, S. (2006). Common vulnerability scoring system. *IEEE Security & Privacy, 4*(6).
4. National Vulnerability Database. nvd.nist.gov/ [online], December, 2016.
5. Liu, Q., & Zhang, Y. (2011). VRSS: A new system for rating and scoring vulnerabilities. *Computer Communications, 34*(3), 264–273.
6. Liu, Q., Zhang, Y., Kong, Y., & Wu, Q. (2012). Improving VRSS-based vulnerability prioritization using analytic hierarchy process. *Journal of Systems and Software, 85*(8), 1699–1708.
7. Spanos, G., Sioziou, A., & Angelis, L. (2013, September). WIVSS: A new methodology for scoring information systems vulnerabilities. In *Proceedings of the 17th Panhellenic Conference on Informatics* (pp. 83–90), ACM.
8. Wang, Y., & Yang, Y. (2012). PVL: A novel metric for single vulnerability rating and its application in IMS. *Journal of Computational Information Systems, 8*(2), 579–590.
9. Spanos, G., & Angelis, L. (2015). Impact metrics of security vulnerabilities: Analysis and weighing. *Information Security Journal: A Global Perspective, 24*(1–3), 57–71.
10. Sharma, R., & Singh, R. K. (2018). An improved scoring system for software vulnerability prioritization. In *Quality, IT and Business Operations* (pp. 33–43). Springer, Singapore.
11. Fruhwirth, C., & Mannisto, T. (2009) Improving CVSS-based vulnerability prioritization and response with context information. In *Proceedings of the 2009 3rd international Symposium on Empirical Software Engineering and Measurement. IEEE Computer Society.*
12. Sibal, R., Sharma, R., & Sabharwal, S. (2017). Prioritizing software vulnerability types using multi-criteria decision-making techniques. *Life Cycle Reliability and Safety Engineering, 6*(1), 57–67.
13. Hwang, C. L., & Yoon, K. (1981). Methods for multiple attribute decision making. In *Multiple attribute decision making* (pp. 58–191). Springer, Berlin, Heidelberg.
14. Opricovic, S., & Tzeng, G. H. (2004). Compromise solution by MCDM methods: A comparative analysis of VIKOR and TOPSIS. *European Journal of Operational Research, 156*(2), 445–455.
15. Zeleny, M. (1982). Multiple criteria decision making, McGraw-Hill Book Company.
16. CVE Details, The ultimate security vulnerability data source. www.cvedetails.com [Online], May 12, 2015.

MIX
Papier aus verantwortungsvollen Quellen
Paper from responsible sources
FSC® C105338
FSC
www.fsc.org

If you have any concerns about our products,
you can contact us on
ProductSafety@springernature.com

In case Publisher is established outside the EU,
the EU authorized representative is:
Springer Nature Customer Service Center GmbH
Europaplatz 3, 69115 Heidelberg, Germany

Printed by Libri Plureos GmbH
in Hamburg, Germany